ANTARCTIC TREATY SYSTEM

An Assessment

Site of the Workshop on the Antarctic Treaty System:
Beardmore South Field Camp, Transantarctic Mountains, Antarctica.

ANTARCTIC TREATY SYSTEM

An Assessment

Proceedings of a Workshop
Held at Beardmore
South Field Camp, Antarctica

January 7-13, 1985

Polar Research Board
Commission on Physical Sciences, Mathematics, and Resources
National Research Council

NATIONAL ACADEMY PRESS
Washington, D.C. 1986

NATIONAL ACADEMY PRESS 2101 Constitution Avenue, NW Washington, DC 20418

NOTICE: The project that is the subject of this report was approved by the Governing Board of the National Research Council, whose members are drawn from the Councils of the National Academy of Sciences, the National Academy of Engineering, and the Institute of Medicine. The members of the committee responsible for the report were chosen for their special competences and with regard to appropriate balance.

This report has been reviewed by a group other than the authors according to procedures approved by a Report Review Committee consisting of members of the National Academy of Sciences, the National Academy of Engineering, and the Institute of Medicine.

The National Research Council was established by the National Academy of Sciences in 1916 to associate the broad community of science and technology with the Academy's purposes of furthering knowledge and of advising the federal government. The Council operates in accordance with general policies determined by the Academy under the authority of its congressional charter of 1863, which establishes the Academy as a private, nonprofit, self-governing membership corporation. The Council has become the principal operating agency of both the National Academy of Sciences and the National Academy of Engineering in the conduct of their services to the government, the public, and the scientific and engineering communities. It is administered jointly by both Academies and the Institute of Medicine. The National Academy of Engineering and the Institute of Medicine were established in 1964 and 1970, respectively, under the charter of the National Academy of Sciences.

Support for the conduct of this workshop was provided under grants from the ARCO Foundation, the Ford Foundation, the William and Flora Hewlett Foundation, the Andrew W. Mellon Foundation, the National Geographic Society, the Tinker Foundation, and the National Science Foundation.

The contributions published in this book were prepared for the Workshop on the Antarctic Treaty System organized by the Polar Research Board of the U.S. National Research Council/National Academy of Sciences that was held in the Transantarctic Mountains, Antarctica, at the Beardmore South Field Camp, January 7-13, 1985.

The views expressed in this book are solely those of each author and are not to be attributed to their respective governments or institutions.

Library of Congress Catalog Card Number 86-60053

International Standard Book Number 0-309-03640-2

Printed in the United States of America

NATIONAL ACADEMY PRESS

The National Academy Press was created by the National Academy of Sciences to publish the reports issued by the Academy and by the National Academy of Engineering, the Institute of Medicine, and the National Research Council, all operating under the charter granted to the National Academy of Sciences by the Congress of the United States.

Organizing Committee for the Workshop on the Antarctic Treaty System

JAMES H. ZUMBERGE, University of Southern California, Chairman
THOMAS A. CLINGAN, University of Miami
W. TIMOTHY HUSHEN, Polar Research Board, National Research Council
LEE A. KIMBALL, International Institute for Environment and Development, Washington, D.C.
ROBERT H. RUTFORD, University of Texas at Dallas
DONALD B. SINIFF, Department of Ecology and Behavioral Biology, University of Minnesota

Staff

SHERBURNE B. ABBOTT, Staff Officer

Polar Research Board

GUNTER E. WELLER, Geophysical Institute, University of Alaska, Chairman
KNUT AAGAARD, Department of Oceanography, University of Washington
MIM HARRIS DIXON, Department of Transportation and Public Facilities, State of Alaska
DAVID ELLIOT, Institute of Polar Studies, The Ohio State University
W. LAWRENCE GATES, Department of Atmospheric Sciences, Oregon State University
RONALD L. GEER, Shell Oil Company
BEN C. GERWICK, JR., Department of Civil Engineering, University of California, Berkeley
DENNIS HAYES, Lamont-Doherty Geological Observatory, Columbia University
ARTHUR H. LACHENBRUCH, U.S. Geological Survey, Menlo Park
LOUIS J. LANZEROTTI, AT&T Bell Telephone Laboratories
GEOFFREY L. LARMINIE, British Petroleum Co. Ltd.
JOHN H. STEELE, Woods Hole Oceanographic Institution
IAN STIRLING, Canadian Wildlife Service, Edmonton, Alberta
CORNELIUS W. SULLIVAN, Department of Biological Sciences, University of Southern California
PATRICK J. WEBBER, University of Colorado
RAY F. WEISS, Scripps Institution of Oceanography, University of California at San Diego

Ex-officio

OSCAR T. FERRIANS, Committee on Permafrost, Chairman
CHARLES F. RAYMOND, Committee on Glaciology, Chairman
JAMES H. ZUMBERGE, U.S. Delegate to SCAR, University of Southern California

Staff

W. TIMOTHY HUSHEN, Staff Director
SHERBURNE B. ABBOTT, Staff Officer
MILDRED L. MCGUIRE, Administrative Secretary

Commission on Physical Sciences, Mathematics, and Resources

HERBERT FRIEDMAN, National Research Council, Chairman
THOMAS BARROW, Standard Oil Company (retired)
ELKAN R. BLOUT, Harvard Medical School
BERNARD F. BURKE, Massachusetts Institute of Technology
GEORGE F. CARRIER, Harvard University
HERMAN CHERNOFF, Massachusetts Institute of Technology
CHARLES L. DRAKE, Dartmouth College
MILDRED S. DRESSELHAUS, Massachusetts Institute of Technology
JOSEPH L. FISHER, Office of the Governor, Commonwealth of Virginia
JAMES C. FLETCHER, University of Pittsburgh
WILLIAM A. FOWLER, California Institute of Technology
GERHART FRIEDLANDER, Brookhaven National Laboratory
EDWARD A. FRIEMAN, Science Applications, Inc.
EDWARD D. GOLDBERG, Scripps Institution of Oceanography
MARY L. GOOD, UOP, Inc.
THOMAS F. MALONE, Saint Joseph College
CHARLES J. MANKIN, Oklahoma Geological Survey
WALTER H. MUNK, University of California, San Diego
GEORGE E. PAKE, Xerox Research Center
ROBERT E. SIEVERS, University of Colorado
HOWARD E. SIMMONS, JR., E.I. du Pont de Nemours & Company, Inc.
ISADORE M. SINGER, Massachusetts Institute of Technology
JOHN D. SPENGLER, Harvard School of Public Health
HATTEN S. YODER, JR., Carnegie Institution of Washington

RAPHAEL G. KASPER, Executive Director
LAWRENCE E. MCCRAY, Associate Executive Director

Participants in the Workshop on the Antarctic Treaty System

Participants in the Workshop on the Antarctic Treaty System

ROLF TROLLE ANDERSEN, Royal Ministry of Foreign Affairs, Norway
JAMES BARNES, Antarctic and Southern Ocean Coalition, Antarctica Project, Washington, D.C.
CHRISTOPHER D. BEEBY, Ministry of Foreign Affairs, New Zealand
ADRIAAN BOS, Ministry of Foreign Affairs, The Netherlands
LEWIS M. BRANSCOMB, IBM Corporation, New York
JOHN H. BROOK, Department of Foreign Affairs, Australia
PETER BRUCKNER, Permanent Mission of Denmark to the United Nations, New York
WILLIAM F. BUDD, Department of Meteorology, University of Melbourne, Australia
ANTONIO CARLOS ROCHA CAMPOS, University of São Paulo, Brazil
KENNETH R. CROASDALE, Croasdale and Associates, Canada
DOMINGO DA-FIENO, Ministry of Foreign Affairs, Peru
HASSAN EL-GHOUAYEL, Permanent Mission of Tunisia to the United Nations, New York
GAO QUINQAN, National Committee for Antarctic Research, National Bureau of Oceanography, China
VLADIMIR GOLITSYN, United Nations Secretariat, New York
JOHN A. GULLAND, Center for Environmental Technology, Imperial College, United Kingdom
GUY G. GUTHRIDGE, National Science Foundation, Washington, D.C.
TREVOR HATHERTON, Ross Dependency Research Committee, New Zealand
JOHN A. HEAP, Foreign and Commonwealth Office, United Kingdom
MARTIN W. HOLDGATE, Department of the Environment and Transport, United Kingdom
W. TIMOTHY HUSHEN, Polar Research Board, National Research Council, Washington, D.C.
ROBERT JONES, Los Angeles Times
ERNST F. JUNG, Federal Foreign Office, Federal Republic of Germany
LEE A. KIMBALL, International Institute for Environment and Development, Washington, D.C.
ALEXANDRE KISS, International Council for Environmental Law, Strasbourg, France
ABDUL KOROMA, Embassy of Sierra Leone, Brussels, Belgium
SACHIKO KUWABARA, United Nations Environment Program, New York

YOON KYUNG OH, Ministry of Foreign Affairs, Republic of Korea
GEOFFREY F. LARMINIE, British Petroleum Co., London, United Kingdom
JULIO CESAR LUPINACCI, Embassy of Uruguay, Santiago, Chile
CRISTIAN MAQUIEIRA, Permanent Mission of Chile to the United Nations, New York
KENTON R. MILLER, International Union for the Conservation of Nature and Natural Resources, Gland, Switzerland
PETER D. OELOFSEN, Department of Foreign Affairs, Pretoria, Union of South Africa
FRANCISCO ORREGO VICUNA, Embassy of Chile, London, United Kingdom (contributed paper but unable to attend)
ANNA C. PALMISANO, University of Southern California, Los Angeles
S. Z. QASIM, Department of Ocean Development, New Delhi, India (contributed paper but unable to attend)
OMAR BIN ABDUL RAHMAN, Office of the Scientific Advisor, Prime Minister's Department, Malaysia
H. P. RAJAN, Department of Ocean Development, New Delhi, India (contributed paper but unable to attend)
ORLANDO R. REBAGLIATI, Ministry of Foreign Affairs, Argentina
FRANÇOIS RENOUARD, Ministry of External Relations, France
E. FRED ROOTS, Department of Environment, Canada
HOLGER ROTKIRCH, Ministry for Foreign Affairs, Finland
ROBERT H. RUTFORD, University of Texas at Dallas
YURI M. RYBAKOV, Ministry of Foreign Affairs, U.S.S.R.
R. TUCKER SCULLY, Department of State, Washington, D.C.
DEBORAH SHAPLEY, Center for Strategic and International Studies, Georgetown University, Washington, D.C.
L. F. MACEDO DE SOARES GUIMARAES, Ministry of External Relations, Brazil
JOSE SORZANO, Permanent Mission of the United States to the United Nations, New York
N. A. STRETEN, Bureau of Meteorology, Melbourne, Australia
BO JOHNSON THEUTENBERG, Ministry of Foreign Affairs, Sweden
ROBERT B. THOMSON, Department of Scientific and Industrial Research, Christchurch, New Zealand
ALEXANDER VAYENAS, Embassy of Greece, Australia
ARTHUR D. WATTS, Foreign and Commonwealth Office, United Kingdom
MITCHELL WERNER, United Nations Secretariat, New York
PETER E. WILKNISS, National Science Foundation, Washington, D.C.

ROGER WILSON, Greenpeace International, Lewes, United Kingdom
RUDIGER WOLFRUM, Institute of International Law, University of Kiel, Federal Republic of Germany
RICHARD A. WOOLCOTT, Permanent Mission of Australia to the United Nations, New York
ZAIN AZRAAI, Permanent Mission of Malaysia to the United Nations, New York
ZHANG KUNCHENG, National Bureau of Oceanography, China
JAMES H. ZUMBERGE, University of Southern California, Los Angeles

Contents

OVERVIEW

1. WORKSHOP ON THE ANTARCTIC TREATY SYSTEM: OVERVIEW 3
   James H. Zumberge and Lee A. Kimball
   Trends in the Debate at the Workshop, 5; Ideas and Suggestions Put Forward, 6; The Antarctic Setting, 9

INTRODUCTION

2. ANTARCTICA PRIOR TO THE ANTARCTIC TREATY—A HISTORICAL PERSPECTIVE 15
   Trevor Hatherton
   Early Notions, 15; The Routes Open, 17; Reduction to Size, 19; Exploitation—The Seals, 19; Science and National Interests, 23; Because It Is There, 26; Exploitation—The Whales, 28; The Modern Era, 29; The International Geophysical Year, 1957-1958, 31

3. JURIDICAL NATURE OF THE 1959 TREATY SYSTEM 33
   Yuri M. Rybakov
   Peaceful Use, 35; Scientific Investigation, 37; Inspection, 38; Consultative Meetings, 38; Recommendations, 40; Additional Conventions, 41

LEGAL AND POLITICAL BACKGROUND

4. ANTARCTICA PRIOR TO THE ANTARCTIC TREATY: A POLITICAL AND LEGAL PERSPECTIVE 49
   Cristian Maquieira

5. ANTARCTIC CONFLICT AND INTERNATIONAL
   COOPERATION 55
   Francisco Orrego Vicuna
   The Early Trends Toward Antarctic Conflict, 55; Localized Territorial Disputes, 56; Generalized Territorial Disputes and International Implications, 58; Strategic Uses and Disputes in Antarctica, 59; Major-Power Rivalry in Antarctica, 59; The Antarctic Treaty: Cooperation as a Factor of Stabilization, 61

6. THE ANTARCTIC TREATY AS A CONFLICT RESOLUTION
   MECHANISM 65
   Arthur D. Watts

7. PANEL DISCUSSION ON THE LEGAL AND POLITICAL
   BACKGROUND OF THE ANTARCTIC TREATY 77

ANTARCTIC SCIENCE

8. SUMMARY OF SCIENCE IN ANTARCTICA PRIOR TO
   AND INCLUDING THE INTERNATIONAL GEOPHYSICAL
   YEAR 87
   Robert H. Rutford

9. THE ANTARCTIC TREATY AS A SCIENTIFIC
   MECHANISM (POST-IGY)—CONTRIBUTIONS OF
   ANTARCTIC SCIENTIFIC RESEARCH 103
   William F. Budd
   Introduction, 103; The Post-IGY International Antarctica Quarter Century, 105; The Profitable Nonrenewable Resources Fallacy, 107; Antarctica as a Global Environmental Science Resource, 117; Antarctic Publications and the Knowledge Explosion, 120; Highlights of Antarctic Discoveries and Research, 128; The Treaty Nations as the United Nations "Antarctic Rangers," 138

10. THE ANTARCTIC TREATY AS A SCIENTIFIC MECHANISM—
    THE SCIENTIFIC COMMITTEE ON ANTARCTIC
    RESEARCH AND THE ANTARCTIC TREATY SYSTEM 153
    James H. Zumberge
    Introduction, 153; The Origin and Growth of SCAR, 154;

SCAR Structure and Procedures, 157; The Interaction of SCAR with the Antarctic Treaty System, 164; A Look at SCAR's Future, 167

11. THE ROLE OF SCIENCE IN THE ANTARCTIC TREATY SYSTEM 169
E. Fred Roots
Background, 169; The Political Role of Science in Antarctica,173; Different Approaches to Science in Antarctica, 174; What Results Can Antarctic Science Deliver?, 175; The Setting of Scientific Priorities in Antarctica, 181; The Future, 183

12. PANEL DISCUSSION ON ANTARCTIC SCIENCE 185

THE ANTARCTIC ENVIRONMENT: MANAGEMENT AND CONSERVATION OF RESOURCES

A. CONSERVATION AND ENVIRONMENT

13. THE ANTARCTIC TREATY SYSTEM AS AN ENVIRONMENTAL MECHANISM — AN APPROACH TO ENVIRONMENTAL ISSUES 195
John A. Heap and Martin W. Holdgate
Introduction, 195; Characteristics of the Antarctic Environment, 195; Human Impacts on the Environment of Antarctica, 198; The Evaluation of Environmental Goals, 199; Environmental Conservation Within the Antarctic Treaty System, 200; The Antarctic Treaty System as a Mechanism for Environmental Conservation, 206

14. PANEL DISCUSSION ON CONSERVATION AND ENVIRONMENT 211

B. LIVING RESOURCES

15. THE ANTARCTIC TREATY SYSTEM AS A RESOURCE MANAGEMENT MECHANISM—LIVING RESOURCES 221
John A. Gulland
Introduction, 221; Marine Resources, 221; International Whaling Commission, 224; The Role of the Antarctic Treaty, 228; Terrestrial Activities, 231; The Role of SCAR, 233

16. PANEL DISCUSSION ON LIVING RESOURCES 235
Biomass, 241; Experimental Fishery, 241; Inspection and Other CCAMLR Measures, 244

C. NONLIVING RESOURCES

17. ARCTIC OFFSHORE TECHNOLOGY AND ITS RELEVANCE TO THE ANTARCTIC 245
K. R. Croasdale
Introduction, 245; Geography and Oil and Gas Resources, 245; The Arctic Offshore Environment, 246; Technology for Arctic Offshore Petroleum Operations, 251; Conclusions, 262

18. DISCUSSION ON TECHNOLOGY AND ECONOMICS OF MINERALS DEVELOPMENT IN POLAR AREAS 265

19. THE ANTARCTIC TREATY SYSTEM AS A RESOURCE MANAGEMENT MECHANISM—NONLIVING RESOURCES . 269
Christopher D. Beeby

20. PANEL DISCUSSION ON NONLIVING RESOURCES 285
Participation, 290; Common Heritage of Mankind, 290; Participation in the Minerals Regime Negotiations, 291; Participation in the Adoption of the Minerals Regime, 292; Participation in Implementation of the Minerals Regime, 292; Participation in Activities and Benefits, 293; Urgency and Timing of Minerals Activities, 294; The Regime, 296; Enforcement and Reporting Requirements, 302

INSTITUTIONS

21. THE ANTARCTIC TREATY SYSTEM FROM THE PERSPECTIVE OF A STATE NOT PARTY TO THE SYSTEM . 305
Zain Azraai
The Response of the Nontreaty Parties (NTPs), 307

22. THE ANTARCTIC TREATY SYSTEM FROM THE PERSPECTIVE OF A NON-CONSULTATIVE PARTY TO THE ANTARCTIC TREATY 315
Peter Bruckner
Introduction, 315; Motives for Accession, 317; Functioning

of the Treaty System, 318; Rights and Obligations of the
NCPs Under the Treaty, 320; The Observer Issue, 325;
Antarctica and the U.N. General Assembly, 327;
Concluding Remarks, 328; Appendix, 333

23. THE ANTARCTIC TREATY SYSTEM FROM THE
PERSPECTIVE OF A NEW CONSULTATIVE PARTY 337
L. F. Macedo de Soares Guimaraes

24. THE ANTARCTIC TREATY SYSTEM FROM THE
PERSPECTIVE OF A NEW MEMBER 345
S. Z. Qasim and H. P. Rajan
Introduction, 345; Background of the Antarctic Treaty,
347; The Antarctic Treaty System, 349; India's Scientific
Expeditions, 365; Political Issues, 367; Conclusions, 371

25. THE INTERACTION BETWEEN THE ANTARCTIC TREATY
SYSTEM AND THE UNITED NATIONS SYSTEM 375
Richard A. Woolcott
Promotion of Principles and Purposes of United Nations
Charter, 376; Links with the United Nations Specialized
Agencies, 379; The Future, 382; Relationship to the
United Nations System in the Future, 388

26. THE EVOLUTION OF THE ANTARCTIC TREATY
SYSTEM—THE INSTITUTIONAL PERSPECTIVE 391
R. Tucker Scully
Introduction, 391; The Antarctic Treaty, 391; The Antarctic
Treaty System—Substantive Content, 395; The Antarctic
Treaty System—Institutional Response, 400; Operation of
the Antarctic Treaty System, 406; Conclusion, 409

27. PANEL DISCUSSION ON INSTITUTIONS OF THE
ANTARCTIC TREATY SYSTEM 413
Legitimacy, 423; Evolution of the Antarctic Treaty System,
427; Concluding Remarks, 431

Participants discuss the Antarctic Treaty System in the Jamesway at Beardmore South Field Camp, Transantarctic Mountains, Antarctica.

OVERVIEW

1. Workshop on the Antarctic Treaty System: Overview
James H. Zumberge and Lee A. Kimball

1.

Workshop on the Antarctic Treaty System: Overview

James H. Zumberge and *Lee A. Kimball*

When Malaysia and Antigua and Barbuda launched their 1983 initiative to bring Antarctica before the United Nations General Assembly (Beck 1985), they could never have predicted that their efforts would help launch the most international undertaking yet attempted in Antarctica (Holdgate 1985).

In September, 1983, the United Nations General Assembly agreed to place "The Question of Antarctica" on its fall agenda. In the same year, the XII Antarctic Treaty Consultative Meeting under the 1959 Antarctic Treaty was taking place in Canberra, Australia. Recalling the highly successful meeting sponsored by the University of Chile at Teniente Marsh, Antarctica, in October, 1982 (Holdgate 1983, Orrego Vicuna 1984), the germ of an idea took shape in Canberra that led to the Workshop on the Antarctic Treaty System. It was held January 7-13, 1985, and was sponsored by the Polar Research Board of the National Research Council. The site was a base camp at 84°03'S 164°15'E on the Bowden Neve in the Transantarctic Mountains, 1700 m above sea level. The camp is called Beardmore South because of its proximity to the head of the great Beardmore Glacier, a huge river of ice flowing 200 km from the interior ice sheet to the Ross Ice Shelf.

The purpose of the workshop was to bring together in Antarctica individuals active in antarctic science and politics with those eager to learn more about the Antarctic Treaty System and eager to play a role in determining the future of Antarctica. In an atmosphere conducive to free interchange, participants were able to gain a deeper appreciation of each others' views on management of potential resources, political and legal regimes, and the evolution of the Treaty System, while experiencing the practical realities of antarctic operations.

Where the Teniente Marsh meeting convened scientists, industrialists, lawyers and government officials from the then 14 consultative parties to the Antarctic Treaty, the Beardmore Workshop at Beardmore South Field Camp specifically sought to expand the circle to include participants from the consultative parties, the nonconsultative parties to the Antarctic Treaty, states not party to the treaty, international and nongovernmental organizations, private industry and the media. All told, 57 individuals from 25 countries participated in his or her personal capacity in the workshop.[1]

The informal nature of the meeting and the fact that participants were invited in their personal capacities allowed them to explore topics of interest off-the-record on a noncommittal basis. (Journalists present could not directly quote participants without their permission.) In most participants' assessment, this fostered a constructive exchange that will serve as a necessary complement to the more politicized discussions taking place in a variety of international forums.

Efforts to plan and organize the workshop fell to a committee chaired by James H. Zumberge, U.S. Delegate to the Scientific Committee on Antarctic Research (SCAR) and the staff of the Polar Research Board, working with National Science Foundation (NSF) officials.[2] Despite some initial skepticism, once the Foundation staff agreed that staging the meeting against an antarctic backdrop would most vividly illustrate the kinds of activities carried out in Antarctica and the difficulties they present, and once they determined that they could move up construction of the Beardmore South Field Camp from the 1985-86 season to the 1984-85 season and thus avoid significant costs to the U.S. Antarctic Research Program, the workshop became a reality. NSF made available transportation between Christchurch, New Zealand, and Antarctica, cold-weather clothing, and facilities at the Beardmore Camp. The Tinker Foundation, Ford Foundation, William and Flora Hewlett Foundation, Andrew W. Mellon Foundation, and Atlantic Richfield Foundation joined the National Geographic Society in granting additional funds to the Polar Research Board to help defray travel costs to New Zealand for many of the participants.

TRENDS IN DEBATE AT THE WORKSHOP

The workshop was divided into five sessions: (1) Introduction and Overview, (2) Legal and Political Background, (3) Antarctic Science, (4) The Antarctic Environment: Management and Conservation of the Environment, of Living Resources and of Nonliving Resources, and (5) Institutions. This volume contains papers and summaries of discussions at the workshop.

The meeting was not designed to produce conclusions or recommendations, but rather to stimulate an open discussion, and in this it succeeded admirably, not only during formal sessions but also in casual conversations at mealtimes, on long walks, and during breaks. The extent to which the inevitable comradery that develops at such an event strengthened personal acquaintances among the variety of antarctic decision-makers represented will alone serve well the determination of future directions for antarctic science and politics. Beyond that, the questions and comments or prepared papers from panelists and audience alike laid the groundwork for continuing dialogue on the positive contributions of the Antarctic Treaty System, criticisms of it by the international and environmental communities, and measures to improve the Antarctic Treaty System that command widespread support.

Statements on the benefits and drawbacks of the Antarctic Treaty System are well recorded elsewhere.[3] These points were all aired at the workshop. In summary, defenders of the Antarctic Treaty System find it a particularly outstanding example of international cooperation today that preserves peace, promotes international scientific collaboration, and protects Antarctica as a 'special conservation area.' They believe that the treaty system is an open one that is evolving to satisfy growing international interest in Antarctica. They oppose any efforts to replace the Antarctic Treaty System or to subject it to a major overhaul.

On the other hand, some members of the international community challenge what they deem as the secret and exclusive nature of the system and have called for wider international cooperation in Antarctica, particularly with respect to potential mineral resources development there. Representing this point of view, Ambassador Zain of Malaysia outlined his concerns at Beardmore as follows:

1. "the assertion of the Antarctic Treaty Consultative Parties...that they--and they alone--have the right to make decisions pertaining to Antarctica ('exclusive');

2. that these would cover all activities in Antarctica ('total'); and
3. that these are not subject to review or even discussion by any other body ('unaccountable')."

The environmental community has questioned the adequacy of environmental protection and conservation measures in Antarctica and whether or not they are being effectively enforced. Some environmentalists have proposed that Antarctica be closed to all mineral resources development activities, while others argue for making Antarctica an international park.

Both environmentalists and members of the scientific and political establishments have begun to explore whether Antarctic management regimes require more directed research, tailored in particular to support the resource management regimes found in the Convention on the Conservation of Antarctic Marine Living Resources (CCAMLR) or in the minerals regime currently being negotiated. They also express concern about the compatibility of science and resources development in Antarctica.

IDEAS AND SUGGESTIONS PUT FORWARD

The participants from consultative parties underscored the importance of Article IV of the Antarctic Treaty on territorial claims as the cornerstone for all antarctic legal/political regimes. Those countries favoring a 'common heritage of mankind' approach to Antarctica that would dismiss the claims heard the claimant states' view that their rights, including rights to offshore zones of jurisdiction, are protected by Arcticle IV. These claims would clearly complicate any effort to fully internationalize Antarctica, as for example, under a United Nations trusteeship. It was equally clear that Article IV was fundamental to the nonclaimant states' support for the Antarctic Treaty System, because it preserves their position as well. In the same vein, participants took note of the consensus requirement for antarctic decision-making, another reassurance against any undermining of either the claimant or nonclaimant position.

Participants also discussed the 'interest' or 'activities' criterion for consultative party status within the Antarctic Treaty System, which according to R. Tucker Scully of the U.S. State Department, "has substituted a functional basis for a political or ideological basis for involvement in decision-making.

Commitment to a particular legal status for Antarctica does not establish eligibility to take part in decisions relating to activities in Antarctica. Demonstration of concrete interest in those activities becomes the standard."

Participants continued to differ over whether Antarctica could or should be internationalized (in the manner of the deep seabed) and over the justification for the 'activities' criterion for decision-making status. But as the workshop evolved, they focused less on challenging the legitimacy of the Antarctic Treaty System per se and more on how to spread its benefits and perfect management mechanisms so as to command widespread international support for the system.

Those participants from the developing countries expressed genuine interest in taking part in antarctic science and pointed out that many of them are directly affected by Antarctica's role in atmospheric and oceanic circulation. This view somewhat surprised others, who believed that the only reason for increased international interest in Antarctica was an interest in resources there. But in the words of Hassan El-Ghouayel of Tunisia, "There won't be any profit from mineral exploitation for many decades. It will be much more valuable for us to share the rewards of science in Antarctica." Several scientists present made the same point, including Lewis M. Branscomb, former chairman of the U.S. National Science Board.

Several participants expressed interest in the possibilities of joint scientific research programs and sharing of program facilities and logistics capabilities, particularly if advanced technology were to become more widely used in Antarctica. Political questions about the extent to which joint scientific activities, and/or supply and transport activities, might be weighed in meeting the 'activities' criterion for consultative party status were also raised. Some participants pressed for clarification or modification of the criteria applied to this determination.

With respect to participation in the Antarctic Treaty System on the political level, those present expressed interest in two emerging trends: (1) participation by the nonconsultative parties in Antarctic Treaty meetings as observers, and the further development of this role; and (2) the further development of working relationships with international and nongovernmental organizations.[4]

The nature of the evolving minerals regime and the role afforded nonconsultative parties and nontreaty states in the negotiation of the regime, its institutions once established, and potential minerals activities under it were discussed in some detail. There were suggestions that funds from potential minerals activities could be made available to foster developing nation participation in antarctic science or other activities, and that developing nations would be able to participate in joint venture arrangements for minerals activities. The idea seemed to persist among several of the workshop participants from states not party to the Antarctic Treaty System that Antarctica has a vast mineral resource potential, even though it has repeatedly been stressed in a number of studies that no minerals of economic worth are known to exist in Antarctica (Behrendt 1983, Holdgate and Tinker 1979, Zumberge 1979).

On the issue of accountability, there was agreement that the consultative parties should persist in increasing the availability of information on antarctic activities and meetings including information on the contributions of antarctic science.[5]

Information alone, however, does not satisfy those who raised the point at the workshop that the Antarctic Treaty System should be accountable in a forum where all states would be able to consider antarctic matters on an equal footing. How the different views on antarctic management would be reconciled to the satisfaction of all concerned, and what role outside interests would have in influencing that process, remains unsettled. That the Antarctic Treaty System would continue to evolve in response to new interests and activities in Antarctica, however, was not in doubt.

In the areas of environmental protection and conservation in Antarctica, participants agreed that the Antarctic Treaty System has provided a valuable mechanism to further scientific research, particularly environmental research, and that it has taken a far-sighted preventive approach to conservation and environmental protection in Antarctica. But they also expressed concern with the increasing level and variety of antarctic activities and their potentially damaging effects. In this context as well, many shared the view that there was some merit in joint station and logistics activities as a means to reduce the buildup and concentration of individual national facilities and to promote efficient use of scarce national resource allocations for antarctic programs.

Different speakers also proposed ways to better coordinate antarctic conservation measures. They suggested the development of a "conservation strategy" for Antarctica, seeking improvements in the institutional arrangements of the Antarctic Treaty System, and drawing on the expertise and experience of international organizations such as the United Nations Environment Program and the International Union for the Conservation of Nature and Natural Resources. The workshop participants from the consultative parties, however, pointed out that under the stewardship provided by the Antarctic Treaty System, Antarctica remains essentially undamaged environmentally except for the localized impacts from logistical support of scientific activities.

The need to improve environmental impact assessment and monitoring of antarctic activities received attention as well. It was pointed out that the Scientific Committee on Antarctic Research was working on assessment procedures for station and logistics activities as requested by the XII Antarctic Treaty Consultative Meeting, and on extending the coverage of protected areas in Antarctica. Suggestions to perfect the application of inspection procedures with respect to compliance with environmental protection measures in Antarctica were also put forward, noting the linkage between inspection, assessment and monitoring arrangements, and reporting requirements under the various antarctic regimes.[6]

Some participants questioned in addition the adequacy of implementation and enforcement under the Antarctic Treaty System, referring in particular to environmental protection measures and to marine living resources conservation under the Convention on the Conservation of Antarctic Marine Living Resources. In their view, these potential weaknesses could be far more problematic if replicated in the minerals regime, which would govern potentially far more damaging activities. Given the problem of lack of data with respect to krill, there was some support for the idea of developing krill as an experimental fishery. Finally, if minerals development were ever to take place in Antarctica, most participants supported the application of stringent safeguards to protect against possible environmental damages from these activities or interference with Antarctica's value as a scientific laboratory.

THE ANTARCTIC SETTING

During the workshop, the National Science Foundation made certain that participants took advantage of the locale by arranging for three fascinating, informative and well-presented lectures by U.S. Antarctic Research Program scientists as well as a one-day trip to the South Pole, a tour of U.S. Amundsen-Scott Station there and brief introductions by researchers at the Pole to their work. The three lectures particularly captured the imagination of participants, who for the most part were not well versed in antarctic science. Dr. Anna C. Palmisano spoke on the ecology of sea-ice microbiological communities in McMurdo Sound, Dr. Imre Friedmann on endolithic microorganisms within antarctic rocks, and Dr. William Cassidy on his study of meteorites discovered in Antarctica.

The trip to the South Pole provided the only real exposure to the extremes of the antarctic climate and environment and to the logistics and structures required to cope with it. The incessant wind, the sunless sky, and the vast expanse of the world's largest ice sheet--interrupted only by the Pole station itself--reminded everyone of the true character of the continent.

Otherwise the setting for the Beardmore South Field Camp and the unusually warm -10°C weather there provided an extraordinarily stunning and welcoming venue for the workshop. The 10-hour flight aboard two Hercules C130, ski-equipped planes from Christchurch, with a refueling stop at Williams Field in McMurdo, afforded a view of sea ice, icebergs, the coastal area, and the Transantarctic Mountains, which not only brought to life the beauty of photographs but also conveyed in a way that photographs cannot the vastness and ice-filled relief that is Antarctica.

The bringing together of scientists and nonscientists proved to be a revealing experience on how differently the two groups seek answers to questions or solutions to problems. The scientists look for factual evidence to support a new hypothesis or refute an older one. The lawyers and diplomats argue from philosophical or legal principles in which precedent plays an important role. The principle of the 'common heritage of mankind' as derived from or embodied in the 1982 Law of the Sea Convention, for example, was the underlying philosophy of those who espoused the replacement of the Antarctic Treaty System by a system of governance in which all nations of the world would have a vote or at least a voice in Antarctica's future.

For those workshop participants who work in the

international political arena, the politics of Antarctica will never again be politics alone. The sense of isolation imparted by Antarctica created an enduring impression that is likely to remain with participants whenever they consider again the management of Antarctica in a far more cosmopolitan setting. Scientists who participated in the workshop learned that their view of Antarctica is certainly not the only one. Nonscientists gained a deeper appreciation of the global and interdisciplinary nature of current antarctic science and of the long tradition of international cooperation in research and exploration of the Antarctic. One can only hope that scientist and nonscientist alike learned from each other during the presentation of prepared papers and the lengthy discussions that followed in the Jamesway hut at Beardmore South Field Camp. While it is certain that no consensus emerged, it is equally certain that the Workshop on the Antarctic Treaty System was an enriching and enlightening interchange of disparate views.

NOTES

1. Individuals from 47 countries were invited, and those attending came from 25. Of the then 16 consultative parties, all were invited and 12 attended, with two last-minute drop-outs; of the then 16 nonconsultative parties 10 were invited and 7 attended; of the nontreaty states 21 were invited and 6 attended, with 3 last-minute drop-outs; and of the international organizations 12 were invited and 10 attended. Two industry representatives and two media representatives also attended.
2. Organizing Committee members in addition to Zumberge were: Thomas A. Clingan, Jr., University of Miami Law School; W. Timothy Hushen, Polar Research Board; Lee A. Kimball, International Institute for Environment and Development; Robert H. Rutford, University of Texas at Dallas; and Donald B. Siniff, University of Minnesota.
3. See United Nations General Assembly records of the First Committee for 1983 and 1984, United Nations Documents No. A/C.1//38/PV. 42-46 and A/C.1/39/PV. 50, 52-55; Lee Kimball, "Antarctica: Summary and Comment:", International Institute for Environment and Development (IIED), April 6, 1984; the United Nations study on Antarctica, U.N. Document No. A/39/583 (Part

I), 31, October 1984, pp. 33-38 and Part II, volumes I-III; discussions in Proceedings of an Interdisciplinary Symposium, June 22-24, 1983, ed. Rudiger Wolfrum (Duncker & Humblot, Berlin, 1984); Proceedings of the Eighth Annual Conference, June 17-20, 1984, ed. Lewis M. Alexander and Lynne C. Hanson (Center for Ocean Management Studies, University of Rhode Island, 1985).
4. Following the Workshop on the Antarctic Treaty System, the nonconsultative parties seemed satisfied with their level of participation in the round of antarctic minerals regime negotiations in Rio de Janeiro in February and March, 1985. On the other hand, the preparatory meeting for the XIII Antarctic Treaty Consultative Meeting, which took place in Brussels in April, 1985, did not act on the decision at the XII meeting to identify international organizations having a scientific or technical interest in Antarctica that could assist in consultative party meeting deliberations and invite them to attend the XIII meeting as observers.
5. The XIII Antarctic Treaty Consultative Meeting in October, 1985, approved a recommendation to enhance public availability of information and documentation on the Antarctic Treaty System.
6. These issues were addressed at the XIII Antarctic Treaty Consultative Meeting.

REFERENCES

Beck, P.J. 1985. The United Nations' Study on Antarctica, 1984. Polar Record, 22 (140): 499-504.

Behrendt, J.C. (ed.) 1983. The Petroleum and Mineral Resources of Antarctica. U.S. Geological Survey Circular 909, Department of Interior, Washington, D.C.

Holdgate, M.V. 1983. Policy for Antarctic Resources. Polar Record, 21 (133): 392-93.

Holdgate, M.V. 1985. International Workshop on the Antarctic Treaty System, 7-13 January 1985. Polar Record 22 (140): 538-39.

Holdgate, M.V. and Tinker, J. 1979. Oil and Other Minerals in the Antarctica. London House of Print.

Orrego Vicuna, F. (editor) 1984. Antarctic Resource Policy. Cambridge University Press.

Zumberge, J.H. 1979. Mineral Resources and Geopolitics in Antarctica. American Scientist, 67-77.

INTRODUCTION

2. Antarctica Prior to the Antarctic Treaty—A Historical Perspective
   Trevor Hatherton

3. Juridical Nature of the 1959 Treaty System
   Yuri M. Rybakov

2.

Antarctica Prior to the Antarctic Treaty—A Historical Perspective

Trevor Hatherton

EARLY NOTIONS

The earliest recorded concept of a southern polar region derived from Aristotle's demonstration of the spherical shape of the Earth, about the middle of the fourth century B.C. Three pieces of evidence were adduced to substantiate that discovery: (1) all matter tends to fall toward a common center, and (2) the more direct observational evidence that the Earth throws a circular shadow on the moon during an eclipse, and (3) as one travels from north to south, familiar stars disappear and new ones come above the horizon. The name of the region reflects its position, i.e., "Antarktos"--opposite the Bear, the northern constellation that contains the pole star, Polaris.

A century later, Eratosthenes was able to make a reasonably accurate estimate of the size of the spherical Earth. The knowledge of the shape and size of the Earth completely changed the problems of geography; not only was the existence of an antarctic region confirmed but the possibility of reaching it could be speculated on. Considerations of symmetry suggested that the Earth could be divided into five zones--the southern temperate and polar regions mirroring the similar known regions to the north, and the equatorial torrid zone as in the map of Macrobius (ca A.D. 410). The Stoic philosophers imposed another symmetry based on continents (Figure 2-1).

Although Eudoxus of Cyzicus is alleged by Poiseidonius to have set out from Cadiz to attempt the circumnavigation of Africa about 100 B.C. (and was never heard of again), the later preoccupation of the Christianized world with spiritual concerns and the uncomfortable implications for theology of other "habitable worlds" on

(a)

(b)

FIGURE 2-1 Grecian notions of the Earth after Aristotle's discovery and Eratosthenes' measurements: (a) the globe as described by Crates, a stoic Philosopher of ca 150 B.C., which satisfied symmetry by inventing, in addition to the known inhabited world (the Oecumene), three other populated continents: Perioeci (peoples around the globe from the Oecumene), Antoeci (peoples below the Oecumene), and the Antipodes (peoples on the opposite side of the globe) (from Raisz, 1948); (b) the Earth as drawn to the ideas of Macrobius (ca 410 A.D.) following Cicero (first century B.C.); the known world, centered on Jerusalem, is balanced by a large southern continent (from Mill, 1905) (Reprinted with permission).

the Earth led to a neglect of physical speculation and
discovery. The antarctic problem was forgotten for almost
a millennium and a half, a period during which mankind
had neither the motivation nor the technology to advance
exploration of any southern land.

THE ROUTES OPEN

The decay of centers of learning in the eastern Mediter-
ranean; the westward transmission of Greek and Arabic
knowledge; the strengthening and centralization of power
in the states of the Iberian Peninsula; the location of
these states on the shores of the Atlantic Ocean; and
developing navigation, shipbuilding and sailing techniques all led, in the early part of the fifteenth
century, to the first long voyages (by Europeans) out of
sight of land. An era of exploration of the west African
coast, passing through the dreaded "perusta," or torrid
zone, initiated by Prince Henry of Portugal 1600 years
after Eudoxus, culminated in 1488 with the rounding of
the southern tip of the African continent by Bartholemeu
Diaz de Novaes. A decade later Vasco de Gama's voyage by
that route to Mombassa and India demolished Ptolemy's
earlier notion of the Indian Ocean as an inland sea
bounded in the south by terra incognita.

Meanwhile, to the west, Columbus made his first, epic
voyage, and exploration of the east coast of South America
proceeded so rapidly that in 1520 Ferdinand Magellan was
able to round that continent by a strait between it and
Tierra del Fuego and emerge into the Pacific Ocean in
about 52°S latitude. Though he was killed in the
Philippine Islands, one of his ships completed the first
circumnavigation of the world.

Strangely, in view of the rapid growth of exploration,
it was more than 50 years before the next circumnavigation. This was led by Francis Drake, who demonstrated
that the Atlantic and Pacific oceans met south of Tierra
del Fuego, in the vicinity of which "we found great store
of foule which could not fly with the bigness of geese,
whereof we killed in lesse than one day three thousand
and victualled ourselves thoroughly," thus exploiting
antarctic resources (penguins) for the first time.

This was still a period of firm belief in the great
southern continent (Figure 2-2), even though both Africa
and South America were now demonstrably separated from
it. Numerous voyagers set out to take possession of this

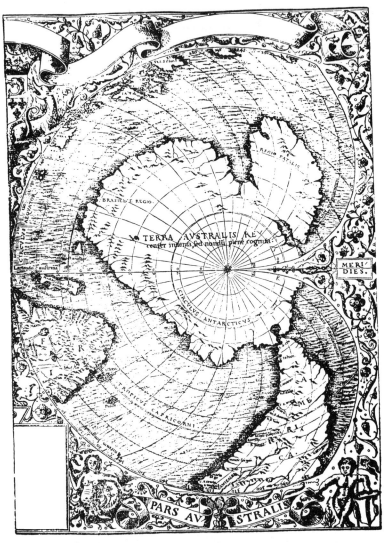

FIGURE 2-2 The Orontius Finaeus Map (Southern Hemisphere) of 1531, which because of the remarkable detail that it shows compared with contemporaneous, or even later, charts has raised speculation (Weihaupt, 1984) that the region was mapped before the Age of Discovery. However, portions of this map stretch north of 30°S latitude and it would be curious if navigators capable of surveying the boundaries of the continent could not make more accurate measurements of latitude, that most easily determined parameter of position (from Hatherton, 1965) (Reprinted with permission).

hypothetical land, most notably Quiros (1605), but after
passing through the Strait of Magellan all were beguiled
by the trade winds into too low latitudes. The most
important voyage in the seventeenth century was that of
Abel Tasman, who swept south of Australia, thus separating
that continent from the antarctic landmass.

Major trading interests now began to display an interest in the hypothetical riches of the great south land,
for the Dutch East India Company (Jacob Roggeveen, 1721)
and the French East India Company (Bouvet de Lozier,
1739) sent out expeditions to discover and annex the
southern lands, and the latter took an ice-clad island in
the South Atlantic, later named Bouvet Island after him,
to be part of it.

REDUCTION TO SIZE

By the middle of the eighteenth century, navigation
methods had greatly improved, and the introduction of the
quadrant gave new precision to determinations of
latitude. This period saw great rivalry between French
and British in the exploration of the Pacific, and the
second expedition of James Cook not only used a reliable
chronometer for the first time but also crossed the
Antarctic Circle (Figure 2-3). This voyage proved beyond
doubt that "habitable" lands did not exist south of the
known continents. It also demonstrated that scurvy could
be prevented by proper diet. Cook's voyage around the
antarctic continent was supplemented during 1819-1821 by
a great Russian encirclement in the ships <u>Mirnyy</u> and
<u>Vostok</u> under Thaddeus von Bellingshausen. This
expedition was superbly planned and executed, and
comparison of tracks (Figure 2-3) shows that
Bellingshausen's vessels sailed over 242° of longitude
south of 60°S latitude, of which 41° were within the
Antarctic Circle, while Cook's vessels made only 125°
south of 60°S latitude and 24° south of the circle.
Bellingshausen's care in crossing all the great gaps left
by his predecessor demonstrated beyond any doubt the
existence of a continuous open sea south of the 60°
parallel.

EXPLOITATION--THE SEALS

Although Dampier had noted the existence of a very large
colony of fur seals at the Juan Fernandez Islands

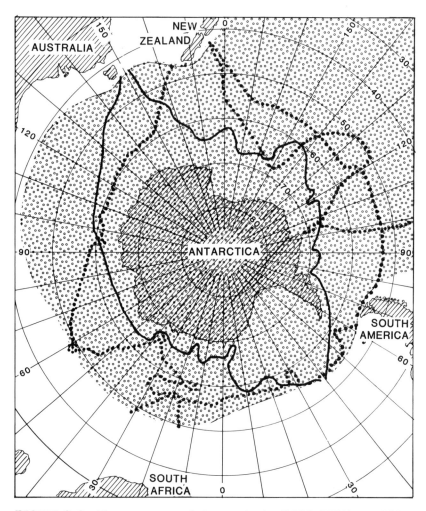

FIGURE 2-3 The voyages of James Cook (1772-1775), solid lines, and Thaddeus von Bellingshausen (1819-1821), dashed lines. The shaded background indicates the unknown area at the time Cook commenced his antarctic voyages.

as early as 1683 while on one of his buccaneering expeditions, the seals remained undisturbed for over a century. However, in seven years in the last decade of the eighteenth century more than 3 million skins were carried from Juan Fernandez to Canton in China, where a good market had been established. This slaughter, which led not surprisingly to the islands being "almost entirely abandoned by the animals," was typical of what was to come. Contemporaneously and farther south, at South Georgia, sealing peaked in 1800-1801 and by 1822 Weddell calculated that at least 1.2 million fur seals (Figure 2-4) had been taken from that island and that the species was virtually extinct there.

The last great refuge of Arctocephalus fur seals was found when the brig Williams was blown off course in 1819 and discovered the South Shetland Islands. Exploitation followed immediately on discovery, and at least 47 vessels worked the islands during the following season. James Weddell complained:

> The quantity of seals taken off these islands, by vessels from different parts, during the years 1821 and 1822, may be computed at 320,000, and the quantity of sea-elephant oil at 940 tons. This valuable animal, the fur seal, might, by a law similar to that which restrains fishermen in the size of the mesh of their nets, have been spared to render annually 100,000 furs for many years to come. This would have followed from not killing the mothers till the young were able to take the water; and even then, only those which appear to be old, together with a proportion of the males, thereby diminishing their numbers, but in slow progression.

Unfortunately, no authority existed that was competent to impose and enforce controlled sealing on a sustainable basis, and Weddell, realizing this, played his own part in hastening the destruction of the seals. On the other side of the Southern Ocean, the sub-Antarctic and other islands in the New Zealand region were divested of their seal population by about 1813.

A brief recrudescence of fur sealing took place between the 1870s and the early twentieth century, though with much smaller catches than in the earlier period. U.S. sealers were active during the same period in the Indian Ocean sector islands of Kerguelen, Crozet, and Heard.

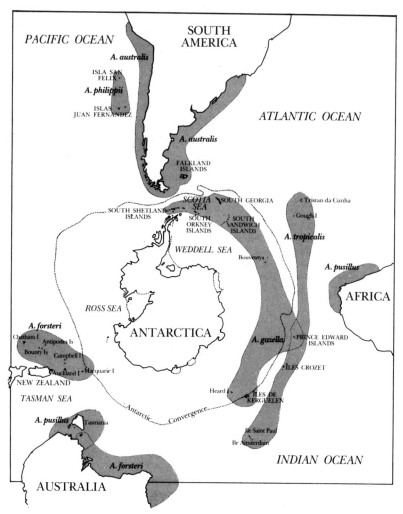

FIGURE 2-4 Distribution of <u>Arctocephalus</u> fur seals in the Southern Hemisphere. These seals are found in their greatest abundance in regions where cool, nutrient-rich waters promote high primary productivity and hence large stocks of the fish and invertebrates on which the seals feed. That these seals thrive in moderate latitudes off South America is due to cool northward-trending currents, such as the Humboldt Current off the coast of Chile (from Bonner, 1982) (Reprinted with permission).

Sealing and discovery were so interrelated in the
region now known as the Antarctic Peninsula, and documentation so sparse, that geographical priorities have
been difficult to establish, with the names of Bransfield,
Smith, Palmer, and Davis being preeminent. But the first
reported sighting of the land inside the Antarctic Circle
(Peter I. Oy) was that of Bellingshausen.

With the depletion almost to extinction of fur seals
in the sub-Antarctic islands of the Southern Ocean, the
small ships ranged more widely. With a brig of 160 tons
and a cutter of 65 tons, James Weddell made a truly
remarkable southing into the sea that now bears his
name. James Biscoe, during 1830-1832, while in command
of another brig from the same firm of Enderby Brothers,
circumnavigated the continent and sighted Enderby Land,
named after his ship's owners. He crossed the whole of
the southern Pacific in high latitudes, discovering the
Biscoe Islands and Graham Land. In 1833 another Enderby
captain, Peter Kemp, discovered Heard Island, and in 1839
Enderby Brothers made their last contribution when John
Balleny discovered the islands named after him.

Throughout all this period and almost to the present
day, the pack ice served to protect the four true
antarctic seals--the Weddell, crabeater, leopard, and
Ross seals--from exploitation. History suggests that
there is no reason to suppose, but for the Convention for
the Conservation of Antarctic Seals of 1972, that these
genera also would not have suffered similar depredation
in more recent times.

SCIENCE AND NATIONAL INTERESTS

Commercial interests declined with the proof that land
found in high latitudes did not enrich the fur seal
industry. The role of antarctic exploration was then
taken up by states, with major French, British, and
American expeditions around the continent during the
years 1838 to 1843 (Figure 2-5). There are about the
Earth two great natural phenomena--a gravitational field
and a magnetic field--both of which have preoccupied the
greatest scientists. The voyage of Halley to the southeast Pacific in 1699 derived from the belief that a
knowledge of variation in magnetic declination and
inclination would enable geographic position to be derived
(this was before the invention of chronometers). However,
by the early nineteenth century it was becoming obvious

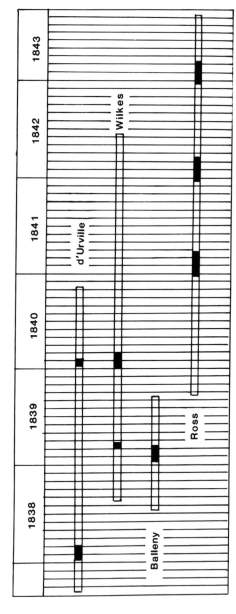

FIGURE 2-5 Synoptic diagram of the almost contemporaneous expeditions of d'Urville, Wilkes, Balleny, and Ross. The time spent south of 60°S latitude is indicated in solid blocks (from Mill, 1905) (Reprinted with permission).

that the geomagnetic field changed in intensity and direction with time, and its study became one of the focal points of contemporary research preoccupying the attention of von Humboldt and Gauss among others. Gauss' "Allgemeine Theorie des Erdmagnetismus" (1839) was one of the great seminal papers of geophysics and predicted the position of the magnetic poles.

All three national expeditions had attaining the south magnetic pole, which Gauss computed to be about 25° south of Tasmania, as one of their objectives, though perhaps in the case of the U.S. and French expeditions it was a subsidiary one. Dispatched by the French Ministry of Marine in the southern summer of 1837-1838, Dumont d'Urville, while trying unsuccessfully to penetrate the Weddell Sea, carried out a season's work near Graham Land before returning to the Antarctic south of Australia in the 1839-1840 season to find his way blocked by ice and land, which he named Terre Adelie.

Charles Wilkes commanded the U.S. Exploring Expedition, the orders of which began, "The Congress of the U.S. having in view the importance of our commerce embarked in the whale-fisheries and other adventures in the great Southern Ocean ...." Though marred from well before its outset by strife and disagreement, this expedition was nevertheless the first to see a major segment of Antarctica. The British expedition, commanded by James Clark Ross, who had considerable Arctic and magnetic survey experience, having learned of the prior journeys of d'Urville and Wilkes, determined to make its high latitudes farther east. On January 10, 1841, Ross' ships Erebus and Terror reached the open waters of a sea that allowed free penetration to a latitude of about 78°S. In November of the same year, Ross returned to that area, and though he did not greatly extend his discoveries, his farthest (south of 78.10°S latitude) was to stand for another 58 years until men grew bold enough to tramp inland across the surface of Antarctica.

The long hiatus can be attributed to a number of factors. British attention was directed to the Arctic in numerous searches for the lost Franklin expedition and, following this, to exploration of Africa. Colonialism was endemic among European powers, and the United States suffered the trauma of its civil war, subsequently concentrating on its internal economic development.

BECAUSE IT IS THERE

The end of the nineteenth century saw the beginning of the first serious exploration of the antarctic continent, conducted largely by the various European powers. Several almost contemporaneous expeditions (Figure 2-6) did startling exploration and scientific work around the offshore islands and coastal regions. The Belgian Antarctic Expedition (1897-1899), led by Adrian de Gerlache, inadvertently became the first party to winter in Antarctica when its ship <u>Belgica</u> was beset by pack ice. The overlapping British <u>Southern Cross</u> Expedition (1898-1900) under Borchgrevink was the first party to winter intentionally on the continent, at Cape Adare. Two other expeditions discovered the dangers of antarctic pack ice--the German Antarctic Expedition under Drygalski, whose ship the <u>Gauss</u> was beset, and the even less fortunate Swedish South Polar Expedition led by Nordenskjold, whose ship the <u>Antarctic</u> was crushed and lost.

But in the first decade of the twentieth century, public attention focused principally on the quest for the South Pole and on four expeditions, all of exceptional interest. Scott's first expedition produced the earliest extensive sledging on the continent and set a pattern for scientific studies, while his second reached the pole only for the party to perish on the return journey, leaving a literary legacy of polar endurance and fortitude. The astonishing ease with which Amundsen reached the pole only 12 years after Borchgrevink's men first set foot on the continent reflects, in Paul Siple's words, "a model of technical performance." But there are reasons for giving pride of place to the exploits of Ernest Shackleton, who among other things pioneered the route up the glacier near the head of which we now meet and which is named after one of his sponsors, William Beardmore. One of Shackleton's geologists, Douglas Mawson, shortly afterward established Australia as an antarctic nation, exhibiting incredible power of individual endurance during one of his journeys from Cape Denison.

From the beginning of the great Age of Discovery in the late 1400s it had taken a century to put Europeans as far as 60°S latitude. It took a further 300 years to breach the pack ice to the continent proper but only a decade to cross the last 12° when the first parties traveled by land. It is a matter of reflection that, in spite of eccentricities of civilizations and demography,

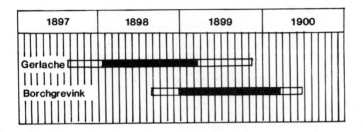
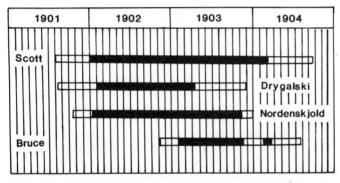

FIGURE 2-6 Synoptic diagram of the earliest wintering parties in Antarctica. The time spent south of 60°S latitude is indicated in solid black (from Mill, 1905) (Reprinted with permission).

the two poles, having otherwise geographically opposed characteristics, were reached for the first time by humans almost simultaneously. The Romantic Age of mid-nineteenth-century Europe motivated citizens to journeys with no material objectives, such as war, trade, acquisition of wealth, and religious conversion. The final technology that enabled the two poles to be reached may well have been liquid fuels and stoves portable and efficient enough to sustain life for several months.

EXPLOITATION--THE WHALES

Whaling has been the major industry by which man has drawn from the great productivity of the cold waters of the arctic and antarctic seas. The oil that is the principal product of the industry is derived from the thick coating of fat or blubber necessary, since whales are warm-blooded mammals, both for insulation and as a reserve store of food and energy during the intervals when whales have to leave for warmer, but less productive, waters to deliver their young.

Modern-style whaling dates from technological developments in the 1860s, when the Norwegian Sven Foyn introduced the harpoon gun with a grenade head and steam-driven catchers became available. Even then, antarctic whale stocks did not become important until the discovery of big packs of the giant Blue Whales by the expeditions at the end of the nineteenth century. In 1904 the first antarctic land station was opened in South Georgia, and whaling rapidly spread to other shore stations and factory ships moored in suitable harbors throughout the Falkland Islands Dependencies. Within three years, antarctic whaling produced more oil than the rest of the world's whaling areas together, and the exploitation of the antarctic regions had moved one step farther south, for with the exception of the South Georgia and South Sandwich Islands region, whales are caught in higher latitudes than those occupied by fur seals.

The British government attempted to control overfishing by issuing licenses for the factories, which were all within its territorial waters. Attempts to avoid such restrictions on catches led eventually (1925) to pelagic whaling using a mobile factory ship with a slipway up which the whales could be hauled for dismemberment and processing. The successive degradation of Blue, Fin, and Sei whale catches (and presumably stocks) since that time can be seen in Figure 2-7. Between 1925 and 1963 a total of 282,903 Blue Whales, the largest of all known animals, living or extinct, was caught. Thus the history of antarctic whaling, like sealing, seems to have been a repeated story of discovery, overexploitation, and collapse.

THE MODERN ERA

Very little exploration was done during the first decade after World War I, but when it resumed it enjoyed the advantage of many developments--including radio, aerial surveys and support, and superior diet, though land travel still depended almost exclusively on dogs until after World War II. The United States was dominant during this period, with both Richard E. Byrd and Lincoln Ellsworth making remarkable contributions. Their achievements and influence encouraged the U.S. government to inaugurate the U.S. Antarctic Service to commence permanent occupation and scientific exploration of parts of Antarctica, but World War II brought this to an untimely end. Nevertheless, the lull was brief, and the winter of 1943 was the last time Antarctica was without at least a transient population. Revenues from whaling licenses assisted the initiation in 1925 of a new era of Southern Ocean studies when the United Kingdom launched the Discovery Investigations to carry out research in the Southern Ocean in support of the whaling industry. At first confined to whales and whaling in the Falkland Islands quadrangle, the program was later expanded to include much broader studies of the Southern Ocean.

A traditional by-product of discovery has been the raising of a national flag by the expedition leader and a proclamation that vaguely describes and delineates the territories that are now the possession of that nation. Such actions can be filed away in a nation's archives and brought out, dusted, and used if an appropriate moment or need arises. Thus, from 1893 when France annexed the Isles de Kerguelen, a steady series of territorial claims has been made by nations based on previous discovery. Ironically, the majority of these claims were made during the period between the two world wars, when antarctic exploration was at a low ebb except for that of the United States, a nonclaimant nation. The analysis of these official actions to establish sovereignty is more properly the sphere of succeeding chapters of this publication.

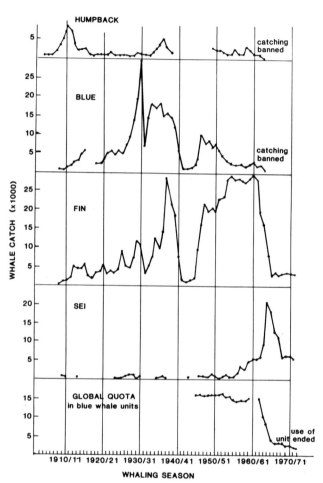

FIGURE 2-7 Baleen whale catches in the Southern Ocean, 1904-1971.

In some cases proclamation was followed by occupation. Although it is outside the Antarctic Circle, Laurie Island in the South Orkneys had been occupied on a permanent basis by Argentina since the base had been handed over to them by the withdrawing Bruce (Scottish) Expedition in 1904. Forty years later, Argentina, Great Britain, and then Chile established a series of bases on the Antarctic Peninsula in support of their national interests. After World War II, France and Australia also moved toward continuous occupation to advance their territorial claims. But scientifically, the most important expedition in the immediate postwar period was the Norwegian-British-Swedish Expedition, which demonstrated for the first time the enormous thickness of the ice cap.

THE INTERNATIONAL GEOPHYSICAL YEAR, 1957-1958

The first two Polar Years of 1882-1883 and 1932-1933 had seen virtually no activity on the southern continent. The burgeoning prosperity of the developing countries and the enhanced prestige of science resulting from World War II ensured that adequate funding would be available for the global synoptic exercise of the International Geophysical Year (IGY), including what was to be its principal showpiece, Antarctica. Technology developed for northern polar regions, including snow and ice runways, ski equipped aircraft, and tracked vehicles, allowed bases to be built and supplied in the interior of the continent. More people were on the continent during that year than had previously set foot on the place during the whole of history; even then the total winter population in the 40 bases (Figure 2-8) established on the continent and its immediate offshore islands by 12 nations* would hardly have peopled a village in many of the countries concerned. Before the IGY was completed at the end of 1958, almost all countries elected to continue occupation of their bases and the conduct of research programs, partly because of the high level of investment but also no doubt for fear that they would lose territorial, political, or strategic advantage to other states should

---

*Argentina, Australia, Belgium, Chile, France, Japan, New Zealand, Norway, South Africa, the USSR, the United Kingdom, and the United States

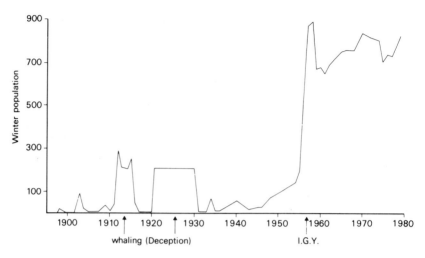

FIGURE 2-8 Total wintering population in the Antarctic south of 60°S latitude from the first overwintering expedition to 1980 (from Sugden, 1982) (Reprinted with permission).

they fail to do so. The scientists happily capitalized on this dilemma, and the scene was set for the formation of the Special Committee for Antarctic Research in 1958--formalizing for the longer term the spirit of cooperation exemplified by the scientists of all 12 nations during the IGY--and for the signing of the Antarctic Treaty in 1959. Easement of potential "international discord" had been quietly sought in diplomatic circles for more than a decade before 1959. It may not be the least of the achievements of the IGY that, by creating unprecedented widespread activity on the continent by 12 nations, both claimant and nonclaimant, it catalyzed that accommodation in the form of the treaty.

REFERENCES

Bonner, W. N. 1982. Seals and Man, Washington Sea Grant, University of Washington (Seattle), p. 170.
Mill, H. R. 1905. The Siege of the South Pole, Alston Rivers (London), p. 455.
Raisz, E. 1948. General cartography, 2nd ed., p. 11.
Sugden, D. 1982. Arctic and Antarctic. Basil Blackwell (Oxford), p. 472.
Weihaupt, J. G. 1984. Eos 65(35): 493-501.

3.

Juridical Nature of the 1959 Treaty System
Yuri M. Rybakov

December 1984 marked 25 years since the signing of one of the most significant international documents of our time--the Antarctic Treaty. Preceding that event was a long era of search, discovery, and the beginning of the exploration of that southern continent.

There had been earlier attempts at theoretically predicting Antarctica's existence and discovering it. As early as 1761 the great Russian scientist, Lomonosov, put forward a bold hypothesis, which later proved to be basically correct, about the possible existence, around the South Pole, of islands and a vast body of land mantled in thick and perennial snows. Later seafarers made numerous attempts to find the southern continent, but for a long time all of them failed.

Today, it is an indisputable fact Russian navigators discovered Antarctica and thus launched an era of scientific research and exploration of the new land. The first Russian Antarctic expedition of 1819-1821, was headed by naval officers Bellingshausen and Lazarev. It belongs in the history of great geographical discoveries, and the scientific research carried out by the expedition was also of importance. In recognition of the Russian navigators' accomplishments, one of the South Pole seas was named after Bellingshausen.

These Russian discoveries and the first scientific research carried out in the Antarctic attracted the attention of many countries to the region, which went beyond purely scientific and exploratory interests. In the early twentieth century a number of states sought a territorial division of Antarctica. Claims were made to several large areas of the continent, covering up to four fifths of its territory. Moreover, three of the claimed areas overlapped, which produced sharp tensions among the claiming states, leading to the demonstration of military force.

These territorial claims were declared unilaterally, and many states, including the United States, Japan, Belgium, the Polish People's Republic, and Brazil, did not accept them. Nor has the USSR ever recognized any territorial claims, especially because it was the Russian sailors who first discovered Antarctica. The USSR's position of principle on this issue is reflected in exchanges of notes and other documents. For example, in response to Norway's statement claiming sovereignty over Peter I island, discovered by the Russian expedition of Bellingshausen and Lasarev, the Soviet government in a note dated January 27, 1939, informed the government of Norway that it did not recognize the Norwegian claim to that island and reserved its position concerning state sovereignty over the lands discovered by Russian navigators.

The end of World War II witnessed a dramatic revival of interest in Antarctica. From July 1957, to December 1958, twelve countries participated in an extensive scientific research program carried out in the region within the framework of the International Geophysical Year (IGY). Most of the 12 states, Argentina, Australia, Belgium, Chile, France, Japan, New Zealand, Norway, South Africa, the USSR, the United Kingdom, and the United States organized expeditions to Antarctica, and together they established 40 scientific research stations. The Soviet Union established six stations, and Soviet scientists made a major contribution to antarctic science. They conducted large-scale exploratory work in the least accessible areas of the continent, in particular in the area of the geomagnetic pole. The IGY was a major global scientific endeavor, and the routine antarctic observations conducted during it continue to this day.

Broad international cooperation in exploring Antarctica within the IGY framework in the 1950s served as an impetus to the conclusion of an appropriate international legal agreement in Antarctica. The position of principle of the Soviet Union as regards the drafting of such an agreement was reflected in particular in a note sent to the U.S. Department of State on May 2, 1958. In the opinion of the USSR, a future agreement would have to be based on the principles of using Antarctica exclusively for peaceful purposes and of freedom of scientific research in the entire region. The note recalled the outstanding achievements of Russian explorers in discovering Antarctica and emphasized that the Soviet Union reserved all the rights arising from discoveries and research done by Russian navigators and scientists, including the right to make appropriate territorial claims in Antarctica.

In 1958 and 1959, preliminary talks were held in Washington, D.C., with active participation of the USSR, which produced the international Antarctic Treaty that was signed on December 1, 1959, and entered into force on June 23, 1961. The drafting and conclusion of the Antarctic Treaty were memorable international events. In its 1984 message of greetings to the members of the antarctic expeditions of the countries that had taken part in drafting the treaty, the Soviet government noted that the treaty would contribute to the further development of cooperation among states in exploring the region and could provide a good example for settling international problems in the interests of universal peace. At the 1960 Meeting of the Political Consultative Committee, the states party to the Warsaw Pact described the document as an important agreement concerning the peaceful use of Antarctica.

The Antarctic Treaty is open for accession by any state. At present, more than 30 states, big and small, situated on every continent of our planet and representing different social and economic systems, have become parties to the treaty; they include the USSR, the German Democratic Republic, Poland, Romania, the United States, Czechoslovakia, the United Kingdom, Argentina, Brazil, Norway, Peru, Papua New Guinea, France, the Federal Republic of Germany, and Japan. India and the Peoples Republic of China acceded to the treaty in 1983, and Hungary, Cuba, Finland, and Sweden did so in 1984.

PEACEFUL USE

The Soviet Union attaches great importance to the Antarctic Treaty as an international legal instrument aimed at curbing the arms race. In accordance with the treaty, a vast continent together with its neighboring islands and adjacent seas is placed totally outside the sphere of military preparations of any form whatsoever, including nuclear-weapon tests. For the first time in the history of international relations and international law, such a region has been established by treaty as a zone for peaceful research and scientific cooperation among states.

Article 1 of the treaty, elaborated with the active participation of the USSR, envisages that Antarctica will be used for peaceful purposes only. Any measures of a military nature, such as the establishment of military bases and fortifications, the carrying out of military maneuvers, and the testing of any type of weapon, are prohibited in the region.

According to the Antarctic Treaty, its provisions apply to "the area south of 60° South Latitude, including all ice shelves, but nothing in the present Treaty shall prejudice or in any way affect the rights, or the exercise of the rights, of any State under international law with regard to the high seas within that area." (Article VI)

Of particular importance today are the provisions of the treaty prohibiting any nuclear explosions and the disposal of radioactive waste material in Antarctica (Article V). Originally there were proposals to permit nuclear explosions subject to prior notification of all parties to the treaty and consultation with them. Adoption of these proposals would have been tantamount to the legislation of nuclear-weapon tests, because technically it is difficult to distinguish peaceful explosions from military blasts. Moreover, nuclear-weapon tests in Antarctica would irreparably damage its unique environment and could lead to unpredictable harmful consequences on a global scale. It was due to the tireless efforts of the Soviet Union that prohibition of any nuclear explosions, both military and peaceful, and of the disposal of nuclear waste was embodied in Article V of the treaty.

The totality of the provisions of Articles I and V, prohibiting in particular any measures of a military nature and any nuclear explosions, bestow on Antarctica a status not only of a demilitarized area of the globe but, for the first time in history, of a zone free from nuclear weapons. This, of course, does not preclude completely the possibility of using nuclear energy, such as the use of nuclear power stations.

Thus the nuclear-free zone and the zone of peace created by the Antarctic Treaty serve as good examples for concluding similar agreements with respect to other regions of our planet and around it. For instance, the Tlatelolco Treaty was signed in 1967. Under that treaty, the Latin American continent became, *de jure* and *de facto*, a zone free from nuclear weapons.

In the present-day international situation, characterized by heightened tensions and a growing threat of nuclear war with all its disastrous consequences, the de-escalation of military presence and the creation of nuclear-weapon-free zones and zones of peace in various regions of the globe is acquiring special significance. Resolutely opposing militarization of ocean expanses, the Soviet Union is in favor of the largest possible part of the world ocean becoming a zone of peace in the near future. In particular, the Soviet Union supports the proposal to create such a zone in the Indian Ocean, a vast geographical region directly adjoining Antarctica and of exceptional importance for world navigation. Turning the Indian Ocean into a zone of peace corresponds to the aspirations of a majority of states in the region, which prefer lowering the level of military activity and dismantling all foreign bases in the area.

SCIENTIFIC INVESTIGATION

The principle of freedom of scientific investigation in Antarctica embodied in Article II of the Antarctic Treaty is highly important. For more than 20 years now it has served as a basis for successfully developing fruitful international cooperation in this inclement and almost inaccessible region of the world. The Antarctic Treaty is rightfully considered to be a unique example of international cooperation among states and with international organizations.

Any state that is a party to the treaty can benefit from scientific data and information obtained by the antarctic expeditions and permanent scientific stations. Even now this information is of considerable practical value for the advancement of knowledge in various fields and for a better understanding of the phenomena and processes taking place on our planet. This undoubtedly is in the interest of all mankind.

The signing of the Antarctic Treaty became an important and effective means to prevent the occurrence of disputes, tensions, and conflicts between states in connection with previously made or potential territorial claims in the region. The 1959 treaty freezes the territorial claims (Article IV).

No less important is the treaty's role in preventing the spread to the Antarctic of crisis situations arising in regions that are in the immediate vicinity of the sixth continent. As practical experience has shown, the 1959 treaty serves as a reliable barrier against extending hostilities to Antarctica and thereby violating its status as a zone of peace. This was borne out by the situation which developed over the Falkland (Malvinas) Islands in 1982.

INSPECTION

In order to promote its objectives and ensure the observance of its provisions, the Antarctic Treaty provides for the possibility of carrying out inspections (Article VII). Each of the consultative parties can designate observers who have complete freedom of access at any time to any or all areas of Antarctica, including all stations, installations, and equipment within those areas. Ships and aircraft at points of discharging or embarking cargoes in Antarctica are also open to inspection.

CONSULTATIVE MEETINGS

In accordance with established international practice, the fulfillment of the provisions of a given international agreement as well as the control over compliance with them and the coordination of states' efforts to translate them into reality are entrusted, as a rule, to a certain body or mechanism, the creation of which is envisaged in such an agreement. Under the Antarctic Treaty, the mechanism in question is the consultative meeting of states that are parties to the treaty.

Under Article IX, the right to participate in consultative meetings belongs to the original contracting parties to the treaty as well as to the contracting parties that conduct substantial scientific research activity in Antarctica, establishing scientific stations and dispatching scientific expeditions there. The latter parties are entitled to participate in the meetings during such time as they demonstrate their interest in research in Antarctica.

After the establishment in 1977 of a permanent scientific station named "Henryk Arctowski," Poland became a consultative party. In 1981, when the station "George von Normaier" became operational, the Federal Republic of Germany also obtained this right. In September 1983, India and Brazil, which set up their scientific stations in Antarctica, acquired consultative status. Thus, at present more than half of the parties to the treaty--16 states--are consultative parties; practice demonstrates that existing procedures envisage real possibilities for any state acceding to the Antarctic Treaty to obtain consultative status.

Consultative meetings deal with matters related to governing the activities of states in Antarctica, including its use for peaceful purposes only, the undertaking to facilitate scientific research and international scientific cooperation in the process of studying the continent, the exercise of the rights of inspection there, and the preservation and conservation of living resources.

As a result of discussions at the consultative meetings, the parties adopt recommendations, which are subject to approval by all states taking part in their elaboration. Thereafter they become standard-setting provisions, building on and complementing the articles of the Antarctic Treaty.

One of the fundamental provisions of the internal regulation of consultative meetings--the Rules of Procedure--is the rule under which the adoption of recommendations requires the unanimous approval by all states participating in the meeting.

The principle of consensus as a method of elaborating and adopting decisions has proved to be effective and efficient. It is of great importance for all consultative parties and reflects the specific nature of regulating the problems of Antarctica in terms of international law. The principle of consensus used in consultative meetings actually means that no decision may be adopted unless it suits every party. This, in turn, not only creates a businesslike atmosphere during the discussion of various issues but it also guarantees the elaboration of well-balanced recommendations and decisions, which reflect the opinions and correspond to the interests of all states participating in consultative meetings.

The additions to the Rules of Procedure adopted by the twelfth consultative meeting, held in September 1983 in Canberra, are of great importance. They determine that the parties to the treaty that do not have consultative status may participate in consultative meetings as observers. Participation of observers in the meetings convened within the framework of the treaty testifies to the open nature of such meetings, the goodwill of the consultative parties and their desire to show the international community the positive nature of the recommendations worked out in the meetings for the benefit of all mankind.

RECOMMENDATIONS

Since the Antarctic Treaty's entry into force, 12 consultative meetings have been held, which have adopted more than 100 different recommendations. The recommendations relate to humans' multifaceted activities in Antarctica. In particular, they regulate questions of radio and telecommunications, tourism and nongovernmental expeditions, information exchange, use of radioisotopes, designation of specially protected areas, use of scientific research rockets, introduction of animals and plants for laboratory research, cooperation in transport, human impact on the antarctic environment, and many others.

Taking into account the unique antarctic environment and its highly sensitive and fragile fauna and flora, the third consultative meeting held in Brussels in 1964 adopted the Agreed Measures for the Conservation of Antarctic Fauna and Flora (Recommendation III-8).

Under Article VI of the Agreed Measures, it is prohibited within the Antarctic Treaty area to kill, injure, or chase any of the native mammals or birds without special permission. Appropriate measures should be adopted with a view to minimizing any harmful interference with normal conditions for mammals and birds. In this context, "harmful interference" means: allowing dogs to run free; flying helicopters or other aircraft in a manner that would unnecessarily disturb birds and seal concentrations; use of explosives or discharge of firearms (Article VII). The Agreed Measures also contain obligations regarding measures to reduce water pollution along the coasts and ice shelves, provide regulations for the introduction and keeping of animals and plants in Antarctica, and establish precautions against accidental introduction of parasites and disease into Antarctica.

Bearing in mind the increase over recent years in the number of nongovernmental expeditions to the Antarctic Treaty area and the number of accidents resulting therefrom, the consultative meetings give careful consideration to the impact of tourism and adopt appropriate recommendations. In this context, great emphasis has been placed on the advance exchange of information on planned activities.

It has traditionally been a principle of antarctic activities to render every possible assistance to antarctic expeditions in emergency circumstances. Unoccupied buildings and refuges exist there that can be used by any expedition in case of an accident. However, in such cases, the authorities that maintain the building or the refuge must be notified of the manner in which those facilities were used. A manual has been drawn up for tourists visiting Antarctica, focusing particularly on measures to preserve the unique environment of the region.

The consultative meetings have also adopted a pattern for information exchange carried out annually before November 30. The information thus exchanged includes the timing of expeditions; routes, types, and equipment of vessels, aircraft, and other means of transportation; the names and positions of bases and support stations; the names of persons in charge of those bases; the functions, numbers, and specialties of station personnel as well as the means of rescue available in case of an accident (medical and transportation services and refuges) together with other data and characteristics. The information provided also covers the intention to use radiosondes and research rockets (including geographical coordinates of the place of launching, the direction of launching, the planned maximum altitude, and the purpose and details of the research program).

Judging by practical experience, all the recommendations adopted by the consultative meetings have an important role in further regulating the antarctic activities of states. They contribute to setting up, developing, and deepening mutually beneficial cooperation in this area of our planet characterized by exacting conditions and difficulty of access on the one hand and extreme vulnerability on the other.

ADDITIONAL CONVENTIONS

In 1978, the Convention for the Conservation of Antarctic Seals (Seals Convention) was worked out within the framework of the consultative meetings. It provides international legal protection for this important species of antarctic fauna, which was almost exterminated at one time.

Further efforts by the consultative parties led them to the formulation and adoption in 1980 of the Convention on the Conservation of Antarctic Marine Living Resources (CCAMLR). Its main objective is to ensure conservation and rational use of marine living organisms to the south of the so-called antarctic convergence line. This line, sometimes referred to as the antarctic polar front, is a composite geophysical border where warm northern waters mix with cold southern waters. This accounts for their high biological productivity.

The CCAMLR is open for accession by any state. At the end of 1984, its membership included 15 states, the USSR being one of them, and the European Economic Community (EEC). Sweden, Spain, Uruguay, the Republic of Korea, and India acceded to it in 1984-1985, and Brazil has initiated the process of doing so. The main working bodies established under it are the Commission on the Conservation of Antarctic Marine Living Resources, the Scientific Committee and the secretariat. The commission's headquarters is located in the city of Hobart on the Australian island of Tasmania. Since the convention's entry into force in 1982, annual sessions of the commission and the scientific committee have been held to work out provisions concerning the commission's secretariat staff, financial regulations and rules of procedures as well as the headquarters agreement and certain other documents.

It must be noted that under Article XII of the CCAMLR, the commission's decisions on matters of substance, like the recommendations of the consultative meetings, are adopted by consensus. Using this method, the commission has already approved an interim agreement between the government of Australia and the commission regarding its privileges and immunities, the budget of the commission, a procedure for appointing an executive secretary, and the commission's financial regulations.

Though it came into force only recently, the CCAMLR has already proved to be an effective instrument of cooperation among states concerned in the conservation, rational use, and study of antarctic marine living resources. This mechanism for international cooperation in the study and rational use of marine living resources in the region, which has been established on the basis of the Antarctic Treaty, serves the interests of all humankind and is proving in practice its efficiency and reliability for the development of such cooperation in the future.

Pursuant to Recommendation XI-1 of the eleventh consultative meeting under the Antarctic Treaty, the consideration of questions related to an antarctic mineral resources regime is now under way. While it is not yet a question of beginning industrial exploitation of antarctic minerals (scientists and experts from various countries regard this as a remote possibility), it is not only feasible but also desirable to establish an appropriate international legal regime before any state considers exploiting the mineral resources in the region. A reliable barrier must be raised against any uncontrolled activity with respect to these mineral resources, and the unique antarctic environment with its dependent ecosystems must be preserved for present and future generations. An international legal regime to further these ends should not be in contradiction with the Antarctic Treaty; it should be fully based on its provisions as their logical continuation, enriching their substance and thus promoting the consolidation of this important international instrument.

The existing geophysical, geological, and geochemical methods as well as powerful drilling equipment enable modern science prospecting and exploration for practically all types of minerals. However, knowledge of Antarctica is insufficient at present to allow fully substantiated estimation of its mineral resource potential.

The question of mineral resources in Antarctica was discussed for the first time at the sixth consultative meeting held in Tokyo in 1970. Later on, a number of recommendations were adopted. The ninth consultative meeting, for example, took a decision establishing a moratorium on any activities relating to the exploitation of mineral resources of Antarctica until an appropriate international legal regime was elaborated that would adequately regulate such activities and provide proper norms to protect the unique antarctic environment and ecological systems dependent on it.

Of special significance were the decisions taken by the eleventh consultative meeting held in Buenos Aires in the summer of 1981, which adopted Recommendation XI-1. This treated the elaboration and establishment of an international legal regime for antarctic mineral resources as a substantial measure that would develop and strengthen the system of the Antarctic Treaty. Moreover, the Buenos Aires meeting recognized the necessity of convening a special consultative meeting to urgently elaborate an appropriate regime, determine its form, including the question of the advisability of setting up an international body, and establish a procedure for conducting further negotiations.

Several sessions of the special consultative meeting have already been held. They have considered questions of the regime's scope of operation, the concept of resources to be regulated by it, and the establishment of stages of future activities in the field of mineral resources, among other issues. Most delegations believe that the regime should be based on the principles and purposes of the Antarctic Treaty and take the form of an international convention. To ensure its successful functioning, an organizational coordinating mechanism should be created along the lines of bodies set up in accordance with CCAMLR.

Actively participating in the elaboration of an international legal regime for the possible development of mineral resources in Antarctica in the future, the Soviet Union seeks to establish in the field a firm international legal order that would preclude any arbitrary action prejudicing the interests of other nations of the world. The Soviet Union would like to see the regime for the development of mineral resources in Antarctica codified in a special international instrument and fully based on the principles of the Antarctic Treaty, which guarantees the use of the region for peaceful purposes only. The elaboration of an appropriate instrument would considerably strengthen the system of the Antarctic Treaty, which serves the interests of all humankind. It should be stressed in this connection that no activities relating to industrial development of Antarctic mineral resources, in accordance with the decisions of consultative meetings, can be carried out until a proper international legal regime adequately regulating such activities is established.

During the period of almost 25 years in which the Antarctic Treaty has been in operation, joint efforts have helped to make a really gigantic leap forward in the scientific studies of Antarctica. The voluminous information obtained and processed in the fields of meteorology, oceanology, physics, and atmospheric phenomena has made it possible, in particular, to better understand the essence of global climatic processes and to forecast them with greater precision. Fundamental discoveries in biology, glaciology, geography, geology, and other natural sciences have been made, which are of great significance not only for identifying laws of Antarctica's nature, but also for understanding the evolution of the planet as a whole. Annually, many states send to the sixth continent scientific research expeditions that have at their disposal up-to-date technology and sophisticated equipment. This development of broad international and mutually beneficial cooperation in Antarctica has become possible due only to the 1959 treaty, a unique international legal instrument whose effectiveness has been tested and borne out by the entire course of its history.

The task of comprehensive strengthening of the Antarctic Treaty has of late acquired particular urgency and relevance. This is explained by the objectively harmful intentions of some states to revise this important international treaty.

The USSR is resolutely opposed to any attempts aimed at revising this important international treaty, no matter what pretexts are used to justify them. Such attempts are fraught with grave negative consequences not only for the countries adjacent to the antarctic region but for humanity as a whole. First and foremost, they can damage the regime of exclusive peaceful use of Antarctica established by the treaty, which undoubtedly, would negatively affect the international situation. The result could be that Antarctica, a zone of peace and fruitful cooperation among states with different socio-economic systems, would turn into a zone of friction and dangerous international conflicts. A new dimension would be added to the acute struggle among states that assert claims to antarctic areas, the issue now frozen by the treaty. It is not the erosion of the Antarctic Treaty System but accession to it by interested states that will guarantee the future continued use of this important region of the globe for the benefit of all humankind.

Moreover, undermining the treaty would mean giving a free hand to those who, to placate their monopolies, seek to develop mineral resources in Antarctica without any prior arrangements, outside any regime. This will inevitably lead to uncontrolled and rapacious exploitation of the above-mentioned resources to the detriment of the continent's unique nature and its fauna and flora.

The significance of the treaty can hardly be overestimated, especially in the present-day tense international situation. The provisions of the treaty prohibiting any measures of a military nature in Antarctica, including nuclear explosions, make it possible to place a reliable barrier to prevent the spread of the arms race to this region.

The Soviet Union is in favor of strengthening the Antarctic Treaty in every possible way as a major international legal instrument of today, aimed at maintaining peace and security both in the Southern Hemisphere and all over the globe.

LEGAL AND POLITICAL BACKGROUND

4. Antarctica Prior to the Antarctic Treaty: A Political and Legal Perspective
    Cristian Maquieira

5. Antarctic Conflict and International Cooperation
    Francisco Orrego Vicuna

6. The Antarctic Treaty as a Conflict Resolution Mechanism
    Arthur D. Watts

7. Panel Discussion on the Legal and Political Background of the Antarctic Treaty

4.

Antarctica Prior to the Antarctic Treaty: A Political and Legal Perspective

Cristian Maquieira

I apologize for the haste with which these general comments had to be compiled, and, therefore for the lack of time for adequate research and reflection. I have the somewhat uneasy sensation of someone who attends a performance at the opera and is plucked out of the audience to sing one of the roles.

Antarctica, the "Terra Australis Incognita" of the ancient maps of the sixteenth century, has been throughout history among many other things a source of theories, fantasy, incredible heroism, commercial activity, and sometimes thinly disguised greed; of enormous scientific knowledge, conflict, and imaginative solutions. It has brought out the best in the human spirit and occasionally the worst.

Political and legal perspectives on Antarctica prior to the Antarctic Treaty--the subject of my presentation--represent an interesting combination of increasing conflict and scientific and commercial activity. This combination stems mainly from the varying decisions of governments to develop an interest in the Antarctic for political, strategic, economic, scientific, or other reasons. Their different emphases led some countries with antarctic interests to establish legitimate territorial claims of sovereignty while others did not.

The first country to establish a claim was the United Kingdom, through the letters of patent of 1908. The final delimitation of the claim took place in 1917. The United Kingdom put under its control some of the best whaling grounds in the Southern Ocean. These included the Antarctic Peninsula and the islands east of it.

The United Kingdom claim was followed by claims from France, Australia, New Zealand, Norway, Argentina, and Chile, which even though latecomers to the quest for antarctic real estate, had been engaged in commercial and other activities in the region since the nineteenth century. As is well known, the claims of Argentina, Chile, and the United Kingdom overlap. The legal bases for all these claims are a combination of discovery, occupation, contiguity, inherited rights, geological affinity, proximity, administrative acts, and also symbolic acts.

At the same time, other states actively engaged in antarctic activities refused to recognize these claims. This refusal set the stage for one type of conflict--those between claimant and nonclaimant states. The overlapping claims of Argentina, Chile, and the United Kingdom set the stage for another type. The immediate post-World War II period, and especially the cold war period between the superpowers, also had its impact on antarctic politics.

Nonrecognition of claims was continually and sometimes not very politely expressed, particularly by the United States and the USSR, every time one of the claimant countries took an initiative to enhance its claim by, among other things, offering facilities for expeditions, announcing the application of conventions in its antarctic territory, or recalling the status of its territory when a decision to overfly was announced by one of the nonclaimant states.

Inevitably, these notes by the claimant states were simply not acknowledged by the nonclaimants. Or they were responded to by the nonclaimant state in the following terms: that the claim was not recognized and that the state in question reserved all its rights in Antarctica and also reserved the right to make a claim of its own. This is particularly true with regard to the position of the United States, which throughout the period prior to the Antarctic Treaty oscillated between nonrecognition of claims and, on at least two occasions, possible assertion of a claim. At the same time, the United States was the country most active in Antarctica and had a basis for a claim as strong as that of any country.

In fact, in 1939 the United States reconsidered its longstanding position of nonrecognition of claims, and the U.S. Department of State handed President Roosevelt a study recommending that the United States seriously consider making a claim that would cover not only Marie Byrd Land but also part or all of the Antarctic Peninsula. This study and its conclusions had a particularly strong impact on the governments of Chile and Argentina and provided an incentive for them formally to define their antarctic territories, which was done the following year.

The formalization of Argentine and Chilean sovereignty in Antarctica gave new impetus to what were increasing complications in the area. Claim overlaps among Chile, Argentina, and the United Kingdom were formally established. In addition, it must be borne in mind that World War II had begun and the Axis powers had begun to use Antarctica in their war operations. Such was the case when the Axis captured three Norwegian whaling factory ships and eleven catcher ships off Queen Maud Land. The United Kingdom, with one eye on the Axis powers and another on Argentina, in order to stop the Axis from seizing control of the south side of the Drake Passage, intensified its naval activity in the Antarctic. So did Argentina.

At the same time, a not very friendly exchange of notes and some waving of flags occurred between Chile and Argentina, because their claims also overlap. Unable to solve this territorial dispute, both countries agreed in 1941 that a South American Antarctic existed and that they had exclusive rights over it.

In 1946, following the war, the United Kingdom issued the Bingham Declaration, by which the United Kingdom indicated that it planned to continue its activities with no regard to the Chilean-Argentine claims. At that time it sent warships, led by the H.M.S. Nigeria, to the area.

In 1947, there was an incident between the United Kingdom and Argentina in Deception Island, where Argentina had decided to build a base. Argentine troop ships arrived and left before the Royal Navy made its entrance. The base still stands.

Also in 1947, the President of Chile visited Antarctica and reaffirmed Chilean titles in the area. Later, in 1948, Chile and Argentina, through the Vergara-La Roza Declaration, recognized the external limits of their claims and agreed to act jointly with regard to antarctic affairs. This agreement, and increasing naval activity by the United Kingdom, fired the intention of both countries to apply the Rio Treaty of 1947 in Antarctica with regard to British activities there.

The fact that the Rio Treaty carried a U.S. reservation designed especially to avoid taking sides in Antarctica did not diminish the positions of Chile and Argentina with regard to the South American Antarctic, which had been established by the Vergara-La Roza Declaration.

Parallel to these developments, the United States in 1947 once again considered establishing a claim in Antarctica. This was the objective of Operation Highjump, the single largest expedition ever dispatched to Antarctica commanded by Admiral Byrd.

In 1948, besides the Vergara-La Roza Declaration, the United Kingdom, Argentina, and Chile signed a tripartite naval agreement, by which they decided that there was no need to send naval ships south of 60°S latitude. This agreement was renewed annually until 1956. A major consideration behind this agreement was the need to solve the problems between Argentina, the United Kingdom, and Chile in the face of common problems vis-a-vis the nonclaimants. Also in 1948, the United States decided to try to put an end to these and other problems by proposing the international administration of Antarctica by the seven claimant nations, with the United States declaring its sovereignty over the unclaimed sector. The U.S. arrangement would have produced an eight-power trusteeship under the United Nations. The objective of this proposal was not only to solve conflicts between claimants and nonclaimants and among the claimants themselves but also, and here the cold war creeps into Antarctica, to prevent a possible Soviet claim. It must be recalled that this initiative was put forward at about the same time that the Berlin blockade was put into effect.

The U.S. trusteeship scheme could not work because there were no inhabitants in what would be the territory held in trust. Moreover, if Antarctica had been placed under the United Nations, the scheme would have made Soviet Union intervention in Antarctica possible. In 1948, the Soviets had not intervened in Antarctica for almost 130 years, except to carry out whaling activities. These reasons led the United States to modify its proposal and move toward a condominium, which is another way of internationalizing the Antarctic. This idea was rejected by Argentina and Chile. The United Kingdom favored it in order to put an end to its disputes with Chile and Argentina, while the other claimants gave the plan a lukewarm reception.

The Chilean response included what is now known as the Escudero Declaration, by which Chile totally rejected the condominium scheme but at the same time suggested a modus vivendi that would welcome scientific cooperation and include a five-year moratorium on territorial claims. Portions of the declaration state that means should be found to alleviate the dangers of conflict and international incidents in Antarctica without prejudice to the individual rights of interested nations in Antarctica.

The significance of the Escudero Declaration, which had no immediate consequence, is that it contained the elements of the final solution to the claims issue found in Article IV of the Antarctic Treaty.

In the early 1950s, mainly because of the rejection of its proposals to internationalize Antarctica, the United States reconsidered establishing a claim as a means to ensure control of Antarctica by the United States and friendly nations. This concept excluded the USSR at a time of mounting cold war tensions.

India became worried by the conflict-prone situation in Antarctica in 1956 and tried to include the item on the United Nations General Assembly (UNGA) agenda for that year. Chile and Argentina, on the basis of their claims, persuaded India not to press the issue.

At the beginning of the International Geophysical Year (IGY) in 1957, heightened problems existed between claimants and nonclaimants, among claimants and between the superpowers of the cold war. Yet the exclusively scientific nature of the IGY and the opening of participation in the process to all interested states were the first steps toward intense international cooperation in Antarctica as it was later embodied in the 1959 Antarctic Treaty. The IGY revealed the possibilities for countries with conflicting interests in the region and elsewhere to work together for a common good.

In 1958, India renewed its efforts to include Antarctica as an item on the UNGA agenda, this time based on the possibility of the exploitation of resources in the area. Its initiative was withdrawn because of the advanced state of preparation for the Washington Conference on Antarctica that produced the Antarctic Treaty.

In conclusion, the period of antarctic activity prior to the Antarctic Treaty was characterized, on the one hand, by enormous scientific development and activity and, on the other hand, by serious and sustained conflicts, which could be summed up in the following way:

Claimant states will not renounce their claims;
Nonclaimant states will not recognize these claims; and
There was conflict between the superpowers.

Finally, I wish to reiterate that I have only tried to indicate in a disorganized way some of the highlights of some of the problems that antarctic countries faced throughout the century before the approval of the Antarctic Treaty. It is only a sketch and not a complete painting of the antarctic situation.

BIBLIOGRAPHY

Auburn, E. F. 1982. Antarctic Law and Politics, C.
    Hurst and Co. (London); Croom-Helm (Cansina), 316 pp.
Pinochet de la Barra, O. 1955. La Antarctica Chilean,
    Editorial del Pacifico, S. A. (Santiago), p. 62.
Quigg, P. W. 1983. A Pole Apart: The Emerging Issue of
    Antarctica, New Press, A Twentieth Century Fund
    Report, McGraw-Hill Book Company (New York), p. 299.

5.

Antarctic Conflict and International Cooperation
Francisco Orrego Vicuna

THE EARLY TRENDS TOWARD ANTARCTIC CONFLICT

Antarctica is singled out today as the paramount example of a region that has been successfully excluded from the trends toward conflict and confrontation that generally characterize the international community at large. It is often thought that the fact that Antarctica is an isolated community or the fact that it is positioned on the margin of world maritime and trade routes can explain the exceptional condition of this region, where the preservation of peace and the maintenance of international security are some of the predominant political features.

However important these facts may be, they do not fully explain the absence of conflict in this region. In this regard it might be useful to draw a comparison with the case of the Arctic, which is also a relatively isolated region of the world because of its climatic conditions and is similarly situated on the fringe of major international routes.[1] Nevertheless, this situation has certainly not prevented the Arctic from being an increasingly conflict-prone region, involving not only the major powers of the world but also various other countries that are geographically related to this northern area.

Although Antarctica is today an area where peace has been effectively preserved, this was not always so. Moreover, it can be argued that in the past the antarctic region was, proportionate to the degree of human activity that took place there, as conflictive as any other area and showing every sign of a progressive deterioration. Had these trends remained unchecked, the region could well be experiencing today a confrontational situation similar to that of the Arctic, be it in terms of boundary disputes or rights of navigation and application of the law of the sea or simply in terms of military strategy and weapons development and deployment. This comparison could apply to other regions of the world as well.

Before attempting to explain why none of this has happened in Antarctica, it might be appropriate to recall some of the basic conflicts that have affected the continent in the past, a situation that is only too often forgotten.

The conflicts that Antarctica has experienced in the past can be grouped into four major categories or types: (1) localized disputes about claims or boundaries, (2) general disputes about claims prompting broader international action, (3) conflicts about strategic uses of the area, and (4) major rivalries between political powers at the general international level.

LOCALIZED TERRITORIAL DISPUTES

The first type of conflict has involved discussions about title to territory, effective occupation, and given activities related to such territories. The Anglo-Argentine-Chilean dispute over the Antarctic Peninsula and associated islands has been one such conflict, although at given points in time in the past the dispute could be seen as escalating into some of the other types mentioned.[2] In this particular case the dispute referred not only to legal and historic title and rights but to every imaginable specific activity undertaken by any of the three countries as well. Such activities included expeditions, stations, personnel, stamps, and other issues, which at the time embittered the relations between these countries in a serious manner.

Another kind of conflict that can generally be included in this category is that related to boundary delimitation in the Antarctic. The United Kingdom and France had a dispute about the precise limits of Adelie Land that was settled by diplomatic means.[3] It is interesting to mention in this regard that at one point it was suggested that this dispute be linked to that relating to the controversy between the same countries about islands in the English Channel. Chile and Argentina had in the past

unsuccessful negotiations about delimitation in the
Antarctic Peninsula.[4]

On various occasions this type of conflict has in turn
been linked with conflicts or disputes that affect sub-
antarctic territories. The linkage between the Anglo-
Argentine dispute over the Falkland Islands and those
countries' respective claims to antarctic areas has been
quite clearly illustrated throughout diplomatic
history.[5] A similar concern emerged in relation to the
controversy between Chile and Argentina over the maritime
delimitation at the southern tip of the South American
continent, insomuch as it was feared that any settlement
in such an area might have implications for antarctic
claims. A specific clause about delinking the two issues
was contemplated, first, in relation to the Beagle
Channel arbitration and, second, in the recent agreement
on maritime delimitation between the two countries.[6]

Problems of maritime delimitation are also beginning
to emerge in Antarctica. The sub-Antarctic French island
of Kerguelen has overlapping maritime areas with the
Australian islands of Heard and Macdonald.[7] The
continental shelf of some of these islands extends into
the Antarctic Treaty area, just as the shelf of antarctic
territories and islands occasionally extends northward of
the treaty area. This poses difficult problems for the
precise definition of the area of mineral resources
regimes. Similarly, the antarctic convergence posed
difficult problems for the area of application of the
Convention on the Conservation of Antarctic Marine Living
Resources, particularly in regard to the question of the
sub-Antarctic French islands.

Generally, this type of dispute has been handled by
means of diplomatic procedures. It is to be noted,
however, that direct diplomatic negotiations have not
been very successful, except for the above-mentioned
Anglo-French settlement of the boundary of Terre Adelie.
In general, all other questions have been left pending.
The judicial approach was attempted by the United Kingdom
on one occasion, but it did not succeed.[8] The difficult
nature of territorial questions in Antarctica has indicated that only through complex international
arrangements, such as that of the Antarctic Treaty, can
the issue be satisfactorily handled without prompting
more acute kinds of crises.

GENERALIZED TERRITORIAL DISPUTES AND INTERNATIONAL IMPLICATIONS

The second type of antarctic conflict refers to disputes about claims that exceed the level of localized controversy and as a consequence lead to broader international action. One such example of a situation is when local conflict eventually escalates into a more profound and conflictive crisis, resulting in the threat or actual use of force. At that point, the countries involved undertake preparations or diplomatic actions of a more serious nature.

The Anglo-Argentine-Chilean conflict resulted in a situation of this kind in the 1950s, when the usual diplomatic protest and other peaceful means gave way to direct confrontation. Installations were reciprocally destroyed, personnel harassment took place on both sides, human relations deteriorated, and there was resort to violence. As Bertram once wrote: "It should not pass unremarked that the first and only shots so far fired in anger in Antarctica were in 1952, when Argentines attempted to allright a British party...."[9]

Naval movements of a military nature were intensified as a result of this escalation, and readiness for further action was quite apparent among the parties concerned. The overall diplomatic relationship between those countries also seriously deteriorated. It was in this environment that the United Kingdom made its attempt to take the case to the International Court of Justice, with regard both to Argentina and to Chile; but it was precisely because of the general deterioration that had taken place that the initiative was aborted. Judicial proceedings require a constructive attitude of mutual trust and cooperation among the parties concerned.

A different approach was then conceived to ease tensions. A tripartite naval declaration was signed in 1949 between Argentina, Chile, and the United Kingdom, undertaking as a measure of detente not to send warships south of 60°S latitude in order to prevent armed engagements.[10] This early form of demilitarization of the continent proved successful, and it later influenced the ideas that led to the signature of the Antarctic Treaty. However, it was only the latter instrument that really changed the political environment in Antarctica, providing a framework to highlight cooperation and minimize confrontation.

STRATEGIC USES AND DISPUTES IN ANTARCTICA

Although there has been much argument over the significance or insignificance of Antarctica in strategic terms, this discussion has been largely theoretical. The fact is that Antarctica has been used in the past for strategic purposes and the conduct of warfare, a situation that typifies the third category of antarctic dispute. In this regard, it should not be forgotten that German submarines operated in antarctic waters during World War II, inflicting heavy damage on the merchant fleets and fishing vessels of a number of countries.[11]

The German and Japanese interests in Antarctica during the war were enormously influential in the development of territorial claims to that continent. The Norwegian claim materialized at the moment when it was felt that a potential German claim had to be stopped.[12] Germany and Japan had been following U.S. policy toward Antarctica very closely, with particular regard to whether the United States was planning to make a claim of its own,[13] an idea that in fact was actively considered at the time. The USSR had occasionally looked into a similar alternative.[14]

It is also interesting to remember that the Chilean decree of 1940, which specified the limits of Chile's antarctic claim, was directly prompted by a diplomatic initiative of President Roosevelt, who was looking for additional ways to prevent a German claim or the establishment of a German base in Antarctica.[15]

It is not an exaggeration to conclude, therefore, that as a consequence of growing interest in the issue of the strategic uses of Antarctica, greater emphasis was placed on sovereign claims. Nor is it mere chance that the provisions of the Antarctic Treaty that freeze the question of sovereignty have been coupled with provisions on demilitarization and peaceful uses. The attainment of one objective necessarily requires the achievement of the other.

MAJOR-POWER RIVALRY IN ANTARCTICA

The fourth major type of conflict in Antarctica has been that arising from superpower rivalry at the general international level. The tensions and difficulties of the cold war began to express themselves in relation to Antarctica just as they became evident in the Arctic region. A number of political incidents that took place between the United States and the USSR during the International Geophysical Year were early indications of how sensitive the power relationship had also become in Antarctica.[16]

The geographical distribution of antarctic stations by the two powers was also to some extent an expression of the interest in establishing a presence throughout the continent, a policy that was not unrelated to strategic interest or to the eventual territorial claims that such powers could ultimately decide to put forward. Both the United States and the USSR actively considered in the past the policy of making territorial claims in Antarctica, and this position has been safeguarded by the Antarctic Treaty in describing the two countries as those having "a basis of claim." It is not difficult to foresee that if for any reason the Antarctic Treaty arrangements were to collapse, and the strategic interests of the major powers revived, a likely consequence might be that these potential territorial claims would be made effective, thereby introducing additional complications in the already complex antarctic scenario.

The nuclear implications that such tensions between the superpowers could have created would have been far more dangerous than the above-mentioned forms of rivalry. The possibility of conducting nuclear explosions in Antarctica had never been explicitly ruled out by either of the major powers, nor had the eventual disposal of nuclear wastes in the continent. While there were continued references to peaceful uses, it is well known that such uses have been interpreted by the major powers as being compatible with the conducting of peaceful nuclear explosions. It was only through an active diplomatic effort undertaken during the negotiation of the Antarctic Treaty that such steps in the domain of nuclear policy were specifically prohibited and remain so until this day.[17]

Just as the situation during World War II prompted diplomatic initiatives to forestall possible moves by the opposite side, so the cold war resulted in similar approaches related to the broader international arrangements sponsored by the major powers. The fact that the American sector of Antarctica was included within the geographical area covered by the Inter-American Treaty of Reciprocal Assistance of 1947 is one such example. To this extent, defense policies in relation to Antarctica became interlinked with treaty arrangements primarily designed for other continents, thereby creating complex legal and political relations.

The Antarctic Treaty came to stabilize in a meaningful way the otherwise conflictive relationship in the area between the powers mentioned. This was partly so because of the specific provisions of the treaty regulating questions such as claims, peaceful uses, and prohibitions of some activities but most importantly because of the positive environment of international cooperation that was established by means of these arrangements, this being in direct contrast with the antagonisms that had prevailed before. Major powers and other countries have learned to work together in a constructive spirit, which has provided the appropriate framework for a real process of detente. Here again the antarctic framework can provide a useful example for similar efforts in other areas of the world.

THE ANTARCTIC TREATY:
COOPERATION AS A FACTOR OF STABILIZATION

The Antarctic Treaty has made the difference between a continent of rivalry and conflict and one of peace and cooperation. As this is a simple legal instrument, with no elaborate provisions on institutions or on settlement of disputes, the question must be asked as to how it has made that significant difference, overcoming and diffusing potential conflicts of a very serious and threatening nature.

The treaty has directly tackled the key issues underlying such disputes by means of concise but effective provisions. The most difficult question of sovereignty, and the related policies of nonrecognition of claims, have been harmonized in the terms of Article IV of the treaty in a very pragmatic and realistic manner. No attempt has been made to try to resolve the problem by means of opting for one or another approach, since this would have proved to be no solution at all and would probably have led to a still more profound crisis. The formulation of this article safeguards all relevant rights while at the same time laying the groundwork for building an effective mechanism of international cooperation. It is on this basis that disputes about sovereignty have come to be effectively controlled.

A similar result has been reached in relation to other types of disputes by means of provisions on demilitarization and the prohibition of nuclear tests and nuclear dumping in Antarctica. These aspects of the treaty, taken in conjuction with those mentioned above on the issue of sovereignty, have disposed of most of the potentially serious causes of tension in the area, thereby also facilitating the gradual development of cooperation.

It should also be borne in mind that such approaches and results were not the products of a secluded deliberation among the countries concerned during the negotiation of the treaty. Since the beginning of the century, every possible idea in relation to the organization of antarctic cooperation has been put forward by authors and governments, ranging from suggestions of subjecting Antarctica to a regime of national sovereignties to those proposing various forms of internationalization, not excluding forms of common property of all nations.[18] From this point of view, the current debate at the U.N. and other proposals are not at all new. What finally emerged as the approach of the treaty was the outcome of detailed consideration of all the alternatives, retaining those that proved to be workable and feasible in the light of antarctic experience.

Another important consideration is that the treaty approach is not a rigid one; it has proved its capacity to evolve and adapt to new circumstances and requirements of antarctic cooperation. The various regimes dealing with the conservation and development of natural resources are paramount examples of this evolution, which, to the required extent, has also included greater institutional development. This evolution has also defused potential conflicts relating to such resources, again not by opting for one or another point of view but through pragmatic formulations that take all necessary interests into account. Sovereignty and cooperation have been integrated with the positive attitudes that have made these developments possible.

It is in this perspective that the consequences of undoing the working cooperation of the antarctic system need to be evaluated and considered. Should the treaty structure be affected in any significant manner, the immediate consequence would be to impede its capacity to control the underlying roots of conflict. The renewal of territorial conflicts and political power struggles and the ensuing arms race and related expenditures in the area would automatically come to the fore, bringing Antarctica in line with what is common in most other areas of the world. This indeed is an entirely unacceptable perspective, particularly for those countries that are closest geographically to the continent and for those that because of their condition as developing countries have limited resources and political influence.

The Antarctic Treaty needs to be upheld as an instrument of peace and cooperation in the area, not excluding of course the capacity of the system to evolve in relation to new requirements and the needs of the international community. The proven effectiveness of the treaty in the light of the many conflicts affecting Antarctica in the past is a fundamental reason for its continuity.

NOTES

1. See generally Sugden, D. 1982. Arctic and Antarctic. A Modern Geographical Synthesis, Basil Blackwell (Oxford), 472 pp.
2. See generally Waldock, C. H. M. 1948. Disputed sovereignty in the Falkland Islands Dependencies. BYBIL:311-353; Pinochet de la Barra, O. 1955. Chilean Sovereignty in Antarctica, Editorial del Pacifico, S.A. (Santiago), 62 pp. Puig, J. C. 1960. La Antarctica Argentina ante el Derecho, R. Depalma (Buenos Aires), 274 pp.
3. For the diplomatic correspondence between France and the United Kingdom on the question of delimitation of Terre Adelie, see Bush, W. M. 1982. Antarctica and International Law, Oceana Publications, Inc. (Dobbs Ferry, N.Y.), Vol. II, pp. 498-506.
4. Pinochet de la Barra, O. 1985. Antecedentes historicos de la politica internacional de la Chile en la Antarctica: Negociaciones Chileno-Argentinas de 1906, 1907 y 1908. In F. Orrego Vicuna and M. T. Infante, eds. Politica Antarctica de Chile.
5. Beck, P. J. 1983. Britain's Antarctic Dimension. Int. Affairs 59:429-444.
6. Chile-Argentina: Treaty of Peace and Friendship, signed on November 29, 1984; Article 15.
7. Bush, W. M. 1982. Antarctica and International Law, Oceana Publications, Inc. (Dobbs Ferry, N.Y.) Vol. II, p. 589.
8. International Court of Justice. 1955. Pleadings, Antarctica cases (United Kingdom v. Argentina; United Kingdom v. Chile).
9. Bertram, C. 1957. Arctic and Antarctic: A Prospect of the Polar Regions, p. 105.

10. See Statements of the Foreign Office of January 18, 1949. In M. M. Whiteman. 1963. Digest of International Law, U.S. Department of State (Washington, D.C.), Vol. 2, p. 1238.
11. See, for example, 1952. German raiders in the Antarctic during the second world war. Polar Rec. 6:399-403.
12. Quigg, P. W. 1983. A Pole Apart: The Emerging Issue of Antarctica, New Press, A Twentieth Century Fund Report, McGraw-Hill Book Company (New York), p. 112.
13. See generally Sullivan, W., 1957-58. Antarctica in a two-power world, Foreign Affairs 36:156.
14. Toma, P. A. 1956. Soviet attitude towards the acquisition of territorial sovereignty in the Antarctic, Am. J. Int. Law 50:611-626.
15. Gajardo Villaroel, E. 1977. Apuntes para un libro sobre la historia diplomatica del Tratado Antartico y la participacion chilena en su elaboracion, Rev. Diffusion 10:41-74.
16. See generally Bullis, H. 1973. The political legacy of the International Geophysical Year, Subcommittee on National Security Policy and Scientific Developments, Committee on Foreign Affairs, U.S. House of Representatives, November.
17. Pinochet de la Barra, O. 1985. La contribucion de Chile al Tratado Antartico. In F. Orrego Vicuna and M. T. Infante, eds. Politica Antarctica de Chile.
18. A forthcoming book on the emerging regime of Antarctic mineral resources by the author of this chapter discusses the broad range of ideas that have been proposed to organize the Antarctic regime.

6.

The Antarctic Treaty as a Conflict Resolution Mechanism
Arthur D. Watts

Any treaty may, in the course of its operation, give rise to conflicting views about its application or interpretation. The Antarctic Treaty[1] is no exception, and it makes provision for the resolution of such disputes.[2] My present focus of attention, however, is not on disputes arising during the course of the Antarctic Treaty's operation, but rather on the underlying and fundamental conflicts of rights and interests that existed before the Antarctic Treaty came into existence--namely, the differences of view regarding territorial sovereignty in Antarctica. The conclusion of the treaty largely neutralized those differences, and it is the way in which it has done so that merits some further consideration.

Before getting into the detail of the subject, however, it may be helpful to see these conflicts in a broader, non-antarctic perspective. Differences over territorial sovereignty underlie many situations of friction and conflict in today's world. Where these differences relate to places of obvious significance, such as those with sizable populations or well-developed economic activity, it is no surprise that they should give rise to serious tensions and even to armed conflict.

What is important to realize is that this can also happen in relation to less obviously significant areas and even to those that are inhospitable. Thus--to give just three examples--in the 1960s two states found themselves engaged in armed hostilities over an area subsequently described by an arbitral tribunal as a "unique geographical phenomenon," bearing "similarity to a desert in the dry season and ... to a lake in the wet season"; the tribunal noted that terms most frequently used to sum up the area were "swamp, marsh, morass, salt marsh, salt water waste, mud and sand, marsh of alluvium."[3] Not surprisingly, in much of the area, inhabitants were few. In another instance, this time in the 1970s, fighting broke out over control of an area forming part of one of the world's great deserts, characterized by sparsity of resources and only spasmodic rainfall, which allowed for only a nomadic population that even then had perpetually to travel across wide areas of the desert in order to survive.[4] Twenty years earlier, in the 1950s, it was a very different kind of deserted area, harsh with cold rather than heat, and this time in the Southern Hemisphere, where different views as to sovereignty led to gun fire by one side to secure the withdrawal of a party from the other.[5] Barrenness, isolation, and the absence of any permanently settled population are no guarantee against serious conflict when the issue concerns territorial sovereignty.

So let us now return to that barren, isolated, and unpopulated location that is Antarctica. The Rann of Kutch (the first example given) is far away, and the conflict there raised issues very different from those relevant to Antarctica. (This was also true with the western Sahara Desert, the second example). However, the third example is of more direct concern: that incident occurred at Hope Bay, which lies within the area to which the Antarctic Treaty now applies.

That incident involved just two states, with differing views as to sovereignty over the place in question. The situation with respect to Antarctica as a whole, however, is not so straightforward: it is not one in which all the interested states exercise territorial sovereignty but find themselves from time to time engaged in controversy over what may be regarded as, in effect, border disputes. That sort of situation would be relatively simple. But Antarctica is much more complex, with states adopting fundamentally irreconcilable positions. There was what has been described as "a massive dispute about sovereignty in Antarctica."[6] Of the twelve states that eventually sat down to negotiate the Antarctic Treaty, seven[7] had for many years asserted sovereignty over areas of Antarctica, although the areas over which three[8] of them did so largely overlapped one another; two other states,[9] while not having made claims to territorial sovereignty, considered themselves as having a basis for doing so, while at the same time not acknowledging the claims made by the seven states previously mentioned; and finally, the remaining three states[10] neither claimed sovereignty nor recognized such claims or bases of claims put forward by others. A final factor to be noted is that a part of Antarctica--the sector between 90°W and 150°W--was unclaimed by any state.

Yet, despite that uniquely complex background, only some half dozen years after the incident at Hope Bay took place, the two states involved, together with the other ten, were able to arrive at a most remarkable agreement. They agreed to the demilitarization and denuclearization of the whole continent. They agreed that their scientists could freely conduct scientific investigations throughout Antarctica, irrespective of assertions of territorial sovereignty by some of those states; furthermore, their agreement did not exclude the use of military personnel or equipment for scientific research, or indeed for any other peaceful purpose; they also agreed that any of them could appoint nationals to carry out inspections in all areas of Antarctica, including inspections of all stations, installations, and equipment there. In terms of traditional attitudes to the exercise of territorial sovereignty, this acceptance by the states concerned of access by other states' scientific and military personnel, and of rights of inspection by nationals of other states, was exceptional. They were just the kinds of activities that in normal circumstances would have all the makings of serious tension and possible conflict.

That this kind of agreement was possible was due to one of the central articles of the Antarctic Treaty, Article IV.[11] Without that article, the Antarctic Treaty would not have been possible. To their credit, the negotiating states got their priorities right: they were determined to secure the benefits of demilitarization, denuclearization, and scientific cooperation for which the treaty was to provide, and in Article IV they established a legal framework within which their respective positions regarding sovereignty could be adequately protected. It is worth looking a little more closely at that article, to see more clearly what it says and also what it does not say.

The article is in two parts, the first paragraph dealing with the possible implications that might flow from the Antarctic Treaty itself, while the second paragraph looks ahead to the implications that might flow from future conduct under the treaty.

By virtue of the first paragraph, nothing contained in the Antarctic Treaty is to be interpreted as a renunciation by any contracting party of previously asserted rights of or claims to territorial sovereignty in

Antarctica. This is essentially a reference to historical facts: the rights or claims that are protected by this provision are those that have been "previously asserted," and such rights are in no way renounced by the treaty. This is a critical provision, since the existence of states asserting territorial sovereignty over parts of Antarctica is a fact that must be taken into account in any workable antarctic regime.

The paragraph also provides that nothing in the treaty is to be interpreted as a renunciation or diminution by any contracting party of any <u>basis</u> of claim to territorial sovereignty in Antarctica that it may have, whether as a result of its activities or those of its nationals in Antarctica or otherwise. This provision primarily protects the positions of the United States and the USSR, both of which maintain that they have a basis for a claim to sovereignty. The treaty leaves their basis of claim as it was.

Finally, the first paragraph of Article IV establishes that nothing in the treaty is to be interpreted as prejudicing the position of any contracting party as regards its recognition or nonrecognition of any other state's right of, or claim or basis of claim, to territorial sovereignty in Antarctica. While by virtue of Article IV itself, all contracting parties have acknowledged the fact that there <u>are</u> claims to territorial sovereignty in Antarctica, and that some parties may not accept those claims, as regards the substance of those claims, the position of each contracting party is protected: those that recognize rights or claims are not prevented by the treaty from continuing to do so; those that do not recognize rights or claims do not have that nonrecognition prejudiced by the treaty. This too is a critical provision, since, just as any workable antarctic regime must take into account the fact that certain states assert territorial sovereignty in Antarctica, so too it must take into account the fact that certain other states do not recognize those assertions.

While the first paragraph of Article IV looks to the consequences that might flow from the Antarctic Treaty itself, the second is concerned with the consequences of future conduct. It provides that no acts or activities taking place while the Antarctic Treaty is in force shall constitute a basis for asserting, supporting, or denying a claim to territorial sovereignty in Antarctica or create any rights of sovereignty in Antarctica.

Two features of this provision should be noted. First, it is not in terms limited to acts or activities taking place pursuant to the Antarctic Treaty: the acts or activities in question are all-embracing. The only express requirement is that they should take place while the Antarctic Treaty is in force.

Second, the all-embracing character of the provision would appear to extend also to the place where acts or activities take place. The provision does not say "no acts or activities in Antarctica."[12]

Paragraph 2 of Article IV also has a further, most important sentence. It provides that no new claim, or enlargement of an existing claim, to territorial sovereignty in Antarctica shall be asserted while the Antarctic Treaty is in force. The broad effect of this provision is apparent. In particular, it does of course prevent those states that maintain that they have a basis of claim from transforming their basis of claim into an actual claim. It also prevents any party to the treaty from making a claim in respect of the hitherto unclaimed sector.

Some interesting questions are, however, left unresolved by this provision. How, for example, does it apply to continental shelf rights or to exclusive economic zones, or to contiguous zones for other purposes? It would seem that for states asserting sovereignty in the area, these are not precluded,[13] since the text prohibits only new or enlarged claims to "<u>territorial sovereignty in Antarctica</u>," and such rights and zones can be plausibly seen as neither involving territorial sovereignty as such (as distinct from sovereign rights or rights of jurisdiction for certain limited purposes) or any "enlargement" of a claim to sovereignty nor as being strictly "<u>in</u> Antarctica"--a term most readily seen as referring to the continent itself rather than to its surrounding maritime areas. Extensions of the breadth of the territorial sea could, however, involve different considerations: this clearly involves sovereignty itself (although there may still be a question over the word "territorial") and, by virtue of its inherent link with adjacent coastal sovereignty, is perhaps to be regarded as "in Antarctica."

Despite a few areas of uncertainty in the scope and interpretation of Article IV of the Antarctic Treaty, it is still a remarkably comprehensive provision, covering a uniquely complicated situation. What is more, it works in practice. Compared with the tensions that existed in the area in the years after World War II, how very different--to take but one of the many recent examples--has been the response of the United Kingdom to the maintenance by Chile since 1969 of a station on King George Island, which the United Kingdom regards as subject to its sovereignty but which Chile considers to be under Chilean sovereignty. In other circumstances this could have been the occasion for a serious conflict, but in practice it has been very different: on March 5, 1984, in answer to a Parliamentary question, a United Kingdom Foreign Office minister was able simply to say "Chilean activities on this island do not prejudice British sovereignty, by virtue of paragraph 2 of Article IV of the [Antarctic] Treaty."[14] Doubtless, for their part, the Chilean government would see the situation in similar terms.

It does not overstate the case to say that Article IV is the cornerstone of the Antarctic Treaty and thus of the whole system that has grown up around it.[15] The effectiveness of that article has, for something like a quarter of a century, kept Antarctica free of the conflicts to which its complex territorial situation would have been most likely to lead and generally has removed it from the usual range of international political tensions.

Yet, however satisfactory the results of Article IV have so far been, there are certain limits to its operation and effectiveness. These limits are sometimes obscured by the very success that Article IV has so far had and by the tendency to get around its complex drafting by summarizing its broad effect by some such phrase as that it "suspends sovereignty claims" in Antarctica or that it has "put sovereignty in abeyance."

What is important always to bear in mind is that the various national claims to and rights of sovereignty in Antarctica are still very much alive--as equally is the opposition to them of those states that do not recognize them. The underlying differences of view remain. In that sense, Article IV has not "solved" the problem. What it has done is to provide a basis on which conflicts arising out of those continuing differences can be avoided.

But this has been achieved only to the extent that Article IV applies. Take Article IV away, and sovereignty rights and claims, and opposition to them, will immediately re-emerge, undiminished in vigor. In an extreme case, involving in some way the Antarctic Treaty or at least Article IV ceasing to be in force, the consequential possibility of a resurgence of conflicts over sovereignty is readily apparent.

Short of that extreme situation, even today the possibilities of conflict once we go beyond the scope of Article IV are real. Being a treaty provision, that article is binding only on states that are parties to the Antarctic Treaty. Nonparties that might wish to be active in Antarctica would not, therefore, be bound by, or have any right to benefit from, Article IV. Since Article IV would not deprive nonparties' activities of possible prejudice to the position of the states asserting sovereignty in Antarctica, the latter states might be expected to react in the usual way to actions that they might see as inimical to their sovereignty.

So Article IV has clearly not buried the sovereignty issue for all purposes. What is less often appreciated is that Article IV does not prevent contracting parties themselves from asserting their sovereignty rights.[16] They in fact do so from time to time, as when they exercise their legislative, curial, or administrative jurisdiction in respect of their antarctic territories. No breach of Article IV results from such actions. This, however, brings us closer to the political heart of the matter. While there would be no breach of a legal obligation if one of the territorial states were to assert its sovereignty against another contracting party, this is not to say that there would be no consequences for it were it to do so in circumstances of any significance: in particular, the other party concerned (at least if not itself a territorial state) might be expected to assert its objections to the territorial state's sovereignty. In that event, conflict could rapidly develop.

If, accordingly, even within the framework of Article IV, conflicts over sovereignty are still possible, what has become of that article's effectiveness as a means of resolving conflicts over territorial sovereignty? Once again, the success in practice of Article IV must not be allowed to obscure its essential limitations: while, in what would otherwise be a potentially serious situation, it takes away the _need_ for the kinds of responses in protection of a state's interests that lead to conflict, the _possibility_ of conflict remains.

What has so far prevented conflicts from arising has been the appreciation by the contracting parties that it is not in their interests to push their various points of view to their logical conclusions. If any contracting party were to do so, it would have to accept the consequences, which would involve other contracting parties' similarly taking their opposing points of view to their logical conclusions. These judgments by the contracting parties are essentially political. Article IV provides a legal framework within which it is at least possible for the contracting parties to have the opportunity of not needing to press their views to their limits. But Article IV itself does not legally avoid the possibility of conflict. The contracting parties have so far avoided that possibility because of politically motivated self-restraint.

In that, they have doubtless had occasion to reflect on certain broad trends in Antarctica in this and the preceding century. As activity grew in the nineteenth century, the situation tended to develop into one of an increasing free-for-all, with all its consequent scope for conflict. Around the turn of the twentieth century this potentially dangerous situation was brought under control by the development by certain states of their respective national jurisdictions, based on their territorial claims, as an effective means of regulating activity in Antarctica. This, however, in turn tended to lead to a degree of international dispute and discord. With the conclusion of the Antarctic Treaty, all could breathe more freely again--within the framework of Article IV of the Antarctic Treaty and on the basis of their mutually valuable self-restraint.

In many respects this self-restraint has been made relatively easy in the last quarter of a century because the range of activities in Antarctica has been limited, and occasions when states will have consciously had to consider whether a policy of self-restraint was justified will have been few. The serious question that arises as activity in Antarctica seems likely to increase is whether that policy of self-restraint will still be seen by each contracting party as the right policy to adopt. Whether this occurs will depend on many factors, some internal to the Antarctic Treaty, some external. Article IV itself will not determine the outcome. Its strength has lain not in solving the underlying differences, nor in precluding their resurrection, but rather in offering an unparalleled opportunity for states, despite fundamental differences as regards so politically sensitive a matter as territorial sovereignty, to cooperate and to coexist in Antarctica. For a quarter of a century the states directly concerned in Antarctica have made full use of that opportunity and have preserved Antarctica as a continent used exclusively for peaceful purposes and free from conflict. Article IV offers them the opportunity--perhaps the only opportunity--to continue on that course.

NOTES

1. For text, see 402 United Nations Treaty Series 71; United Kingdom Treaty Series No. 97 (1961); 12 U.S. Treaties 794. The treaty was signed at Washington on December 1, 1959, and entered into force on June 23, 1961.
2. Article XI.
3. Seventeen United Nations Reports of International Arbitral Awards 1, at pp. 24, 527.
4. I.C.J. Rep., 1985, 41-2.
5. See United Nations Secretary-General's Study on the Question of Antarctica, U.N. Document Number A/39/583 (1984) paragraph 42 (although that paragraph incorrectly puts the place in question--Hope Bay--in the South Orkney Islands: it is on the Antarctic Peninsula). See paragraphs 25-27 generally for a summary of various other incidents occasioning tension or conflict in the Antarctic between 1945 and the conclusion of the Antarctic Treaty in 1959.
6. See Chapter 19 in this volume by C. D. Beeby "The Antarctic Treaty System as a Resource Management Mechanism--Nonliving Resources."
7. Argentina, Australia, Chile, France, New Zealand, Norway, and the United Kingdom.
8. Argentina, Chile, and the United Kingdom.
9. The USSR and the United States.
10. Belgium, Japan, and South Africa.
11. Article IV is as follows:

    1. Nothing contained in the present Treaty shall be interpreted as:
    (a) a renunciation by any Contracting Party of previously asserted rights of or claims to territorial sovereignty in Antarctica;
    (b) a renunciation or diminution by any Contracting Party of any basis of claim to territorial sovereignty in Antarctica which it may have whether as a result of its activities or those of its nationals in Antarctica, or otherwise;

(c)   prejudicing the position of any Contracting
      Party as regards its recognition or
      nonrecognition of any other State's right of
      or claim or basis of claim to territorial
      sovereignty in Antarctica.

2. No acts or activities taking place while the
present Treaty is in force shall constitute a basis
for asserting, supporting or denying a claim to
territorial sovereignty in Antarctica or create any
rights of sovereignty in Antarctica. No new claim,
or enlargement of an existing claim, to territorial
sovereignty in Antarctica shall be asserted while
the present treaty is in force.

As to the "gentlemen's agreement" that operated
during the International Geophysical Year
(1957-1958) and that was a forerunner of Article IV
of the Antarctic Treaty, see F. M. Auburn, <u>Antarctic
Law and Politics</u>, C. Hurst and Co. (London),
Croom-Helm (Canberra), 1982, pp. 89-93.

12. There may, however, be some uncertainty in this
respect. It could be argued that in the context of
the Antarctic Treaty as a whole, the acts or
activities in question are implicitly limited to
those taking place in Antarctica. Furthermore, the
provisions of Article VI of the treaty (which
provides that the treaty applies to the area south
of 60°S latitude) might suggest that the geographical scope of paragraph 2 of Article IV is
similarly limited. A consequence of this
uncertainty has been that, whereas the existence of
Article IV has effectively put a stop to the
exchanges of diplomatic protest notes in respect of
things being done by contracting parties in
Antarctica, it has not completely put a stop to the
sort of "paper war" that takes place, for example,
within the framework of international organizations
in which statements and counterstatements are made
outside the Antarctic Treaty area [see, e.g., with
respect to an exchange in relation to the Universal
Postal Union in 1979, British Year Book Int. Law,
53:425-426 (1982); and in relation to the
International Telecommunications Union in 1981,
British Year Book Int. Law, 53:458-459 (1982).]

13. See, e.g., the proclamation establishing an Australian 200-mile fishing zone: Bush, <u>Antarctica and International Law</u>, Oceana Publications, Inc. (Dobbs Ferry, N.Y.) 1982, Vol. 2, pp. 202-204, 208-209; as to France, see pp. 586-588.
14. Parliamentary Debates (Commons), March 5, 1984, written answers, col. 413. See also ibid., December 13, 1984, col. 599, regarding the use by Chile of the former British station on Adelaide Island.
15. The two separate antarctic conventions concluded since 1959 concern the Convention for the Conservation of Antarctic Seals 1972 [United Kingdom Treaty Series No. 11 (1973)] and the Convention on the Conservation of Antarctic Marine Living Resources 1980 [United Kingdom Treaty Series No. 48 (1982)]. Both include provisions to secure the application in those contexts of Article IV of the Antarctic Treaty: see, respectively, Articles I and IV of those conventions. As to the current negotiations for an antarctic minerals regime, see paragraph 5 of Recommendation XI-I adopted at the Antarctic Treaty consultative meeting in Buenos Aires in 1981, requiring that the minerals regime safeguard the principles of Article IV of the Antarctic Treaty: see also Chapter 19 in this volume by C. D. Beeby.

    It may also be noted that Article IV of the Antarctic Treaty served as an obvious model for a provision in the United Kingdom/Argentina exchange of notes of August 5, 1971 [United Kingdom Treaty Series No. 64 (1972)], which provided a "sovereignty umbrella" for bilateral talks between those two states about the Falkland Islands, whereby neither state's legal position concerning sovereignty over the Falkland Islands was prejudiced by the talks.
16. Other articles of the Antarctic Treaty may, however, do so. Thus, Article V prevents a state from exercising in respect of its antarctic territory what would otherwise be its right to use it for the disposal of nuclear waste; Article VIII imposes restrictions on the territorial state's rights of jurisdiction over observers and exchanged scientific personnel.

7.

Panel Discussion on the Legal and Political Background of the Antarctic Treaty

The panel consisted of François Renouard (moderator), Rudiger Wolfrum, and Felipe Macedo de Soares Guimaraes.

SUMMARY

The papers that form the chapters of this section provided historical background on the conflicts affecting Antarctica before the 1959 Antarctic Treaty and on the continuing potential for dispute and conflict in the area. They discussed the origins of the treaty's modus vivendi, which, by neutralizing contracting parties' differing positions on the territorial status of Antarctica, preserved the demilitarization and denuclearization of Antarctica and international cooperation in scientific research there.

Arthur Watts' paper focused on Article IV as the cornerstone of the Antarctic Treaty, and many participants agreed that the treaty's objectives would not have been achieved absent the spirit of compromise and cooperation exhibited by the contracting parties in dealing with their respective positions on territorial sovereignty.

Watts' final question led to a discussion of the future evolution of the Antarctic Treaty System (ATS) in relation to new requirements and the interests and needs of the international community as a whole: In the face of increasing activity in Antarctica, will the states directly concerned in Antarctica continue to find it the right policy to exercise self-restraint and avail themselves of the opportunity presented by Article IV to cooperate in Antarctica? This question was explored both in the context of the existing ATS and in relation to a more radical restructuring of the system of governance for Antarctica.

In the context of the existing ATS, panel participants explored the implications of Article IV with respect to claims to offshore zones of sovereignty and jurisdiction.

(See below.) They also discussed the fact that Article IV does not resolve the territorial status of Antarctica and that the claims are still alive, as is the continuing possibility of conflict over territorial sovereignty.

At the same time, there was some debate over the fundamental legitimacy of the ATS and the possibility of revising or modifying the Antarctic Treaty after 1991. (The Antarctic Treaty provides in Article XII that at any time after the treaty has been in force for 30 years, a consultative party to the treaty may request a conference of all contracting parties to review the operation of the treaty and possibly to amend it. The Antarctic Treaty entered into force on June 23, 1961.) Several participants suggested that Antarctica be "internationalized" in the sense of developing a new agreement to be negotiated by all states or through the U.N. They argued that Antarctica should be internationalized in order to preserve the benefits achieved by the Antarctic Treaty to date and to extend them to the wider international community, taking for granted that internationalization would be compatible with preserving the benefits of the present system.

The opposing view held that Antarctica is already internationalized in the sense that the ATS is an open one. Those holding this view argued that it would be impossible to renegotiate the disarmament and other provisions of the Antarctic Treaty in today's world. An attempt to do so could disrupt the compromise enshrined in Article IV, because it would not take account of claimant states' positions; this in turn could make it impossible to carry out the purposes of the Antarctic Treaty in securing peace and stability in the area and international cooperation in scientific research.

Many participants believed that the Antarctic Treaty had worked well so far; some indicated that they would be willing to consider realistic proposals for improvement and encouraged their presentation.

REMARKS BY RUDIGER WOLFRUM

Wolfrum addressed the legitimacy of the ATS by stating his view that the international community has acquiesced to the Antarctic Treaty because for 25 years it has accepted the treaty as a valid vehicle for the preservation of peace in the area. He believed that acquiescence has taken place even though countries not party to the treaty might have accepted the treaty as one applicable only to states' parties.

Others challenged Wolfrum's contention that the Antarctic Treaty had been accepted in the past and noted that since the claims were disputed even among parties to the treaty, it would be presumptuous to say that non-participants in the treaty had acquiesced to them.

Wolfrum also took the view that the claimant states have always had continental shelf rights because these are inherent in the continental claim of a coastal state. But he distinguished a shelf claim from an exclusive economic zone (eez) claim, stating that an eez claim must be declared if a state is to exercise the rights related to it; it is not inherent in the continental claim. Nevertheless, he argued that an eez declaration would not enlarge a claim to territorial __sovereignty__ because an eez claim represents no more than a series of __jurisdictions.__

Another speaker questioned these conclusions, arguing that because continental shelf and eez rights flow from territorial sovereignty on land, they do involve sovereignty and could be construed as an enlargement of a claim to sovereignty under the Article IV(2) stipulation that "no new claim, or enlargement of an existing claim, to territorial sovereignty in Antarctica shall be asserted while the present treaty is in force."

This provoked a wider exchange of views, with some participants clearly regarding any such claims as a violation of Article IV, while others were of the opinion that Article IV does not preclude states with long-standing claims to Antarctica from now asserting their interests in the nature of eez or continental shelf rights.

One participant agreed with the principle that all offshore jurisdictions derive from the continental claim but added that states not recognizing land jurisdiction obviously do not recognize maritime jurisdictions either. He also noted that some countries had claimed offshore jurisdictions before the conclusion of the Antarctic Treaty. (Chile claimed offshore a zone in 1947 and Argentina 1946 decreed its claim extending from the South Pole to 60°S latitude.)

Several speakers acknowledged Watts' point that the extension of a territorial sea claim, since it is in fact an extension of __territorial sovereignty,__ could be seen as an enlargement of an existing claim within the terms of Article IV quoted above. (The 1982 U.N. Convention on

the Law of the Sea permits coastal states to claim
territorial seas up to 12 nautical miles in breadth.
Before agreement on this convention, there was no
universally recognized breadth for the territorial sea.)

On a final point, Wolfrum addressed the effect of
Article IV on states not party to the Antarctic Treaty.
He noted that claims in Antarctica and Article IV operate
on two levels: on one level between claimant and
nonclaimant states that are party to the Antarctic Treaty
and on another level between claimant states and states
that are not party to the Atlantic Treaty. He reiterated
his view that nonparties to the treaty have in effect
acquiesced to the treaty regime.

This point was further developed by another participant, who noted that Article IV(2) prohibits "acts or
activities taking place while the present treaty is in
force... [from] constitut[ing] a basis for asserting,
supporting or denying a claim to territorial sovereignty
in Antarctica or creat[ing] any rights of sovereignty in
Antarctica" <u>with respect to states parties to the
treaty</u>. For those not party to the treaty, activities
that take place while the Antarctic Treaty is in force
can be regarded by the claimant state carrying them out
as enforcing its claim vis-a-vis countries that do not
object.

REMARKS BY FELIPE MACEDO DE SOARES GUIMARAES

Soares stressed the value of Article IV in the context of
the internationalization of Antarctica, since Article IV
in his view not only internationalizes Antarctica but
effectively puts to one side the alternative: the
nationalization of Antarctica. He pointed out that this
Article preserves the opportunity for all countries,
including Brazil, to express an interest in the whole of
Antarctica, not just a slice of it. He believed that,
because Antarctica is already internationalized, those
proposing this option are in effect proposing a kind of
ultranationalization of it.

He also noted the special characteristic of the
Antarctic Treaty that identifies a category of states,
the United States and the USSR, which have a "basis for
claim" in Antarctica as a result of historic involvement
in the area. Contrary to usual practice, whereby a state
either has a claim or does not, this provision maintains
U.S./USSR interest in all Antarctica. He stated that

likewise Brazil, even without a basis for claim enshrined in the treaty, has an interest in conducting scientific research and other activities in all of Antarctica.

One participant doubted that the United States and the USSR, which today refuse to accept the justifications for claims expounded by the claimant states, could in fact argue at some future date that some of these same justifications could underpin their own claims. He noted that the possibility of this happening presumes the abrogation of the Antarctic Treaty as now construed. Another speaker responded that if the Antarctic Treaty no longer applied, circumstances would revert to the status quo ante, and both claimant states and those with strong arguments for a basis for claim would be free to pursue these positions.

Finally, Soares expressed concerns about the relationship of the continental shelf in the Antarctic Treaty area to the deep seabed beyond the shelf and about the possibility of conflict between the authority of the ATS and that of the International Seabed Authority (ISA) provided for in the 1982 Law of the Sea (LOS) Convention. (According to the LOS Convention, mineral activities in the seabed and ocean floor beyond the limits of national jurisdiction are governed by the ISA in accordance with the convention.)

DISCUSSION

In discussing the evolution of the ATS, participants focused on the role of consensus in decision making under the Antarctic Treaty and the position of the nonconsultative parties to the treaty. The minerals regime negotiations were also considered.

Consensus

That consensus might become more difficult to obtain as more countries acquired decision-making rights under the Antarctic Treaty was acknowledged, but in general participants supported consensus as a workable procedure in multilateral and international forums. One participant noted that it is better than alternative voting systems because it does not create a disgruntled minority that feels discriminated against.

Nonconsultative Parties to the Antarctic Treaty

Three points were raised about the role of NCPs under the Antarctic Treaty as follows.

First, the decision that an NCP has met the requirements of Article IX(2) of the treaty and is entitled to appoint representatives to participate with full decision-making rights in meetings of the consultative parties is taken by consensus. [Article IX(2) permits CP status for states other than original signatories to the treaty "during such time as that contracting party demonstrates its interest in Antarctica by conducting substantial scientific research activity there, such as the establishment of a scientific station or the despatch of a scientific expedition."]

Second, any state that is a member of the U.N. may accede to the Antarctic Treaty, and no decision by the states party to the treaty is called for. Nevertheless, there is an anachronism in the 1959 Antarctic Treaty, which applies special circumstances to states that are not members of the U.N. These states may accede only if invited with the consent of all the CPs. Today this circumstance applies to countries such as the Democratic Peoples Republic of Korea and the Republic of Korea, and Switzerland; it applied to the German Democratic Republic and the Federal Republic of Germany before they became U.N. members in 1973.

Third, if several countries joined forces to undertake substantial scientific research in Antarctica and then individually sought CP status, the determination that each state had individually met the criterion in Article IX(2) for CP status would have to be made by each of the CPs. Whether these states might jointly exercise CP decision-making rights was not specifically addressed, although the mechanics of doing so in a consensus decision-making process might prove difficult. (The possibility of jointly conducted activities in Antarctica is considered further in Chapters 12, 14, 21, and 25.)

Minerals Regime Negotiations

The bearing of the differing positions on antarctic claims on the minerals regime negotiations was discussed. One participant noted that whether the provisions of Article IV(2) apply to continental shelf rights or to eezs has implications for the minerals regime negotiations. Another added that because there is nothing in the treaty relating to minerals, it cannot be assumed that countries not party to the treaty have acquiesced to the jurisdiction of a small group of states over this issue. (This subject is considered in more depth in Chapters 20 and 27.)

The possibility that the unresolved status of territorial claims could produce conflicts in the future between claimant and nonclaimant states was also raised in this context because the minerals negotiations deal for the first time with questions directly related to territorial sovereignty in Antarctica. These include issues such as title to minerals; responsibility and liability; and authority to issue permits, receive fees and royalties, and enforce regulations. Other speakers countered that the CCAMLR negotiations had already dealt successfully with similar issues of territorial sovereignty in Antarctica.

Participants concluded that because the minerals negotiations are complicated in addition by issues of East/West and North/South relations, and by the imperative of protecting the antarctic environment, it will take much goodwill, imagination, and compromise to achieve a workable and satisfactory result.

Compliance with the Antarctic Treaty System

In response to a question whether there had been any incidence of problems or possible violations of the Antarctic Treaty stipulation that "Antarctica shall be used for peaceful purposes only," it was pointed out that none of the inspections conducted under Article VII of the treaty has turned up any violations. Moreover, there have been no incidents of this sort, even during the period of United Kingdom/Argentine dispute over the Falklands/Malvinas Islands.

One participant questioned a point in Watts' paper, that there would be no breach of Article IV if one of the claimant states should exercise its sovereignty rights or claims while the treaty is in force. The questioner drew attention to Article VIII, which specifies that each contracting party exercise sole jurisdiction over its nationals designated to carry out inspections and those conducting scientific research and their staffs. He asked whether the right of a claimant state to exercise jurisdiction in such situations would not be contrary to its obligations under the treaty. Watts pointed out that Article VIII imposes restrictions on the territorial states' rights of jurisdiction over observers and exchanged scientific personnel; because such exercise of jurisdiction is expressly provided for in the treaty, it does not involve any derogation from sovereign rights. He added, however, that in relation to a number of subjects not expressly covered in the treaty, actions by states could cause controversy over the exercise of sovereignty in Antarctica.

ANTARCTIC SCIENCE

8. Summary of Science in Antarctica Prior to and Including the International Geophysical Year
    Robert H. Rutford

9. The Antarctic Treaty as a Scientific Mechanism (Post-IGY)—Contributions of Antarctic Scientific Research
    William F. Budd

10. The Antarctic Treaty as a Scientific Mechanism—The Scientific Committee on Antarctic Research and the Antarctic Treaty System
    James H. Zumberge

11. The Role of Science in the Antarctic Treaty System
    E. Fred Roots

12. Panel Discussion on Antarctic Science

8.

Summary of Science in Antarctica Prior to and Including the International Geophysical Year

Robert H. Rutford

A summary of antarctic science up to and including the International Geophysical Year (1957-1958) is not an easy undertaking. It is difficult to separate geographic exploration from science, and it is clear that for most of the history of antarctic science the two have been closely tied. This review is based on summaries that have been prepared for other purposes, and includes liberal use of those documents listed at the end of this article.

The legendary southern continent, now known as Antarctica, has long intrigued geographers and cartographers. Pythagoras in 600 B.C. postulated a spherical world and convinced his students that large land masses would be found in the Southern Hemisphere to balance those already known in the Northern Hemisphere. In the second century A.D., Ptolemy drew a map with a huge southern land mass called "terra incognita": a name that is appropriate even today. Ptolemy linked southern Africa with the Malay Peninsula, making the Indian Ocean a closed basin, and it was not until the end of the fifteenth century when European sailors reported rounding the Cape of Good Hope and sailing into the Indian Ocean that this concept was proven wrong. Magellan's transit from the Atlantic into the Pacific through the Straits of Magellan in the early part of the sixteenth century lead to the conclusion that the land to the south, Tierra del Fuego, was part of the great southern continent, Terra Australis. By 1531 a southern continent was shown on a map drafted by Orantius, a map copied by Mercator in 1538. There is shocking similarity between the shape of the postulative land mass and the actual land mass as determined several centuries later.

In the late fifteenth century, Sir Francis Drake was blown well south of Tierra del Fuego and Cape Horn on his voyage from the Atlantic to the Pacific, thus proving that if a land mass did exist to the south it was not connected to Tierra del Fuego and must lie much farther south than had been postulated.

Continued exploration and attempts to establish trade routes led to additional voyages in the Drake Passage area. It has been suggested that South Georgia was sighted by a British merchant vessel captained by de la Roche in 1675. As a result of exploration in the Pacific and the discovery of New Zealand, it was suggested by some that if a southern continent did exist, it was not the cold inhospitable continent we know it to be today.

In 1739 the French Captain Bouvet discovered the island that now bears his name. He continued to sail south, traversed along the edge of the pack ice, reported large icebergs, but saw no land mass. His travels gave the first real impression of the nature of the land mass if in fact one did exist.

The circumpolar navigation by Captain James Cook during the period 1772-1775 established the fact that there was no land connection to the legendary southern continent. Cook, although he apparently never sighted land, crossed the Antarctic Circle three times, charted the north coast of South Georgia, and landed at Possession Bay. His ship penetrated to 71°10' in the Bellingshausen Sea. He reported the rich seal fauna present, and a century of intense sealing followed. Cook had with him astronomical consultants from the Greenwich Observatory, who in addition to assisting with navigation, made seawater temperature observations, noting a warm layer of water below the cold surface layer.

In the early 1800s, exploration in the Southern Ocean was largely related to exploitation of the seal resource. The South Shetland Islands were discovered by Captain William Smith in 1819, and apparently the first seals were taken from those islands. The first overwinter sealing party wintered over on King George Island during 1820-1821. U.S. sealers discovered the Biscoe Islands in the same time period, and in 1821, Captain Palmer (U.S.) and Powell (U.K.) discovered and charted the South Orkney Islands. Powell measured water temperature, again noting the warmer water below the cold surface water.

Captain Thaddeus Bellingshausen, leading a Russian exploration team during the period 1819-1821, discovered Peter I Island and Alexander I Island and charted the

South Shetland Islands, South Sandwich Islands, and the south coast of South Georgia. Bellingshausen towed nets behind his ships, noting that there was a difference between daylight and darkness trawl catches; he postulated a vertical migration of the animals to avoid daylight. Deacon notes that "he may have been the first to mention Euphasia superba," krill, the staple food of the whales and penguins.

It was during this period in the early 1820s that the antarctic continent itself was apparently first sighted. There is some debate as to whether Palmer from the United States, Bransfield from the United Kingdom, or Bellingshausen from Russia should be recognized as the "discoverer." Incomplete records and less than accurate navigational equipment make it difficult to resolve this matter. It is clear that another captain, John Davis of the United States, first entered the words in his log, "I think this Southern Land to be a Continent." Davis on this voyage landed members of his crew at Hughes Bay.

Captain James Weddell, sponsored by the Enderby Brothers, made the deepest penetration south into the sea that now bears his name, reaching 74°15'S in February 1823. Strong winds prevented him from reaching the edge of the Filchner Ice Shelf. Weddell reported on magnetic variations, ice movements, winds, and ocean currents. Captain Brisbane, whose ship sailed with Weddell, worked in the South Orkney Islands and made rough charts of their southern coasts.

In 1829-1831, Captain Foster, sponsored by the British government, made magnetic observations and measured gravity by pendulum measurements at Deception Island. John Biscoe, also from the United Kingdom, made a circumnavigation of the continent. Sailing eastward from the Falklands, he discovered land at Cape Ann in Enderby Land. After wintering at Hobart, Tasmania, Biscoe continued eastward, discovering Adelaide Island and landing on Anvers Island before returning to the Falklands.

James Eights, a naturalist from Albany, New York, accompanied a U.S. expedition led by Captain Benjamin Pendleton to the peninsula and then on a cruise to the west. Following the voyage, Eights wrote a report on the natural history of the South Shetland Islands, making excellent observations on geology, fauna, and flora. He is credited with finding the first fossils in Antarctica, with discovering the ten-legged marine spiders called Decolopoda, and with deducing from his observations of icebergs on the western cruise that there must be a large land mass close by. Eights may well have been the first trained scientist to visit the Antarctic.

In the late 1830s and early 1840s a flurry of scientific activity occurred in the southern polar regions. Captain Dumont d'Urville, sponsored by the French Ministry of Marine, set out to sail farther south than Weddell. He was forced by unfavorable ice conditions to turn back and worked in the South Shetlands and Antarctic Peninsula before sailing into the Pacific. Following a year in the Pacific, he sailed south from Tasmania in a futile attempt to reach the south magnetic pole. In the process, however, he discovered the Adelie and Clarie coasts and noted that the magnetic pole lay inland from the coast.

The U.S. Exploring Expedition in 1838-1842 was led by Lieutenant Charles Wilkes, who was sponsored by the U.S. Navy with a Congressional Appropriation. Wilkes' vessels explored both sides of the Antarctic Peninsula, then headed westward in the Pacific. Later, they returned to antarctic waters and sailed westward along the Wilkes Coast for over 1500 miles. It was on this voyage that Wilkes established the continental dimensions of Antarctica, and thus provided evidence that Antarctica was, indeed, a continent.

In 1839-1843, an expedition led by James Clark Ross and sponsored by the British Admiralty and the Royal Society ventured south in an attempt to reach the south magnetic pole. Ross penetrated deep into the Ross Sea, discovered the Ross Ice Shelf, Victoria Land, and eventually sailed eastward to the Falklands, and then south in an attempt to penetrate the Weddell Sea.

The three expeditions led by Wilkes, Ross, and d'Urville all had planned scientific programs, and while attempts to reach the south magnetic pole failed, they did bring back a great deal of information about magnetic variations, winds, currents, water temperatures, and fauna and flora of the areas they visited. Ross made numerous biological collections but had problems with preservation. Deacon notes that following the voyage, 234 species were described, of which 145 were new. Wilkes also brought back a great deal of biological materials and despite better preservation, it appears that little was published on these materials. One monograph that was published was written by James Dwight Dana, who was apparently the first to describe <u>Euphasis superba</u> in his excellent work on crustaceans.

In 1841-1842, Captain William Smyley, a sealer from

the United States, recovered a minimum thermometer on
Deception Island that had been left there by Foster in
1829. The temperature of -5°F was the lowest reported
temperature in Antarctica until 1898.

During the 1850s Captain John Heard discovered the
Heard Islands, Captain John McDonald discovered the
island that now bears his name, and in 1857 a group of
U.S. sealers first wintered on Heard Island.

In 1872 the Challenger voyages began under the sponsorship of the British Admiralty and the Royal Society.
The scientific leader of the 1872-1876 voyage of the
Challenger, under the command of Captain George Naves,
was Professor C. Wyville Thompson. The studies carried
out, primarily in the subantarctic islands, were part of
a larger oceanographic cruise. Challenger dredged continental type rocks from the seafloor, proving the
existence of Antarctica as a true continent, and noted
both a rich fauna and flora in the antarctic marine
environment. The studies also included observation on
depth, chemical composition, temperature, and currents of
these southern waters.

During this same time period the ship Gronland,
commanded by Captain Edward Dallman and sponsored by
Albert Rosenthal and the German Society for Polar
Navigation, sailed south along the west side of the
Antarctic Peninsula, pressing south of Biscoe Island and
discovering the Bismark Straits in an area Biscoe in 1832
had presumed was land. A second German party, led by von
Rubritz, sailed to Kerguelan and Heard Island to investigate possible sites for a base to observe the transit
of Venus.

Almost 20 years lapsed before activity resumed in the
exploration of Antarctica. In the early 1890s, a
Scottish whaling expedition of four vessels visited the
peninsula area. Two surgeons, W. S. Bruce, who later
returned to Antarctica, and C. W. Donald, accompanied
this cruise and, despite the concentration on commercial
activities, were able to make some observations on the
fauna of coastal zones. One of the ships discovered
Dundee Island, the strait between it, and the Jooinville
Islands, and explored both Erebus and Terror gulf to the
south.

The Norwegian effort began in earnest at this time
also. Captain C. A. Larsen combined whaling, sealing,
and exploration, collecting plant and animal fossils from
Seymour Island and exploring the east side of the
peninsula along the Larsen Ice Shelf. In 1894-1895,

Captain Leonard Kristensen, while in search of right whales, made significant collections of rock specimens, lichen, and seaweed during a landing at Cape Adare. A landing was also made on Possession Island, where rock specimens, lichen, and penguins were collected for later study. It was one of these landings that the first mummified seals in Antarctica were discovered and reported.

The now-famous cruise of the Belgica under command of Lieutenant Adrien de Gerlache of Belgium was initiated as a truly scientific expedition. The Belgica sailed south from Tierra del Fuego to the South Shetlands, then along the west side of the peninsula, penetrating south through Gerlache Strait and eventually into the Bellingshausen Sea. The ship became locked into the ice, and became quite unintentionally the first scientific expedition to winter over in the Antarctic. The ship broke free after a year and was able to return home with observations on fauna and flora, geology, glaciology, and the first set of continuous winter temperature observations from the Antarctic.

In 1898 a German expedition led by Professor Karl Chan provided additional information about the form of the ocean bed in the southern Atlantic and Indian oceans and reported an accurate location of Bouvet Island.

C. E. Borchgrevink, who had accompanied Kristensen to Antarctica in 1894-1895, returned in 1898 as leader of the privately funded British Antarctic Expedition. He established a land base on the continent at Cape Adare. His group was the first to land at a base actually on the continent, and they obtained continuous meteorological and magnetic observations as well as geologic and biologic collections. An important contribution was the description of the shallow water marine life in the area, observations that set aside once and for all the notion that these waters were barren of life. Borchgrevink's party was picked up in 1900, sailed south and then along the edge of the Ross Ice Shelf, landed at the Bay of Whales, from where Borchgrevink and two others sledged south to 78°50'S, man's farthest penetration south to that time.

The first decade of the twentieth century has been called the "Heroic Age" of antarctic exploration. More important to this discourse is the fact that the increased activity included major emphasis on the acquisition of scientific data as part of the expeditions. During the 1890s, using reports from the Challenger cruise and evidence from the ocean floor sediments and drop-stones from melting icebergs, Murray attempted a reconstruction of the geology, meteorology, and glaciology of this still unknown continent. Murray pleaded with the Royal Society to fund scientific research on the continent. Clement Markham, president of the society, picked up the theme and lobbied for support. His efforts finally resulted in support for the National Antarctic Expedition of the United Kingdom. Utilizing support from the government, the Royal Geographical Society, the Royal Society, and private donors, the ship <u>Discovery</u>, under the command of Robert Falcon Scott, sailed to Antarctica in 1902.

At the Bay of Whales, Scott utilized a captive balloon ascent to make observations of the Ross Ice Shelf topography and extent. He then sailed west and established his base at Hut Point on Ross Island. The three summers spent on Ross Island resulted in many new scientific observations. The discovery of the ice-free Taylor Valley, two sledge trips up the Ferrar Glacier to the Polar Plateau that reached as far south as 77°59'S, and observations on life in the area were the highlights of scientific efforts, efforts that served as the basis for additional work in the area for the next decade and that are still referred to today.

On the other side of the continent a German party led by Professor Erich von Drygalski was working along Wilhelm II Coast at about 90°E. Their ship <u>Gauss</u> was frozen in the pack ice and drifted for over a year. During this period, stations were established on the ice for magnetic, geodetic, climatological, and tidal data. Drygalski also used a captive balloon to view the local terrain. The extinct volcano Gaussberg was discovered and named by a sledge party that visited the site. Glaciologic and geologic studies and collections were carried out, and a biologic program including descriptions and collections of birds, lichen, mosses, and marine fauna was completed.

In the Antarctic Peninsula area the Swedish Antarctic Expedition under the leadership of Dr. Otto Nordenskjold set out to explore the east side of the peninsula. Extreme ice conditions made it impossible to get to the area, and as a result, the party split up, with a group including Nordenskjold establishing a station on Snow Hill Island, while the ship and the remainder of the party returned to winter on South Georgia. Difficult times beset the ship as it attempted to return to Snow Hill Island the next spring. The ship could not reach the island and set a party ashore in an attempt to reach the Snow Hill Island group. The ship, Antarctic, was then crushed by ice in the Erebus and Terror Gulfs and the crew landed on Poulet Island. Eventually, all groups rejoined and were rescued by the Argentine ship Uruguay.

The Nordenskjold expedition produced significant scientific studies. Continuous records of meteorology and magnetics came from the Snow Hill Island party, geologic and cartographic work came from the sledge parties, and observations on glaciology were recorded also.

Dr. William Bruce returned to the Antarctic as an expedition leader in 1902, concentrating his scientific efforts in the area of the Weddell Sea. In addition to meteorologic, oceanographic, and magnetic observations, Bruce established a meteorology station on Laurie Island, a station that was taken over by the Argentines in 1904 and has provided continuous observations since that time.

The French expedition led by Dr. J. Charcot set out to assist Nordenskjold but arrived after his rescue by the Argentines. His work was concentrated along the west side of the peninsula, and the contributions included studies of tides, sea ice, magnetics, geology, and biology, including major collections of marine organisms.

Shackleton arrived in the Ross Sea with the stated intent of conducting scientific studies and reaching both the south magnetic and geographic poles. Unable to reach Hut Point and re-occupy Scott's base, he established a new camp at Cape Royds. Important studies included studies of freshwater lakes and the living organisms within them, geologic studies, and oceanographic observations. A motor-driven vehicle was used to establish caches prior to the attempts to reach the poles. Shackleton's group reached 88°23'S in a vain attempt to reach the geographic pole, while Professor David's group was successful in reaching the south magnetic pole, then located at 72°25'S, 155°16'E.

Charcot returned during the 1908-1910 period to continue his work on the peninsula. The scientific results were significant with glaciology, geology, and biology the major efforts. Important collections were made, especially a wide variety of species of birds, marine invertebrates, and marine mammals.

The start of the second decade of the 1900s saw greatly increased activity with the conquest of the south geographic pole a major stimulus. Amundsen, beaten in his attempt to be the first to reach the North Pole,

sailed south and established a base camp at the Bay of
Whales at the east end of the Ross Ice Shelf. At the
same time, Scott returned to Ross Island and set up a
station at Cape Evans. In the fall of 1911 both parties
set out for the South Pole, Amundsen arriving and
establishing the location on December 16, 1911. Scott
arrived January 17, 1912, and to his disappointment found
he had been beaten. Amundsen, having successfully used
dogs to make the journey, arrived back at the Bay of
Whales in late January. Scott's party of five had
man-hauled, and were unable to complete the return trip.
All perished by late March of 1912.

The contrast between those two expeditions has often
been cited. Amundsen's sole goal was to reach the pole
and return safely. Scott combined scientific work with
his effort, and some 35 pounds of geologic specimens were
still on the sledges when the ill-fated party was found
the following season. In addition, other members of the
Scott expedition made studies of the Cape Crozier Emperor
Penguin rookery, completed important geologic studies in
the areas to the west and south of McMurdo Sound and on
Ross Island, made extended continuous meteorological
observations including the use of balloons for upper air
studies, and also included magnetic and auroral research
as part of the overall effort. The party left at Cape
Adare did geologic work primarily. The results of this
scientific work, published as the Terra Nova Reports,
include also the oceanographic work done on-board the
ship on the various legs of its voyages in antarctic
waters.

Lieutenant Shirase, leading a Japanese party,
initially set out to reach the South Pole. Hearing of
the plans of Amundsen and Scott, the party worked instead
along the east end of the Ross Ice Shelf, where they
named Kainan Bay and sledged south from the Bay of Whales
to 80°S. A second group visited Edward VII Peninsula and
the Alexandra Mountains.

Dr. Bruce had suggested from his earlier visits to
Antarctica that there was a possibility that the Weddell
and Ross seas were connected. Dr. Wilhelm Filchner
proposed that parties travel south from the edges of the
sea and meet someplace in the middle of the continent.
Unable to raise sufficient funds, his German expedition
was able only to start the approach from the Weddell Sea
side. He established a base on what is now the Filchner
Ice Shelf, but was again set back by a calving of the
edge of the shelf and loss of equipment. The goal of the
expedition was not accomplished.

Sir Douglas Mawson, leader of the Australasian Antarctic Expedition, established a base at Cape Denison on the Adelie Coast in 1912. Five parties worked in the area under the most extreme weather conditions. The first radio communication link was established with the outside world via a station on Macquarie Island, and the scientific accomplishments were most productive in geology, glaciology, and terrestrial biology. The expedition ship *Aurora* established a western base after charting portions of the Davis Sea and Queen Mary Coast. The ship carried out oceanographic work between Australia and Anarctica and discovered specimens of the Ross Seal. This was the last of the expeditions of this era that provided significant scientific results. Schackleton's 1914-1916 and 1921-1922 expedition produced little science but much adventure, and the southern oceans were dominated by whalers during the next decade.

In the mid-1920s, scientific work began again. The cruises of the German ship *Meteor* introduced the use of the echo sounder to provide details of the topography of the seafloor, the South Sandwich Trench was discovered, and the presence of four distinct water masses in the South Atlantic was identified.

At the same time the first of the Discovery voyages was undertaken. These voyages, carried out over 15 years using three different ships, were sponsored by the Discovery Committee of the British Government to provide scientific information related to the various conditions that influence the distribution and number of whales. A scientific station was established at Grytviken on South Georgia, and summer research programs were conducted each whaling season from 1925 to 1931. The importance of krill to the diet of baleen whales was substantiated, the relationship between the areas of upwelling, phytoplankton, krill, and whales was studied, whale markings were made, and significant additions to the understanding of the physical, chemical, and biological components of the waters around Antarctica resulted. These voyages established an admirable record of scientific accomplishment, charted a great number of important areas adjacent to Antarctica, and developed techniques used by later expeditions in the design of their scientific research.

A series of Norwegian expeditions during the 1926-1937 period made significant observations to the east of the Weddell Sea along the coastal fringe of Queen Maud Land as far east as 80°. The major contributions were mapping, use of airplanes for aerial reconnaissance, and the meteorological and oceanographic data recorded.

In 1929-1930, Mawson led an international party on two voyages; the first worked along the Enderby Coast, the second along the coast of Wilkes Land. In addition to the further definition of the coast lines in these areas, valuable research was carried out in geology, oceanography, and biology, including marine mammals as well as small marine organisms.

The U.S. interest in antarctic exploration and science had been dormant for almost a century until the first Byrd Expedition established a base on the Ross Ice Shelf in 1929. This expedition is best known for the first flight over the South Pole, but important geologic observations and discoveries were made as well. Rocks were collected from the Queen Maud Range, and the Rockefeller Mountains were discovered.

The second Byrd Antarctic Expedition in 1933-1935 included a major scientific component, and the utilization of the best technology available to accomplish the goals was a significant factor. Meteorologic records from Little America included both surface and upper air observations, seismic techniques were used to measure ice thickness and subice topography, cosmic ray studies were initiated, significant biologic work was carried out, and again geologic parties worked in Marie Byrd Land and the Queen Maud Range. Aerial reconnaissance flights fairly well established that there was no surface expression of a connection between the Ross and Weddell seas, and the general outline of Roosevelt Island was defined.

In the mid-1930s, Rymill (U.K.) worked in the Antarctic Peninsula area, made important geographic discoveries including surveys that established that there was not a strait separating the peninsula from the continent. Long-term meteorologic observations were obtained, and an important study of antarctic seas was completed.

The expedition of Ellsworth provided little scientific data, but of some significance was the discovery of the Ellsworth Mountains, now known to contain the highest peak in Antarctica. The German party led by Captain Alred Ritscher made a quick visit to the coastal areas just east of the Weddell Sea with the expressed intent of establishing a claim and mapping the areas. A lack of geodetic control made the photos taken of little value, and the claim of New Schwabenland was never established.

The Antarctic Service Expedition led by Byrd in 1939-1941 again brought a large group with a great deal of equipment to the continent. Two bases were established, one at Little America on the Ross Ice Shelf, a second on Stonington Island on the west side of the peninsula. A wide variety of scientific studies were carried out from each base camp. The usual geologic, meteorologic, biologic, and oceanographic studies were carried out along with one of the first studies of the physiologic and psychologic reactions by man to the cold.

Between 1942 and 1955 both Argentina and the United Kingdom carried out numerous voyages that had components of scientific research related to hydrographic, biologic, and meteorologic studies. Bases were established by both nations, not only for scientific purposes, but to establish their claims as well.

Admiral Byrd returned to Antarctica during the 1946-1947 austral summer as the leader of Operation Highjump, the largest antarctic expedition to date. The intent was a massive effort to photograph the coastal regions of a large portion of the continent. The use of airplanes required ground meteorological data, and the use of synoptic maps helped identify characteristics and movement of air masses and fronts. Operation Windmill followed the next year with the intent of obtaining ground control for the photos taken during Operation Highjump. In addition, both geologic and biologic work were carried out. Helicopters were first used in support of research in the Antarctic.

The pace and scope of research activities picked up during this period. The privately funded Ronne Expedition to the west side of the Antarctic Peninsula included scientists, and the Chilean government supported numerous expeditions to the Antarctic Peninsula and adjacent islands. The Norwegian-British-Swedish Antarctic Expedition conducted glaciologic, geologic, and meteorologic studies along the Princess Martha Coast. French parties from 1949 to 1953 worked in the area south of the Adelie Coast. The Australian party led by Law established Mawson Station at 64° on the Mawson Coast with a strong scientific component as part of the planning effort. A broad spectrum of observations was carried out, additional areas were explored, and new Emperor Penguin rookeries were discovered.

All of this renewed scientific interest helped set the stage for the International Geophysical Year (IGY) activities in Antarctica. In 1882-1883, the first International Polar Year (IPY) had been held; 12 nations and 14 stations were involved. The two Southern Hemisphere stations were established, one on South Georgia by the Germans and one at Cape Horn by the French. Emphasis was placed on meteorology, geomagnetism, and the aurora. The climatology of the polar regions was greatly enhanced, and the value of a coordinated synoptic network of stations was demonstrated.

The second International Polar Year of 1932-1933 again concentrated on meteorologic, geomagnetic, and auroral observations. The first IPY had shown that high-latitude meteorologic observations were helpful in the understanding of low-latitude processes. The introduction of the radiosonde for upper air soundings was a key part of the program. Despite very difficult financial times, the second IPY was held. Forty-four countries participated, but still no antarctic station was established.

Following World War II, the idea of a Third Polar Year began to be discussed. The technology that had been developed during the war and the increased understanding of the third dimension was a necessary ingredient to the understanding of both the atmosphere and the solid earth. The Third Polar Year soon became the International Geophysical Year. The Antarctic was included as an essential element, and 12 nations agreed to participate in the establishment of stations on and around the continent. Forty-eight new stations were established in addition to the seven already in operation on the peninsula. Eleven programs, aurora, cosmic rays, geomagnetism, glaciology, gravity, ionospheric physics, meteorology, international weather control, oceanography, seismology, biology, and medicine were included as part of the scientific program. The antarctic continent was crossed for the first time, and inland stations were established at the South Pole, close to the south magnetic pole (Vostok), and in the middle of Marie Byrd Land (Byrd Station).

The scientific results of the IGY are voluminous, and an attempt to summarize them here would be frivolous. The explosion of knowledge about the Antarctic that took place during this 18-month period provided substantial results and benefits. What is often neglected is the fact that the IGY research did not include geology, and it was not until after the completion of the IGY that geologic studies became an integral part of research.

A second fact that is often forgotten is that despite the great increase in research activity during the IGY, at the end of that effort there were still large areas of the continent that remained unexplored and unknown scientifically. The interior areas of both east and west Antarctica except for those limited areas traversed by geophysical or oversnow supply vehicles remained blank on the available maps. Third, the IGY certainly can be credited as the international activity that spawned the Scientific Committee on Antarctic Research (SCAR) and the Antarctic Treaty. These two international organizations, each with a special mission, have provided a framework for continued scientific activity in a peaceful and essentially apolitical environment that has been marked by exchange of scientists and scientific data, coordinated international research efforts, and the continued growth in our knowledge of the Antarctic and the surrounding seas.

BIBLIOGRAPHY

American Geographical Society. 1975. History of Exploration and Scientific Investigation; in Folio 19, Antarctic Map Folio Series, V. C. Bushnell, editor, American Geographical Society, NY, 15 plates with text.

Anderson, J. J. 1965. Bedrock Geology of Antarctica: a summary of exploration 1831-1962; American Geophys. Union Ant. Res. Series, vol. 6, pp. 1-70.

Bertand, K. J. 1971. Americans in Antarctica 1775-1948; Special Publication No. 39, American Geographical Society, Lane Press, Burlington, VT, 55 pp.

Corby, G. A. 1982. The First International Polar Year (1882/83); in WMO Bulletin, World Meteorological Organization, pp. 197-214.

Crary, A. P. 1982. International Geophysical Year: Its Evolution and U.S. Participation; in Antarctic Journal of the United States, National Science Foundation, Washington, D.C., Vol. XVII, No. 4, pp. 1-6.

Duncan, Sir George E. R. 1977. The Southern Ocean: History of Exploration; in Adaptations Within Antarctic Ecosystems, George A. Llano, editor, Proceedings of the Third SCAR Symposium on Antarctic Biology, Gulf Publishing Co., Houston, TX, pp. XV-XXXVIII.

Gould, L. M. 1978. The Emergence of Antarctica; in Polar Research: to the Present and the Future, M. A. McWhennie, editor, AAAS Selected Symposium #7, Western Press, Boulder, CO, Chapter 1, pp. 9-26.

Laursen, V. 1982. The Second International Polar Year (1932/33); in WMO Bulletin, pp. 214-222.

Nicolet, M. 1982. The International Geophysical Year 1957/58; in WMO Bulletin, pp. 222-231.
United States Board on Geographic Names. 1956. Geographic Names of Antarctica, Gazeteer No. 14, Office of Geography, U.S. Department of the Interior, U.S. Government Printing Office, Washington, D.C., 332 pp.
U.S. Navy Hydrographic Office. 1943. Sailing Directions for Antarctica; H. O. Publication No. 138, U.S. Government Printing Office, Washington, D.C., 312 pp.

9.

The Antarctic Treaty as a Scientific Mechanism (Post-IGY)—Contributions of Antarctic Scientific Research

William F. Budd

INTRODUCTION

The main thesis of this chapter is that <u>humankind's quest for knowledge needs to be recognized as the primary motivation for the high level of continued interest and activity in the Antarctic</u>. The treaty nations, through the Antarctic Treaty System, have supported this objective. Their fundamental basic tenets include peaceful cooperation among nations, freedom of exchange of results, and preservation of the antarctic environment. The basic quest for knowledge was also the primary motivation for the First International Polar Year 1882-1883 (Corby, 1982), the Second International Polar Year 1932-1933 (Laursen, 1982) and the International Geophysical Year (IGY) 1957-1958 (Nicolet, 1982). Much of the new development activity in the Antarctic initiated by the IGY has been continued and expanded in the following period. This great post-IGY period of scientific research in Antarctica has led to a knowledge explosion producing an order of magnitude more information than available pre-IGY. This extensive antarctic information data base has become pervasive through a large part of other basic scientific disciplines.

The importance of antarctic science in the general scheme of scientific knowledge was early recognized by the International Council of Scientific Unions (ICSU) in a 1957 agreement to form a Special Committee for Antarctic Research, later to evolve into the Scientific Committee on Antarctic Research (SCAR), of ICSU. Since the IGY, SCAR has provided the international forum for exchange of information on scientific research plans, activities, results, logistics, management, and cooperation.

The Antarctic Treaty System formed after SCAR provided an international political framework for continued peaceful cooperation in antarctic research. Much of this antarctic research is essentially international in character and global in significance. For example, the atmosphere and the oceans are not restrained by national boundaries, nor are the marine nutrients or the biomass. Consequently, other international bodies are also interested in the antarctic region; for example, the World Meteorological Organization (WMO) and the Intergovernmental Oceanographic Commission.

In recent years, increased attention has been focused on the possible resource potential of Antarctica (cf. Wright and Williams, 1974; Holdgate and Tinker, 1979; Zumberge, 1979; Lovering and Prescott, 1979). More detailed assessments, however, have revealed that economically viable exploitation in the foreseeable future is not a likely prospect and could therefore not provide the rationale for supporting the expensive "big science" inherent in the operation of continuing antarctic expeditions (see, e.g., Behrendt, 1983a; Quilty, 1984; Tingey, 1984).

On the other hand, the scientific information is invaluable. For example, the global information set required to extend weather forecasts to several days or to understand interannual climatic fluctuations cannot exclude data from such a large and influential region of the Earth as the Antarctic. The impact of interannual climatic fluctuations on agricultural production and the global economy could be considered reason enough for a continuing antarctic program aimed at increasing our understanding of the global climate system. Antarctic research is much broader than this, however, and has impacted extensively on the wide spectrum of science. A few examples of the key role antarctic research has played in the advancement of science are given below.

The treaty nations represent that group of nations sufficiently interested in Antarctica to cover the expense of antarctic field activities. One subgroup of the members of SCAR includes the Southern Hemisphere nations most nearly adjacent to Antarctica: Argentina, Australia, Chile, New Zealand, and South Africa. The remainder tend to be technologically advanced, high-northern-latitude nations with strong traditional polar interests: Belgium, France, the Federal Republic of Germany, the German Democratic Republic, Japan, Norway,

Poland, the United Kingdom, the USSR, and the United States. Some nations, although perhaps similarly classified, have extensive Arctic activities, for example, Denmark and Canada.

The treaty nations have been prolific in their research and publications. These publications have become universally available and extensively disseminated in international scientific literature. This means that any nation can now become active in antarctic research through the analysis of an immense amount of data. This is particularly relevant when it is realized that the antarctic region provides an excellent resource for global monitoring of the environment. It is a mark of the importance of Antarctica on the global scene that, now, tropical nations such as India and Brazil have joined SCAR, and other nations of the U.N. have expressed interest in Antarctica.

THE POST-IGY INTERNATIONAL ANTARCTICA QUARTER CENTURY

The continuation and expansion of many of the antarctic activities initiated during the IGY has lead to a quarter-century accumulation of antarctic knowledge. In addition, the advancement of technology and science generally has resulted in a greatly increased capacity for antarctic data collection and research. Some examples of these advances include satellites for remote sensing, geodetic location, and communications, automatic stations and drifting buoys, aircraft remote sensing and ice thickness sounding, deep ice core drilling, and sophisticated ice core analyses.

In particular, the polar-orbiting satellites have been greatly improved and now provide a data bank of many years' complete mosaic coverage of the globe, including the polar regions, as illustrated in Figure 9-1. The cloud imagery depicts the high concentration of intense cyclones around the edge of the Antarctic. These large systems play a major role in the global weather and climate system.

The antarctic stations as shown in Figure 9-2 form an extensive coverage of both surface and upper-air observations essential for extended weather prediction. The already archived data are invaluable for testing models and theories of global circulation and the causes of climatic change.

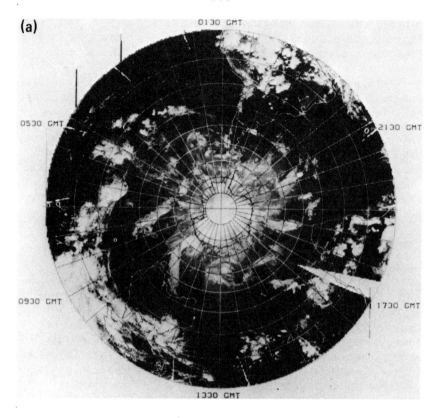

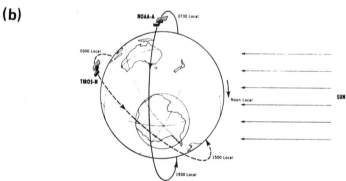

FIGURE 9-1 Examples of: (a) a satellite thermal infrared mosaic of Antarctica and the Southern Hemisphere produced from National Oceanic and Atmospheric Administration (NOAA) Nimbus 6 data on March 19, 1980; and (b) typical orbits of a polar-orbiting satellite system.

The antarctic continent has also been gradually more fully covered by oversnow traverses, as shown in Figure 9-3. These traverses collect a wide range of data, including snow accumulation and ice thickness, crucial for understanding the ice sheet mass balance and its implications for global sea level changes.

The ice thickness distribution has been more extensively determined from aerial radio echo sounding, as shown by a compilation of flight lines in Figure 9-4. This type of work has resulted in construction of detailed maps of the major features of the Antarctic. Earlier versions post-IGY include the U.S. Antarctic Map Folio Series Folio 2 by Bentley et al. (1964) and the Soviet Antarctic Atlas (Bakayev, 1966). A more recent folio, including extensive aerial sounding data, has been produced by the Scott Polar Research Institute (Drewry, 1983).

Other landmark results include the features of the ocean, particularly from the voyages of the Ob (USSR) and the Eltanin (USA; cf. Gordon and Goldberg, 1970); the marine life (Balech et al., 1968; and Be and Hedgpeth, 1969); the bedrock geology, land morphology, and sediments (Craddock et al., 1970; Heezen et al., 1972; and Goodell et al., 1973); the sea ice from satellite sensing (Zwally et al., 1983); the upper atmosphere (Penwdorf et al., 1964, and Waynik, 1965); and the climate (see, e.g., Weyant, 1967; and Schwerdtfeger, 1970, 1984).

Many comprehensive reviews of the progress in antarctic science have been produced (see, e.g., Quam, 1971; Washburn, 1980; Polar Group, 1980; and McWhinnie and Denys, 1978).

The extensive resulting publications are considered further below. For the present it is apparent that scientific achievements of the period have been immense. Many data centers around the world provide an extensive service for archiving and accessing data at low cost to other users. In regard to the provision of scientific information for the global community, antarctic activities have been very successful.

THE PROFITABLE NONRENEWABLE RESOURCES FALLACY

In recent times, consideration has been given to the possibility that the potential of the Antarctic for economic resources could lead to international conflict (see, e.g., Auburn 1977, 1982, 1984; and Sollie 1983).

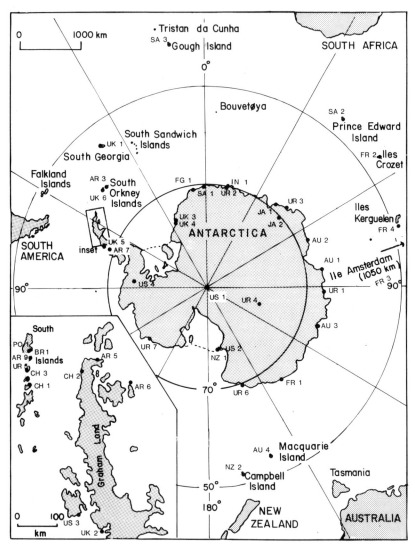

FIGURE 9-2 Location and nationality of the network of antarctic meteorological stations involved in routine surface and upper air observations (modified from SCAR, 1984) (Reprinted with permission).

KEY FOR FIGURE 9-2

Argentina

AR1 Belgrano II, 77°51' S, 34°33' W
AR3 Orcadas, 60°45' S, 44°43' W
AR5 Esperanza, 63°24' S, 56°59' W
AR6 Marambio, 64°14' S, 56°38' W
AR7 San Martin, 68°07' S, 67°08' W
AR8 Primavera, 64°09' S, 60°57' W
AR9 Jubany, 62°14' S, 58°38' W

Australia

AU1 Davis, 68°35' S, 77°58' E
AU2 Mawson, 67°36' S, 62°52' E
AU3 Casey, 66°17' S, 110°32' E
AU4 *Macquarie Island, 54°30' S, 158°56' E

Brazil

BR1 Comandante Ferraz, 62°05' S, 58°23' W

Chile

CH1 Capitán Arturo Prat, 62°30' S, 59°41' W
CH2 General Bernado O'Higgins, 63°19' S, 57°54' W
CH3 Tenient Rodolfo Marsh, 62°12' S, 58°54' W

Federal Republic of Germany

FG1 Georg von Neumayer, 70°37' S, 8°22' W

France

FR1 Dumont d'Urville, 66°40' S, 140°01' E
FR2 *Alfred-Faure, Iles Crozet, 46°26' S, 51°52' E
FR3 *Martin-de-Vivies, Ile Amsterdam, 37°50' S, 77°34' E
FR4 *Port-aux-Francais, Iles Kerguelen, 49°21' S, 70°12' E

India

IN1 Dakshin Gangotri, 70°05' S, 12°00' E

Japan

JA1 Syowa, 69°00' S, 30°35' E
JA2 Mizuho, 70°42' S, 44°20' E

New Zealand

NZ1 Scott Base, 77°51' S, 166°45' E
NZ2 *Campbell Island, 52°33' S, 169°09' E

Poland

PO1 Arctowski, 62°09' S, 58°28' W

South Africa

SA1 Sanae, 70°18' S, 02°24' W
SA2 *Marion Island, Prince Edward Islands, 46°53' S, 37°52' E
SA3 *Gough Island, 40°21' S, 09°53' W

United Kingdom

UK1 *Bird Island, South Georgia, 54°00' S, 38°03' W
UK2 Faraday, Argentine Islands, 65°15' S, 64°16' W
UK4 Halley, Caird Coast, 75°35' S; 26°40' W
UK5 Rothera, Adelaide Island, 67°34' S, 68°07' W
UK6 Signy, South Orkney Islands, 60°43' S, 45°36' W

United States of America

US1 Amundsen-Scott, 90°S
US2 McMurdo, 77°51' S, 166°40' E
US3 Palmer, 64°46' S, 64°03' W

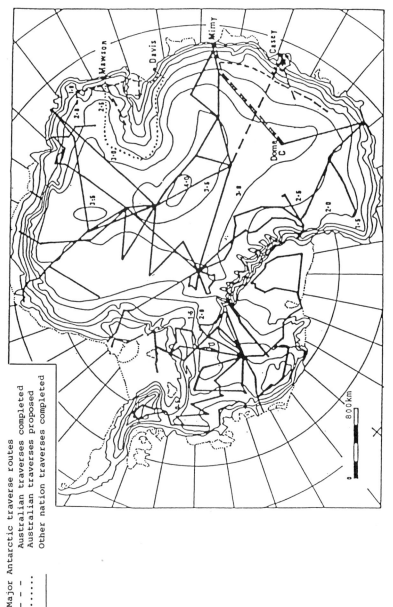

FIGURE 9-3 Major antarctic oversnow traverse route compilation (from Budd, 1984) (Reprinted with permission).

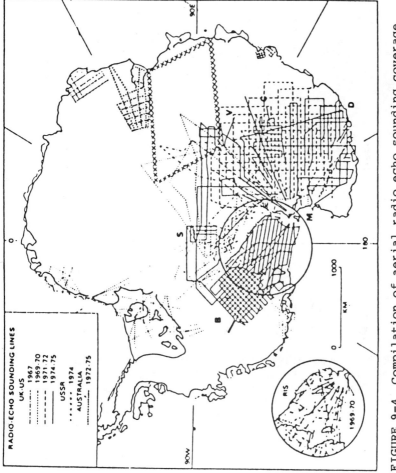

FIGURE 9-4 Compilation of aerial radio echo sounding coverage (modified from Drewry, 1983; Radok, 1977; and Budd, 1984) (Reprinted with permission).

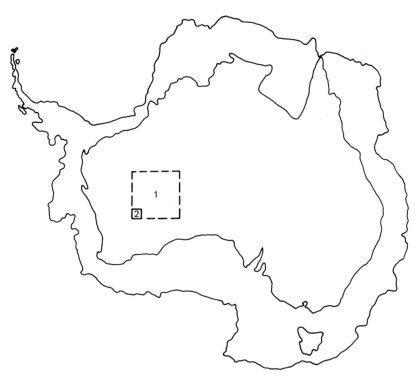

FIGURE 9-5 Relative areas of exposed rock in the Antarctic (Area 1: $0.33 \times 10^6 km^2$ or 2.4%) compared with that in the Australian Antarctic Territory (Area 2: $.011 \times 10^6 km^2$).

Such possibilities have lead to extensive activities in

(1) Evaluation of resource potential estimates,
(2) Evaluation of possible environmental consequences of exploitation,
(3) Consideration of legal and political implications of resource-oriented activities,
(4) Increased interest of the wider international community in the Antarctic as a resource prospect, and
(5) The diversion of antarctic science programs toward resource-oriented topics.

It is a thesis of this chapter that these activities are heavily premised on the idea that Antarctica may have significant economically exploitable resources within the foreseeable future.

It is here contended that

(1) If resources are being sought, there are far more prospective sources elsewhere, and
(2) The redirection of antarctic activities toward resource-oriented questions may detract from the more important objectives of basic and applied scientific research.

To support these contentions, four examples are discussed: (1) land-based minerals, (2) offshore minerals, (3) marine living resources, and (4) ice as a water resource.

(1) The antarctic rock exposures are small in total area by continental standards, they are widely dispersed over the continent, and they are generally in highly inaccessible locations. Working in the Antarctic for mineral extraction would be very costly. To place this in perspective, Figure 9-5 shows the antarctic continent compared with Australia. The small area of total exposed antarctic rock in the Australian Antarctic Territory represents a small fraction of western Australia, where a wide range of easily accessible minerals is relatively abundant. Similar comparisons with the other, larger continental land masses are also valid. The rest of the antarctic continent is covered in ice, averaging about 2.5 km in thickness. This makes the rock underneath largely inaccessible. The ice, however, does offer a greater real prospect as a renewable resource.

(2) The prospects for offshore hydrocarbon resources around Antarctica have been frequently alluded to (see, e.g., Holdgate and Tinker, 1979; and Lovering and Prescott, 1979). The most prospective continental shelf region is also the region of most highly concentrated large icebergs, which, as shown by Figure 9-6, move around the coast in the eastwind drift current like a conveyor belt. A technique for extracting offshore hydrocarbons in regions of large icebergs (which may be kilometers in extent) has not yet been devised, let alone proven. Even then the cost of extraction would be high and would have to be competitive with known large resources from other sources, for example, shale oil, tar sands, coal conversion, and renewable fuel crops. The prospects for commercial ventures in the Antarctic, therefore, must be rated very low (cf. Quilty, 1984; and Behrendt, 1983b). Nevertheless, it needs to be emphasized that antarctic research into offshore sediments is invaluable for studying the processes and conditions of formations of the continental margin sediments in general, particularly in relation to those of neighboring continents. For future generations, when other, lower cost sources of nonrenewable hydrocarbons have been depleted, advances in technology may make any antarctic prospects more feasible.

(3) With regard to the resource potential of the marine biomass around Antarctica, it is well recognized that antarctic waters are highly productive (see, e.g., Laws, 1983; Chittleborough, 1984). The mechanisms involved in productivity are not well known and are very complex (Tranter, 1982, 1984). Furthermore, there have been gross imbalances caused to the marine life by past human exploitation activities, as shown in Figure 9-7 (from Chittleborough, 1984). Unbridled harvesting of species such as krill could lead to loss of an important renewable resource. Therefore, a long-term project of measurement and monitoring is required to understand interrelations among the biota before the real resource potential can be determined. Such a program is being addressed by the SCAR-sponsored program entitled Biological Investigations of Marine Antarctic Systems and Stocks (cf. Laws, 1983). The intergovernmental Commission for the Conservation of Antarctic Marine Living Resources also carries out this function.

(4) Since renewed interest in the prospects for using icebergs as a water resource was raised in the early 1970s (see, e.g., Weeks and Campbell, 1973a, 1973b), a

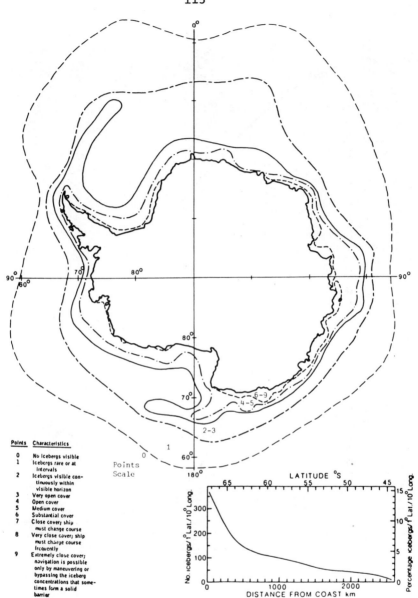

FIGURE 9-6 Average concentrations of icebergs around Antarctica from the Soviet Antarctic Atlas (from Bakayev, 1966; and Morgan and Budd, 1978) (Reprinted with permission).

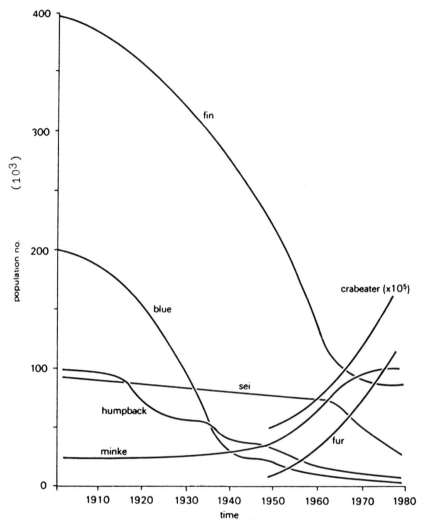

FIGURE 9-7 Major changes in the antarctic oceans biomass associated with past harvesting activities, as given by (Chittleborough, 1984) (Reprinted with permission).

great deal of new research on icebergs has been carried out. This work has resulted in several international conferences (see, e.g., Husseiny, 1978 and the Annals of Glaciology, Vol. 1), an international monitoring program, and publication of an international news magazine, Iceberg Research.

The more recent reviews of problems and prospects have indicated that, judging from the results obtained so far, the topic is worthy of further consideration, and icebergs as useful resources could be beneficial to some Southern Hemisphere cities such as Perth and Adelaide (cf. Schwerdtfeger, 1979; Budd et al., 1980; Russell-Head, 1980; and Lawson and Russell-Head, 1983). A practical scheme for utilizing icebergs requires an iceberg several hundred meters in horizontal extent, such as those depicted in Figure 9-8. Much larger icebergs are difficult to tow, while much smaller icebergs could be expected to have largely melted before reaching the destination. To test the scheme further, a series of well-instrumented trial towing and melt-rate experiments would be useful. In the interim, it appears that the iceberg water could be quite competitive with other sources of fresh water, particularly in times of drought.

ANTARCTICA AS A GLOBAL ENVIRONMENTAL SCIENCE RESOURCE

Although little direct commercial exploitation of the Antarctic can be expected in the medium term, it needs to be recognized that the antarctic region is invaluable as a resource for scientific research, particularly for the environmental sciences that help us better to manage our planet. It is generally understood that the well-being of advanced technological societies is largely due to the scientific research and development of previous generations. One important question that needs to be addressed is, "What kind of planet shall we be leaving for following generations?"

This question needs to be examined with reference to the impacts already made on the environment and the contributions that can be made by antarctic research toward alleviating future problems, such as

(1) The change in balance of the antarctic marine biota from past harvesting regimes (Figure 9-7, Chittleborough, 1984),

FIGURE 9-8 A typical antarctic iceberg of volume about 10^8 m^3 (R. Wills), (from Budd et al., 1980) (Reprinted with permission).

(2) The continuing impact on long-term climatic trends from continued CO_2 production,
(3) Possible effects on the antarctic mass balance and future sea level changes from climatic warming, and
(4) Possible changes in the ocean and marine biosphere from reduced sea ice cover following climatic warming.

In addition, there are many ways in which antarctic data are invaluable in contributing to global problem-solving and research. For example:

(1) Because of the increased environmental stress caused by growing human populations, climatic fluctuations even in the present era, such as the El Nino and Southern Oscillation episodes, can have major social and economic impacts. The antarctic region is an integral part of this global climate system and, particularly with regard to sea ice, needs to be studied along with changes in the atmosphere and ocean.
(2) Antarctic rock and sediments provide the unique resource of an additional continent to further better understanding of the development and origin of rocks, the history of the formation of the continents, and the development of deep sediments.
(3) It is one of the great discoveries in the Antarctic that certain parts of the ice sheet are the world's greatest sources of meteorites. Already there is an extensive literature base on antarctic meteorites (cf. Nagata, 1983).
(4) The ice sheet offers a unique resource for studying past climate from the analysis of deep ice cores. Information is also obtained on other features, such as volcanic and anthropogenic fallout, atmospheric gases, cosmic activity, global ice volume, and sea level.
(5) The antarctic region is important as a location for monitoring the upper atmosphere, the ionosphere, and solar terrestrial relations. This information is of fundamental importance for high-frequency telecommunications.
(6) Finally, it needs to be re-emphasized that the antarctic stations provide a unique resource for meteorological measurements, particularly upper air soundings to provide the initial states required by numerical modeling for weather prediction. The Southern Hemisphere is very

sparsely covered compared to the Northern Hemisphere. The antarctic and Southern Ocean island stations play an important role in contributing to hemispheric and global coverage.

ANTARCTIC PUBLICATIONS AND THE KNOWLEDGE EXPLOSION

Although antarctic scientists may be well aware of the wide impact of antarctic science within the general science literature, it may be of interest to try to quantify the impact since the IGY. This task is worthy of a much more thorough analysis than is attempted here, but this review can be regarded simply as a very preliminary survey of the publications. In the first case, only the numbers of publications for the different fields over time and from different nations are considered. This leaves quite open the types of publications and their depth. No doubt this approach has inherent biases and therefore should not be used to infer relative merit without more detailed study.

Techniques of citation analysis also provide a means of gauging relative interest in various topics as well as other relative measures of impact. This section therefore summarizes some of the results of publications analysis in a general way, and the next section gives some examples of the specific highlights resulting from antarctic research. As a first indication of the extent of antarctic publications and how they have changed over time, various bibliographies and data bases are useful. For example, the U.S. Library of Congress' Antarctic Bibliography is one such source. This series has 14 volumes from 1961 to July 1984, containing 30,097 abstracts. Another single volume with an additional 4,773 abstracts extends the series back in time to cover the period 1951-1961. About one third of this group could be regarded as post-IGY. The full series, therefore, serves to show the large order-of-magnitude increase in antarctic publications during the quarter century following the IGY. Figure 9-9 illustrates this continuing growth of antarctic publications as well as a tendency toward increasing growth rates.

In order to compare the post-1950 publications numbers with those from earlier times, reference is also made to The Antarctic, 1739-1957, Vol. III of the Bibliography of Regional Meteorological Literature, produced by the South African Weather Bureau and compiled by Venter and Burdecka (1966). Although not directly comparable with the Library of Congress bibliography, it provides an

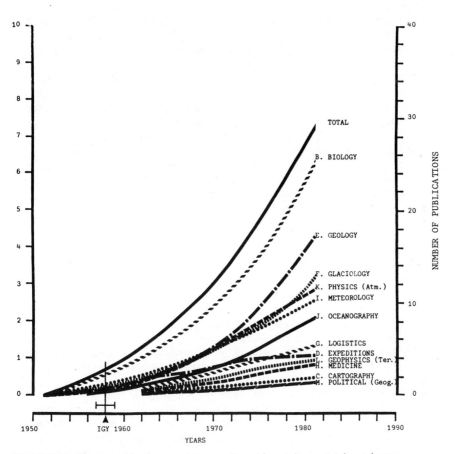

FIGURE 9-9 Growth in numbers of antarctic publications from the U.S. Library of Congress' Antarctic Bibliography.

independent listing with a breakdown into some of the important environmental disciplines. Table 9-1 shows the numbers of publications in these different categories from 1739 to 1957. It can therefore serve as a pre-IGY compilation, going back to the earliest times for antarctic publications. It should be noted that its limitation to topics related to meteorology omits the two large discipline areas of biology (apart from bioclimatology) and geology. Nevertheless, the total numbers of publications and the breakdown into the other categories are useful for comparison with the Library of Congress series in later years, as shown in Table 9-1. An independent analysis of recent publications from the national reports to SCAR given by Budd (1984) is shown in Table 9-2. Although the categories again are somewhat different, because they were based on the SCAR disciplines, the relative numbers in the different disciplines are similar. This table also shows national preferences for some disciplines in antarctic activities, and the totals for the different nations for this recent period are compared with the national origins of literature in the Library of Congress compilation for the earlier period in Table 9-3. The average rate for the total publications of about 1,200/year compares with an average of 1,500/year from the Antarctic Bibliography.

Evidence that this now large body of information obtained from antarctic research has made some impact on science generally has been provided in citation analysis by Cozzens (1981), Cozzens and Small (1982), and Guthridge (1983). From publications appearing between 1961 and 1978, the citation analysis was used to identify topics for which antarctic research had a significant role. From an analysis of 314 different journals, highly cited documents were identified in 189 different specialities. Of these highly cited specialities, 52 included citations of antarctic documents. A total of 2,942 papers appearing in the Antarctic Bibliography was cited by the journals at least once between 1961 and 1978. The total number of citations to antarctic literature was 28,974, giving an average citation rate per paper of about 10 over the period. From the list of 52 topics of the highly cited documents for which antarctic work was referenced, as given by Guthridge (1983), Table 9-4 lists the first 26. Guthridge reported that 84 antarctic papers published between 1961 and 1978 are regarded as citation classics (defined as having been cited more than 50 times during that period). Of the top 12 reported by Guthridge,

TABLE 9-1

Numbers of Antarctic Publications over Time and in Different Disciplines (%)

South African Meteorological Antarctic Bibliography	1739-1957 (%)	Discipline	Library of Congress Antarctic Bibliography											National Reports to SCAR[a]	1979-1983 (%)
			1951-1961 to: 1966	1968	1970	1971	1973	1974	1976	1977	1979	1980		
			Vol. 2	3	4	5	6	7	8	9	10	11		
General	4.6	General	15.0	9.6	7.7	6.1	7.6	11.8	7.0	5.5	6.1	7.7	9.4	Other	5.4
Bioclimatology	1.0	Biological sciences	20.6	19.5	19.4	22.1	21.8	19.1	15.5	26.7	29.7	22.6	28.8	Biology	32.6
Expeditions	33.6	Cartography	2.0	2.3	2.3	1.6	1.1	1.6	1.1	1.5	1.2	.9	.7	Geodesy and cartography	2.3
		Expeditions	11.2	4.4	2.9	4.0	3.6	4.5	1.8	0.9	0.9	.6	.7	Geology	18.7
Glaciology	7.3	Geological sciences	6.6	11.3	13.7	13.7	14.4	15.1	20.1	19.6	20.0	17.3	18.4	Glaciology	12.8
Logistics	1.8	Ice and Snow	8.4	11.3	11.0	12.3	8.7	9.4	10.0	9.7	9.1	15.2	10.7	Logistics	1.6
		Logistics, equipment, and supplies	5.3	4.4	4.5	7.0	2.4	4.0	5.6	3.5	2.7	3.7	2.4		
Medical	1.9	Medical sciences	2.0	4.2	3.2	2.7	3.8	3.1	5.4	3.5	2.4	3.4	1.8	Medicine	2.9
Meteorology	24.9	Meteorology	11.2	10.9	10.4	7.4	9.6	7.0	8.2	8.8	8.5	7.7	8.3	Meteorology	5.0
Climatology	13.0														
Oceanography	3.6	Oceanography	5.2	6.7	6.1	7.6	8.7	8.9	8.5	7.6	9.4	8.0	6.0	Oceanography	5.6
Atmospheric physics and cosmic rays	3.7	Atmospheric physics	7.0	10.5	13.5	11.7	14.2	10.4	10.9	8.8	6.7	8.7	7.8	Physics	9.6
		Terrestrial physics	3.5	4.4	4.5	3.3	3.6	3.8	4.1	2.1	2.1	2.5	2.0	Geophysics	4.7
Geography	5.2	Political geography	2.2	0.6	0.9	0.4	0.7	0.5	0.9	1.8	1.2	1.5	2.7		
Total	3,170	Totals	4,773	2,000	2,000	2,000	2,000	2,244	2,202	2,452	2,348	2,479	2,361	Total number of publications	3,478

[a] After Budd (1984).

TABLE 9-2

National Origin of Antarctic Publications (%)

Nation	Library of Congress[a] (%)				Reports to SCAR[b]
	1962-1966 Vol. 2	1968 Vol. 3	1970 Vol. 4	1972 Vol. 5	1980-1983 (+)
Argentina	2.3	2.2	1	1	0.8
Australia	2.9	2.8	3	1.3	5.8
Chile			1	1	1.9
Federal Republic of Germany	2.0	1.9	1	1	6.3
German Democratic Republic					2.0
France	7.3	3.4	3	2.1	6.6
Japan	3.5	4.1	11	2.4	12.9
New Zealand	2.8	3.2	4.5	6.1	6.7
Norway		1.1		2.2	1.2
South Africa				2.4	5.9
United Kingdom	14.1	10.9	10.5	11.0	16.9
USSR	27.8	30.2	19	21.7	15.0
U.S.	28.3	34.3	41	39.4	18.0
Others	9.0	5.9	11.5	8.4	

[a]The Antarctic Bibliography editor (G. A. Doumani) noted that these statistics represent the numerical proportions included in the volume and not the world's output of Antarctic literature.
[b]From Table 3 (after Budd, 1984) from the SCAR members' annual national reports.

TABLE 9-3

Relative Publication Frequencies in Different Disciplines[a]
for a Three-Year Period from National Reports to SCAR after 1979

Nation	Biology	Medicine	Ocean-ography	Meteo-rology	Glacio-logy	Geo-physics	Geology	Geodesy and Cartography	Physics	Logistics	Other	Total
Argentina	11	4			1		4	1	2		1	4	29
Australia	73	9		22	26	13	47	1	9				200
Chile	56		2	3	3		3						67
F.R.G.	109		18	2	21	2	29	13	10		5	9	220
G.D.R.	14	1		1	9	1	29	1	6		5	8	71
France	119	3	4	8	31	(7)	(21)	3	29		3		228
Japan	(83)	(8)	4	(24)	(40)	(46)	(150)	5	83		4		447
New Zealand	52	6	4	4	24	8	91		1		10	34	234
Norway	16		5	1	14	4	1	2					43
Poland	(not available)												
S. Africa	131		23		(6)	(2)	(8)	15	35		7		205
U.K.	238	17	8	16	65	16	82	1	80		7	44	588
USSR	106	42	72	(22)	95	54	42	1	(43)			43	520
U.S.	125	9	53	69	111	12	143	2	35		23	45	626
Totals[b]	1133	99	194	172	446	165	650	44	333		54	188	3,478
Percent of total	32.6	2.9	5.6	5.0	12.8	4.7	18.7	1.3	9.6		1.6	5.4	

[a] In some cases subdivision into categories is not provided or categories are used differently to the SCAR disciplines used here. Articles listed as "in press" have not been counted.
[b] For the totals subdivisions have been apportioned as for the averages of the others.

TABLE 9-4

The First 26 of 52 Specialties in Which Citation
Analysis Demonstrates that Antarctic Research Had
a Significant Role Between 1961 and 1978

Specialty	Specialty
Atmospheric pollution	Stratigraphy of deep-sea carbonates
Magnetospheric physics	Models of climatic change
Aurora and electric-fields	Genetic variation in marine invertebrates
Biological antifreeze mechanisms	Geomagnetic exclusions and reversals
Reproductive strategies in arctic and antarctic birds	Carbon dioxide and climate
Thermospheric dynamics	Radionuclides and suspended matter in seawater
Stable isotope geochemistry	Life on Mars: Viking results
Asymmetric sea-floor spreading	Biochemical adaptations to temperature
Geothermal dynamics of ocean ridges	Paleoclimatology of the Antarctic region
Glaciation models: arctic data	Magnetospheric wave phenomena
Very-low-frequency waves in the magnetosphere	Hydromagnetic waves and geomagnetic pulsations
Trace-element and isotope geochemistry	Tectonics of the Scotia Sea
Midocean ridges	Stratospheric warmings

[a] From Guthridge (1983).

TABLE 9-5

The Top Six of the 84 Antarctic "Citation Classics" Papers Cited More Than 50 Times between 1961 and 1978

No. of Citations	Papers
546	Heirtzler, J. R. 1968. Marine magnetic anomalies, geomagnetic field reversals and motions of the ocean floor and continents. J. Geophys. Res. 73:2119.
432	Morgan, W. J. 1968. Rises, trenches, great faults, and crustal blocks. J. Geophys. Res. 73:1959.
313	Bullard, E. 1965. Fit of the continents around the Atlantic. Phil. Trans. R. Soc. London 258:41.
312	Carpenter, D. L. 1966. Whistler studies of the plasmapause in the magnetosphere. 1. Temporal variations in the position of the knee and some evidence on plasma motions near the knee. J. Geophys. Res. 71:693.
205	Smith, R. E. 1962. Metabolism and cellular function in cold acclimation. Physiol. Rev. 42:60.
168	Smith, A. G. 1970. Fit of the southern continents. Nature 225:139.

Table 9-5 lists the first six. In spite of the reservations associated with citation analysis, Guthridge concludes that "...the masses of literature evaluated in this study point inescapably to the conclusion that antarctic research has advanced understanding on many of the world's frontiers of science."

HIGHLIGHTS OF ANTARCTIC DISCOVERIES AND RESEARCH

From the large number of antarctic topics that are important on the global scene, only a brief mention of a few examples can be attempted here.

In marine biology, some understanding of the changes in the balance among the species shown in Figure 9-7 is being developed. The interactions within the biomass are illustrated by Figure 9-10 (from Laws, 1983). The antarctic waters are remarkable for their high productivity (Knox, 1983; and Chittleborough, 1984), and the Antarctic Ocean's biomass is large on the world scale. Already, antarctic research has indicated some of the controlling factors on biomass productivity (Tranter, 1982).

The physical and chemical oceanographic studies of the Antarctic Ocean have shown the important role that the antarctic region plays in global ocean circulation, the control of our climate, the circulation of atmospheric gases, and the distribution of nutrients (Gordon and Goldberg, 1971; DeWitt, 1971; and El-Sayed, 1975). The antarctic bottom water has been referred to as the "lung of the ocean" for its role in transferring oxygen and nutrients to other regions of the global ocean system (cf. Edmond, 1975).

Figure 9-11 (from Gordon, 1971, as modified by Budd, 1980; and from Knox, 1983) shows the main water-mass transfers and the prominent role of antarctic sea ice in driving the deep mixing by salt rejection.

High on the list of topics for which antarctic research has been a key factor is the concept of continental drift and the reconstruction of the past locations of the continents. Figure 9-12 (from the U.S. Antarctic Journal cover picture of May-June 1970 and Colbert, 1970) illustrates the crucial role of Antarctica in Gondwanaland reconstruction and that the fossil flora and fauna found in Antarctica provide crucial evidence to support the theory.

Further crucial evidence of the processes of separation of the continents has come from the matching patterns of

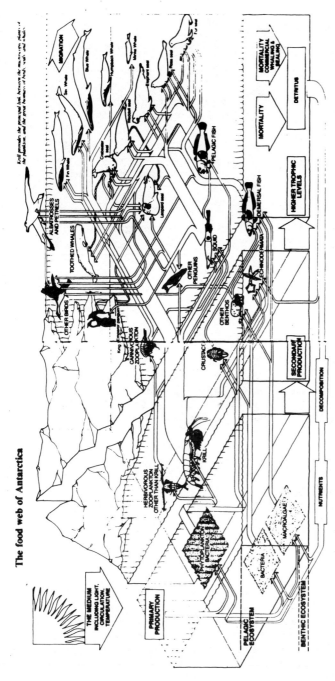

FIGURE 9-10 Schematic diagram of antarctic biomass interactions (from Laws, 1983) (Reprinted with permission).

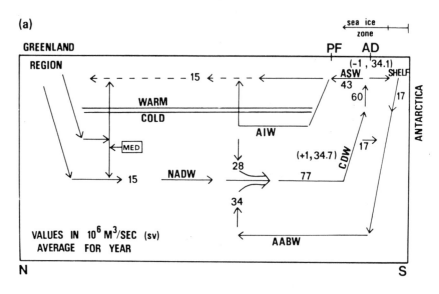

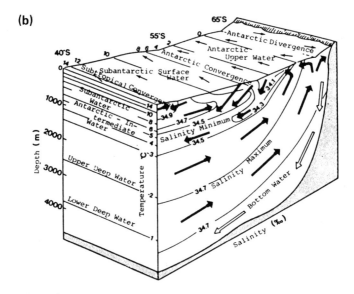

FIGURE 9-11 Antarctic and global ocean water masses with estimated characteristics and fluxes: (a) estimated gross meridional annual fluxes between world ocean water masses (modified by Budd, 1980; after Gordon, 1971); and (b) meridional and zonal flow in the Southern Ocean under summer conditions (from Knox, 1983) (Reprinted with permission.

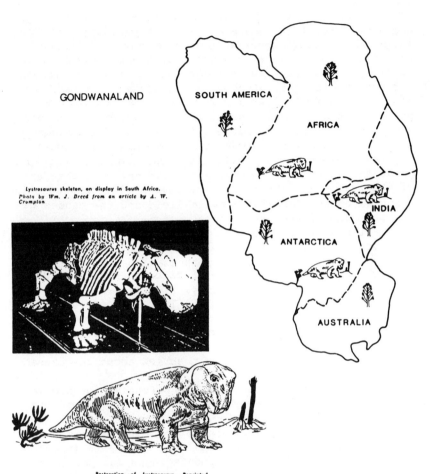

FIGURE 9-12 Gondwanaland reconstruction with representations of reptile fossils found in Antarctica as well as the other continents [modified from U.S. Antarctic Journal (Vol. 5, No. 3) and Colbert (1970)] (Reprinted with permission).

magnetic variations in the sediments on opposite sides of the midocean ridge, as shown in Figure 9-13 (from Weissel and Hayes, 1972). Measurements of the paleomagnetic locations of the south magnetic pole from antarctic rocks have provided valuable evidence that the average position of the magnetic pole in recent millennia clusters around the geographic pole, as shown in Figure 9-14 (from Funaki, 1984). This concept is a cornerstone for determining the paleocontinental locations of the long geological history of the Earth. As a supplement to the paleodata, measurements in both polar regions have shown how the magnetic dip poles are currently well displaced from the geographic poles and have a well-established movement over the period of observations. These data are invaluable for developing theories on the Earth's magnetic field and the causes of its past and future variations.

Studies of deep ocean sediments around the world have provided clear evidence of the large fluctuations of sea level and climate associated with glacial and interglacial variations. Figure 9-15 (from an Antarctic Ocean sediment study presented by Hays, 1978) shows clearly the typical variations over about the last 130,000 years, which are well recognized from sediments in other parts of the world. The Southern Ocean data have been crucial for comparing the time scales of the changes in the different hemispheres and thereby theories about the causes of these changes.

The antarctic ice sheet has the potential for providing a much clearer record of climate and ice changes of the last few glacial cycles, along with other factors that changed at the same time, such as atmospheric gases, volcanic fallout, dust, and sea salt. Figure 9-16 (from Budd and Young, 1983) shows computed ages and particle paths for the ice sheet in the region of Vostok Station. Results from the Vostok deep core given by Gordienko et al. (1983) and Grosswald (1983) already show part of the similar variations seen in the deep sea sediments as shown by Figure 9-17. Other cores from Antarctica and Greenland also show the same characteristic features but much more compressed near the base (cf. Robin, 1983). The deep ice in East Antarctica is useful for going back much further in time.

The basic dynamics of the present-day antarctic ice sheet, determined from many years' observations, has provided the basis for computing the behavior of the ice age ice sheets that covered North America and Europe some 20,000 years ago (Budd and Smith, 1981). A more recent

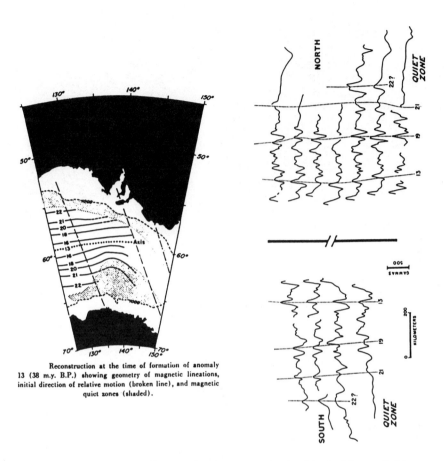

FIGURE 9-13 Magnetic variations on opposite sides of the midocean ridge between Australia and the Antarctic showing the similar magnetic patterns with distance from the ridge (from Weissel and Hayes, 1972) (Reprinted with permission).

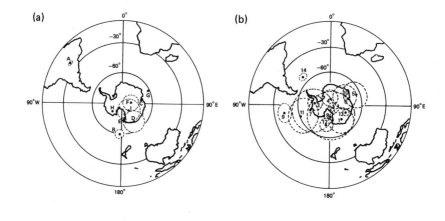

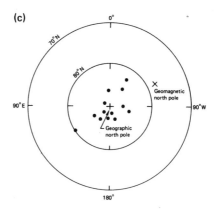

FIGURE 9-14 Plots of geologically recent paleo virtual geomagnetic poles (VGP's) determined in the Antarctic and the Arctic: (a) Previous Cenozoic VGP positions for Antarctica, modified from Funaki (1984); (b) VGP positions from McMurdo volcanics; and (c) Pleistocene and recent Northern Hemisphere VGP. Note the clustering of VGP's around the modern geographic pole rather than around the present geomagnetic pole (redrawn from Cox and Doell, 1960) (Reprinted with permission).

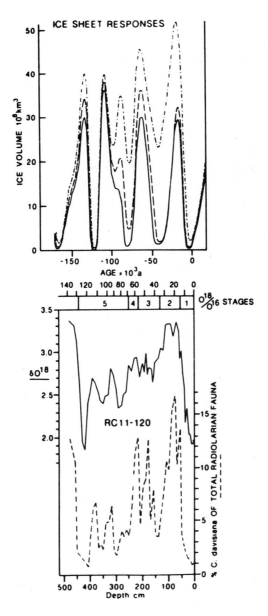

FIGURE 9-15 Comparison of sea sediment paleoclimatic indicators (from Hays, 1978) with computed ice volume changes from ice sheet modeling in response to the Earth's orbital radiation changes (from Budd and Smith, 1981, 1985) (Reprinted with permission).

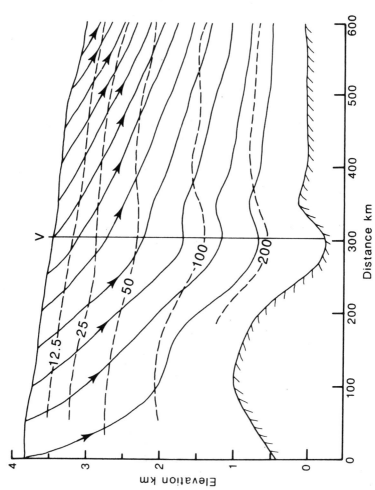

FIGURE 9-16 Computed antarctic ice sheet particle paths and ages from steady state with the present regime along a flowline near Vostok (from Budd and Young, 1983) (Reprinted with permission).

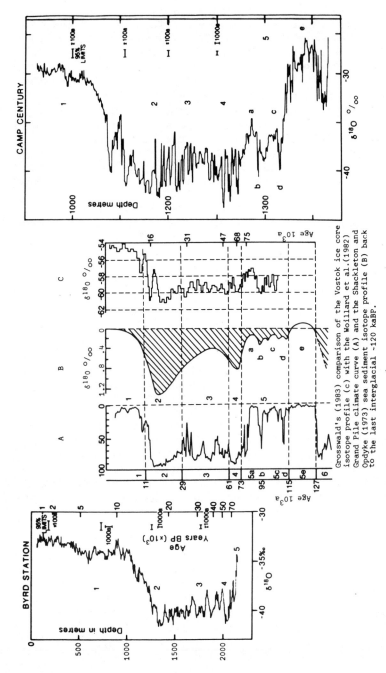

FIGURE 9-17 Comparison of ice sheet isotope profiles as paleoclimate indicators with the deep sea sediment patterns [from Grosswald (1983) for the Vostok core and Robin (1983) for the Byrd and Camp Century core]. The similarities are highlighted by the numbered stages 1 to 5e (Reprinted with permission).

computation shows that the ice sheet and climate changes derived from the Earth's orbital radiation changes give rise to ice volume variations that closely match the sea sediment data over the past 400,000 years, Figure 9-18. The deep antarctic ice cores will be crucial in confirming these results.

The discovery that radio waves of certain frequencies can travel large distances through ice has led to the development of the radio echo sounder. Figure 9-19 (from Robin, 1983) shows clearly how the bedrock and ice thickness can be determined and also important subsurface layering within the ice. This subsurface layering may be able to be tied in the ice core data to serve as clear historical markers of the ancient layers preserved in the ice.

Finally, it needs to be reiterated that the Antarctic is invaluable for its role in global environment monitoring. The long series of measurements of CO_2 from the South Pole Station are classic and have been instrumental in the formulation of theories of the global circulation of the gas (cf. Mook et al., 1983). Similarly, the measurements of CO_2 at the coast of Antarctica have been instrumental in determining the role of the antarctic sea ice in transferring the CO_2 to the deep ocean (cf. Budd, 1980; Fraser et al., 1980; and Milne and Smith, 1980).

Other monitoring of global importance in Antarctica includes that of weather and climate, the upper atmosphere and ionosphere, cosmic rays, turbidity, volcanic fallout, anthropogenic fallout and pollution, radioactivity, atmospheric composition, solar-terrestrial phenomena, earthquakes, magnetic fields, polar motion, and earthtides. This list is meant to give some examples rather than to be comprehensive. It is quite clear that antarctic data form an essential part of the global environment monitoring system and are needed as a regular part of the global coverage.

THE TREATY NATIONS AS THE UNITED NATIONS "ANTARCTIC RANGERS"

From the foregoing review it is clear that the Antarctic is a valuable global resource for scientific research. It is also an important natural habitat for wildlife, particularly around the coastal and oceanographic zone. Antarctica is unique for its scenery of unsurpassed

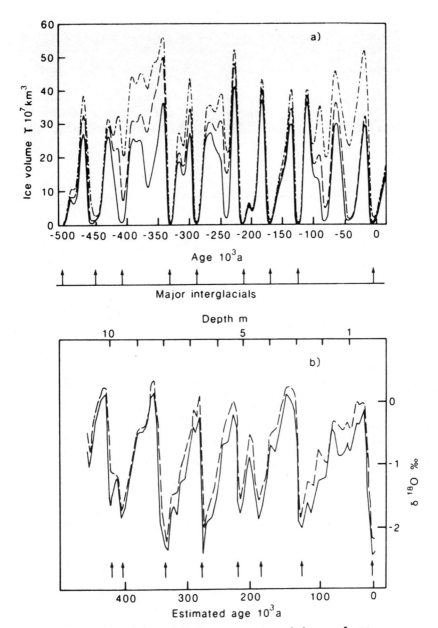

FIGURE 9-18 (a) Longer term changes of ice volume computed by Budd and Smith (1985) compared with (b) deep sea sediment isotope changes over 400,000 years (from Broecker and van Donk, 1970) (Reprinted with permission).

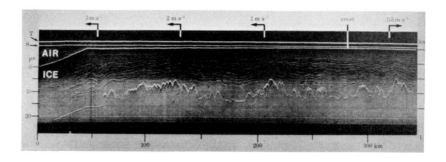

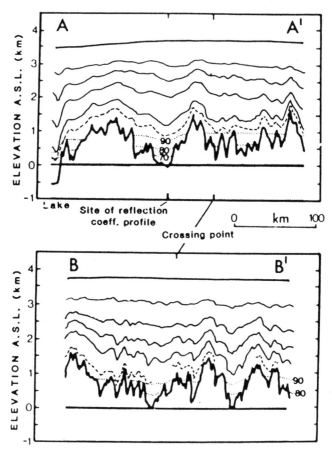

FIGURE 9-19 Radio ice thickness sounding records given by Robin (1983) showing clear bedrock reflections and systematically varying internal echos (Reprinted with permission).

beauty. Part of the attractive character of the Antarctic is its large area of unspoiled natural wilderness. There is no doubt that the Antarctic is one of the Earth's natural wonders, which can be appreciated by all people today and by the generations to come.

The Antarctic Treaty System provides for the preservation of the Antarctic and for its use as a region of peaceful international cooperation and scientific research. The treaty nations may be regarded as those nations most interested in antarctic research and most prepared to provide the support to undertake it. The data, information, and material obtained from the Antarctic are made freely available to the rest of the global community, and by using these data, other nations can also become involved in antarctic research.

Antarctic expeditionary activities, however, are rather costly because of the expensive logistics required for transport, living, and communication in the harsh environment. Table 9-6 provides some indications of the major transport logistics in use by the nations currently working in Antarctica. The number of stations and wintering personnel, along with some indication of the order of magnitude of the number of summer visitors, is given in Table 9-7. The high cost of logistics facilities required means that only the more technologically advanced nations can be expected to be able to afford to mount regular, large-scale expeditionary activity.

Cooperation among nations has been most effective. Many cases of exchange of expeditioners occur among the treaty nations. In addition, the treaty nations have arranged for members of other nations to participate in antarctic field work. With the growing international interest in the Antarctic and discussions within the U.N. of current antarctic matters, the promotion of greater awareness of the region is most appropriate. Participation in antarctic activities for a wider cross section of the global community should be encouraged, but this should be done with care to preserve the principles of the Antarctic Treaty, particularly regarding preservation, peaceful cooperation, and free exchange of scientific research results.

In this regard the treaty nations can play a major role, somewhat similar to the concept of park rangers working to understand the region better, to publicize the value and features of the region, and to preserve it well for future generation. Thus, if the Antarctic is

TABLE 9-6

Logistics in Use by Various Antarctic Treaty Nations (approximate)

Nation	Ships	1976/1977 Aircraft				1983/1984 Aircraft		
		Intercontinental	Within Antarctica Fixed Wing	Rotary[a]	Ships	Intercontinental	Within Antarctica Fixed Wing	Rotary[a]
Argentina	5	C-130	3	5	2	C-130	2	4
Australia	2	Twin Otter	1	5	3	Twin Otter		4
Chile	5		2	2	2			2
F.R.G.	1			1	3			8
France	1			3	1			1
Japan	1		1	3	1			3
New Zealand		C-130				C-130		
Norway	1							
U.K.	3	Twin Otter	2	2	3	Twin Otter	2	2
USSR	7	Il-18	7	4	10	Il-18	6	8
		C-141				C-130		
U.S.	6	C-130	5	4	6	C-141	6	11

[a] Or on ships.

SOURCES: Webster (1977), CIA (1978), Antarctic Treaty exchange information, National Reports to SCAR.

TABLE 9-7

National Populations at Antarctic Stations (approximate)[a]

	1955		1960		1965	1970	1975	1980		1983		Typical Summer Peak 1976-1979
	S	P	S	P	P	P	P	S	P	S	[S]	P
Argentina	(5)	68	(5)	76	120	116	126	(8)	179	(9)		300
Australia	(1)	15	(3)	59	52	56	68	(3)	76	(3)	[1]	150
Chile	(4)	35	(4)	35	39	40	43	(3)	52	(3)		100-200
D.D.R.			(0)	3	2	4	4	(0)	5	(0)		7
F.G.R.					0	0		(1)	5	(1)		18
France			(1)	14	22	26	35	(1)	26	(1)	[3]	60-70
Japan			(1)	15	18	32	32	(2)	34	(2)		60-70
New Zealand			(2)	31	12	18	10	(1)	11	(1)	[1]	100-200
Norway										(0)		20
Poland								(1)	(19)	(9)		50
South Africa			(1)	13	14	20	18	(1)	16	(1)	[2]	20
USSR			(3)	129	145	220	244	(7)	288	(7)		600
United Kingdom	(6)	51	(7)	71	86	59	54	(4)	55	(5)	[1]	100-200
U.S.			(3)	169	290	243	90	(4)	111	(4)		900
Totals	(16)	169	(30)	614	800	834	724	(36)	878	(38)	[11]	2,486-2,805

[a]S, stations; ( ), antarctic; [ ], sub-antarctic (north of 60°S); P, personnel.

SOURCES: Dubrovin and Petrov (1971), Webster (1977), CIA (1978), U.S. Antarctic Journal (March 1981), Holdgate (1984), National Reports to SCAR, Treaty exchange information.

regarded as an international park--perhaps the greatest natural park in the world--then the treaty nations could be regarded as working for the U.N., and the global community as "antarctic rangers" through the SCAR-ICSU system, to ensure that the laudable principles of the Antarctic Treaty are maintained.

BIBLIOGRAPHY

Auburn, F. M. 1977. Offshore oil and gas in Antarctica, In German Yearbook of International Law, Vol. 20 (Berlin), 173 pp.

Auburn, F. M. 1982. Antarctic Law and Politics, C. Hurst and Co. (London), 361 pp.

Auburn, F. M. 1984. The Antarctic minerals regime: sovereignty, exploration, institutions and environment. In S. Harris, ed. Australian Antarctic Policy Options, CRES Monograph 11, Australian National University (Canberra).

Bakayev, V. G., ed. 1966. Soviet Antarctic Expedition Atlas of Antarctica. Main Administration of Geodesy and Cartography of the Ministry of Geology USSR (Moscow).

Balech, E., S. Z. El-Sayed, G. Hasle, M. Nueshul, J.S. Zaneveld. 1968. Primary productivity and benthic marine algae of the Antarctic and sub-Antarctic, Folio 10, Antarctic Map Folio Series, American Geographical Society (New York), 12 pp. and 15 maps.

Be, A. W. H., J. W. Hedgpeth, eds. 1969. Distribution of selected groups of marine invertebrates in waters south of 35°S latitude, Folio 11, Antarctic Map Folio Series, American Geographical Society (New York), 44 pp. and 29 maps.

Behrendt, J. C., ed. 1983a. Petroleum and mineral resources of Antarctica, Geological Survey Circ. 909, U.S. Geological Survey (Washington, D.C.), 75 pp.

Behrendt, J. C. 1983b. Are there petroleum resources in Antarctica? In J. C. Behrendt, ed. Petroleum and Mineral Resources of Antarctic, Geological Survey Circ. 909, U.S. Geological Survey (Washington, D.C.), pp. 3-24.

Bentley, C. R., R. L. Cameron, C. Bull, K. Kojima, A. J. Gow. 1964. Physical characteristics of the Antarctic Ice Sheet, Folio 2, Antarctic Map Folio Series,

American Geographical Society (New York), 10 pp. and 10 maps.

Brodie, J. 1965. Oceanography. In T. Hatherton, ed. Antarctica, Methuen and Co. Ltd. (London), pp. 101-127.

Broecker, W. S., J. van Donk. 1970. Insolation change, ice volume and the 18O record in deep sea cores. Rev. Geophys. Space Phys. 8:169-198.

Budd, W. F. 1980. The importance of the Antarctic region for studies of the atmospheric carbon dioxide concentration. In G. I. Pearman, ed. Carbon Dioxide and Climate: Australian Research, Australian Academy of Science (Canberra), pp. 115-128.

Budd, W. F. 1984. Scientific research in Antarctica and Australia's effort. In S. Harris, ed. Australia's Antarctic Policy Options, CRES Monograph 11, Australian National University (Canberra), pp. 219-253.

Budd, W. F., I. N. Smith. 1981. The growth and retreat of ice sheets in response to orbital radiation changes. In I. Allison, ed. Sea level, Ice and Climatic Change, IAHS Pub. No. 131, pp. 369-409.

Budd, W. F., I. N. Smith. 1985. A 500,000 year simulation of the North American Ice Sheet and Climate. In T. H. Jacka, ed. Australian Glaciological Research 1982-83. Australian Antarctic Research Expedition (ANARE) Research Notes, Antarctic Division, (Hobart) (in press).

Budd, W. F., N. W. Young. 1983. Application of modeling techniques to measured profiles of temperatures and isotopes. In G. de Q. Robin, ed. The climate record in polar ice sheets, Cambridge University Press (Cambridge), pp. 150-177.

Budd, W. F., T. H. Jacka, V. I. Morgan. 1980. Antarctic iceberg melt rates derived from size distributions and movement rates, Ann. Glaciol. 1:103-112.

Central Intelligence Agency. 1978. Polar Regions Atlas, National Foreign Assessment Center (Washington, D.C.).

Chittleborough, R. G. 1984. Nature, extent and management of Antarctic living resources. In S. Harris, ed. Australia's Antarctic Policy Options, CRES Monograph 11, Australian National University (Canberra), pp. 135-161.

Colbert, E. H. 1970. The fossil tetrapods of Coalsack Bluff, V. Antarc. J. 5(3):57-61.

Colbert, E. H. 1977. Cynodont reptiles from the Triassic of Antarctica, Antarct. J. U.S. 12(4):119-120.

Corby, G. A. 1982. The first International Polar Year (1982/83), WMO Bull. 31:197-214.

Cox, A., R. R. Doell. 1960. Review of Paleomagnetism. Geo. Soc. Am. Bull. 71:734.

Cozzens, S. E. 1981. Citation analysis of Antarctic research, Antarct. J. U.S. 4:233-235.

Cozzens, S. E., H. Small. 1982. Citation analysis of Antarctic research: final report on NSF grant DPP 79-24011, Inst. for Scientific Information (Philadelphia), 87 pp.

Craddock, C., R. S. Adie, S. J. Carryer, A. B. Ford, H. S. Gain, G. W. Grindley, K. Kizaki, L. L. Lackey, M. G. Laird, T. S. Laudon, V. R. McGregor, I. R. McLeod, A. Mirsky, D. C. Neothline, R. L. Nichols, P. M. Otway, P. G. Quilty, E. F. Roots, D. L. Schmidt, A. Sturm, T. Tatsumi, D. S. Trail, T. Van Autenboer, F. A. Wade, G. Warren. 1970. Geology of Antarctica, Folio 12, Antarctic Map Folio Series, American Geographical Society (New York), 6 pp. and 45 maps.

DeWitt, H. H. 1971. Coastal and deep-water benthic fishes of the Antarctic, Folio 15, Antarctic Map Folio Series, American Geographical Society (New York), 10 pp. and 5 maps.

Drewry, D. J., ed. 1983. Antarctica: Glaciological and Geophysical Folio. Scott Polar Research Institute (Cambridge), 9 map sheets with text.

Dubrovin, L. I., V. N. Petrov. 1971. Scientific stations in Antarctica 1882-1963, Hydrometeorological Service (Leningrad), 1967. English translation: Polar Information Service, National Science Foundation (Washington, D.C.).

Edmond, J. M. 1975. Geochemistry of the circumpolar current, Oceanus 18(4):36-39.

El-Sayed, S. Z. 1975. Biology of the Southern Ocean, Oceanus 18(4):40-49.

Fraser, P. J., P. Hysen, G. I. Pearman. 1980. Global atmospheric carbon dioxide space and time variability: an analysis, with emphasis on new data for model validation. In G. Pearman, ed. Carbon Dioxide and Climate: Australian Research, Australian Academy of Science, Australian National University (Canberra), pp. 33-40.

Funaki, M. 1984. Paleomagnetic Investigation of McMurdo Sound region, Southern Victoria Land, Antarctic, Mem. Nat. Inst. Polar Res. Ser. C, No. 16, 81 pp.

Goodell, H. G., R. Houtz, M. Ewing, D. Hayes, B. Naini,
R. J. Echols, J. P. Kennet, J. G. Donahue. 1973.
Marine sediments of the Southern Ocean, Folio 17,
Antarctic Map Folio Series, American Geographical
Society (New York), 18 pp. and 9 maps.

Gordienko, F. G., V. M. Kotlaykov, E. S. Korotkevich,
N. I. Barkov, S. D. Nikolayev. 1983. New results of
oxygen-isotope studies of the ice core from Vostok
Station down to the depth of 1412 m (in Russian), Data
of Glaciological Studies, Chronicle, Discussion, Pub.
No. 46, Academy of Sciences of the USSR (Moscow), pp.
168-171.

Gordon, A. L. 1971. Oceanography of Antarctic waters.
In J. L. Reid, ed. Antarctic Oceanology, Antarctic
Research Series Vol. 15, American Geophysical Union
(Washington, D.C.), pp. 169-203.

Gordon, A. L., R. D. Goldberg. 1970. Circumpolar
characteristics of Antarctic waters, Folio 13,
Antarctic Map Folio Series, American Geographical
Society (New York), 6 pp. and 19 maps.

Grosswald, M. G. 1983. On the interpretation of a new
oxygenisotope curve from Vostok station (in Russian),
Data of Glaciological Studies, Chronicle, Discussion
Pub. No. 46, Academy of Sciences of the USSR (Moscow),
pp. 171-174.

Guthridge, G. G. 1983. Citation of research
literature, Antarctic J. U.S. 18(2):12-13.

Hays, J. D. 1978. A review of the Late Quaternary
climate history of Antarctic Seas. In E. M. van
Zinderen Bakker, ed. Antarctic Glacial History and
World Paleoenvironments, Balkema (Rotterdam), p. 57-71.

Hays, J. D., J. A. Lozano, N. J. Shackleton, G. Irving.
1976. Reconstruction of the Atlantic and Western
Indian Ocean. In R. M. Kline and J. D. Hays, eds.
Investigations of Late Quaternary Paleoceanography and
Paleoclimatology. Geol. Soc. Am. Memoir (Boulder),
145:337-372.

Heezen, B. C., M. Tharp, C. R. Bentley. 1972.
Morphology of the Earth in the Antarctic and
Sub-Antarctic, Folio 16, Antarctic Map Folio Series,
American Geographical Society (New York), 16 pp. and 8
maps.

Holdgate, M. W. 1984. The use and abuse of polar
environmental resources. Polar Rec. 22(136):25-48.

Holdgate, M. W., J. Tinker. 1979. Oil and other
minerals in the Antarctic, House of Print (London),
51 pp.

Husseiny, A. A., ed. 1978. Iceberg utilization, In Proceedings of the First International Conference, Ames, Iowa, 1977. Permagon (New York), pp. xi-xiii, xvii, 731-732.

Judson, S. 1971. Physical Geology, 4th ed. Prentice-Hall (Englewood Cliffs, N.J.), 687 pp.

Knox, G. A. 1983. The living resources of the Southern Ocean: a scientific overview. In F. Orrego Vicuna, ed. Antarctic Resources Policy: Scientific, Legal and Political Issues, Cambridge University Press (Cambridge), pp. 21-60.

Laursen, V. 1982. The Second International Polar Year (1932/33) WMO Bull. 31:214-222.

Laws, R. 1983. Antarctica: a convergence of life, New Sci. 99:608-616.

Lawson, J. D., D. S. Russell-Head. 1983. Augmentation of urban water by Antarctic icebergs, In Proceedings of the Eighteenth Coastal Engineering Conference, ASCE/Cape Town, South Africa/Nov. 14-19, p. 2610-2618.

Lovering, J. F., J. R. V. Prescott. 1979. Last of Lands: Antarctica, Melbourne University Press (Melbourne), 212 pp.

McGregor, P. M., A. J. McEwin, J. C. Dooley. 1983. Secular motion of the south magnetic pole. In R. L. Oliver et al., eds. Antarctic Earth Sciences, Australian Academy of Science (Canberra), pp. 603-606.

McWhinnie, M. A., C. J. Denys. 1978. Antarctic marine living resources with special reference to krill Euphausia superba: Assessment of adequacy of present knowledge, Report to the National Science Foundation (10B-21492), DePaul University (Chicago), 209 pp.

Milne, P., J. D. Smith. 1980. Measurement of dissolved carbonate parameters in Antarctic coastal waters. In G. I. Pearman, ed. Carbon Dioxide and Climate: Australian Research, Australian Academy of Science, Australian National University (Canberra), pp. 137-142.

Mook, W. G., M. Koopmans, A. F. Carter, C. D. Keeling. 1983. Seasonal, latitudinal and secular variations in the abundance and isotopic ratios of atmospheric carbon dioxide, 1, Results from land stations, J. Geophys. Res. 88:915-920, 933.

Morgan, V. I., W. F. Budd. 1978. The distribution, movement, and melt rates of Antarctic icebergs. In A. A. Husseiny, ed. Iceberg Utilization, Proceedings of the First International Conference, Pergamon (London), pp. 220-228.

Nagata, T., ed. 1983. Proceedings of the Eighth Symposium on Antarctic Meteorites. Mem. Nat. Inst. Polar Res., Spec. Issue No. 30, 457 pp.

Nicolet, M. 1982. The International Geophysical Year 1957/58, WMO Bull. 31:222-231.

Penndorf, R., T. M. Noel, G. F. Rourke, M. A. Shea. 1964. Aeronomical Maps for the Antarctic, Folio 1, Antarctic Map Folio Series, American Geographical Society (New York), 6 pp. and 9 maps.

Polar Group (D. J. Baker, U. Radok, G. Weller). 1980. Polar atmospheres-ice-ocean processes: a review of polar problems in climate research, Rev. Geophys. Space Phys. 18:525-543.

Quam, L. O., ed. 1971. Research in the Antarctic. American Association for the Advancement of Science (Washington, D.C.), 768 pp.

Quilty, P. G. 1984. Mineral Resources of the Australian Antarctic Territory. In S. Harris, ed. Australia's Antarctic Policy Options, CRES Monograph, Australian National University (Canberra), pp. 165-203.

Radok, U. 1977. International Antarctic Glaciological Project: past and future, U.S. Antarct. J. 12(1 and 2):32-38.

Robin, G. de Q., ed. 1983. The climatic record in polar ice sheets, Cambridge University Press (Cambridge), 212 pp.

Russell-Head, D. S. 1980. The melting of free-drifting icebergs, Ann. Glaciol. 1:119-122.

Schwerdtfeger, P. 1979. On icebergs and their uses, A report to the Australian Academy of Science, Cold Regions Sci. Technol. 1:59-79.

Schwerdtfeger, W. 1970. The climate of the Antarctic. In S. Orvig, ed. World Survey of Climatology, Vol. 14, Climates of the Polar Regions, Elsevier (New York), pp. 253-361.

Schwerdtfeger, W. 1984. Weather and Climate of the Antarctic, Developments in Atmospheric Science 15, Elsevier (New York), 261 pp.

Scientific Committee on Antarctic Research. 1984. SCAR Bulletin, No. 78, September, pp. 354-356.

Serson, P. H. 1980. Tracking the north magnetic pole. GEOS (Winter):15-17.

Shackleton, N. J., N. D. Opdyke. 1973. Oxygen isotope and paleomagnetic stratigraphy of equatorial Pacific core V28-238: oxygen isotope temperatures and ice volumes on a 10^5 and 10^6 year scale. Quat. Res. 3:39-55.

Shil'nikov, V. I. 1969. Icebergs. In Antarctic Atlas, Vol. II (in Russian), Hydrometeorological Service (Leningrad), pp. 455-465.

Sollie, F. 1983. Jurisdictional problems in relation to Antarctic mineral resources in political perspective. In F. Orrego Vicuna, ed. Antarctic Resources Policy, Cambridge University Press (Cambridge), pp. 317-335.

Thuronyi, G. T. 1981. Recent review papers in Antarctic literature, Antarct. J. U.S. 16:232-233.

Tingey, R. R. 1984. Comments on Mineral Resources. In S. Harris, ed. Australia's Antarctic Policy Options, CRES Monograph 11, Australian National University (Canberra), pp. 204-215.

Tranter, D. J. 1982. Interlinking of physical and biological processes in the Antarctic Ocean, Oceangr. Mar. Biol. Ann. Rev. 20:11-35.

Tranter, D. J. 1984. Comments on: Nature, extent and management of Antarctic living resources. In S. Harris, ed. Australia's Antarctic Policy Options, CRES Monograph 11, Australian National University (Canberra), pp. 161-163.

Venter, R. J., J. M. Burdecka. 1966. Bibliography of Regional Meteorological Literature, Vol. III, The Antarctic Weather Bureau (South Africa), 485 pp.

Washburn, A. L. 1980. Focus on polar research, Science 209:643-652.

Waynik, A. H., ed. 1965. Geogmagnetism and Astronomy. Antarctic Research Series Vol. 4, American Geophysical Union (Washington, D.C.).

Webster, J. J. 1977. Antarctica. An information paper presented by Senator the Honorable J. J. Webster, Minister for Science, Australian Government Publishing Service (Canberra), 34 pp. plus appendixes.

Weeks, W. F., W. J. Campbell. 1973(a). Icebergs as a fresh water source: an appraisal, Symposium on the Hydrology of Glaciers 1969, IAHS Pub. No. 95, p. 255.

Weeks, W. F., W. J. Campbell. 1973(b). Icebergs as a fresh water source: an appraisal. J. Glaciol. 12:207-233.

Weissel, J. K., D. E. Hayes. 1972. Magnetic anomalies in the southeast Indian Ocean. In D. E. Hayes, ed. Antarctic Oceanology II, Vol. 19, Antarctic Research Series, American Geophysical Union, (Washington, D.C.), pp. 165-196.

Weyant, W. S. 1967. The Antarctic Atmosphere: climatology of the surface environment, Folio 8,

Antarctic Map Folio Series, American Geographical Society (New York), 4 pp. and 13 maps.

Woillard, G. 1981. Grand Pile pollen records (NE France) and correlations with the northwestern European stratigraphy: brief report, In IGCP Project 73/1/24 Quaternary glaciations in the Northern Hemisphere, Rep. 6, pp. 285-289.

Woillard, G. M., W. G. Mook. 1982. Carbon-14 dates at Grand Pile: Correlations of land and sea chronologies. Science 215:159-161.

Wright, N. A., P. L. Williams. 1974. Mineral Resources of Antarctica, U.S. Geological Survey Circ. No. 705, 29 pp.

Zillman, J. W. 1977. The first GARP Global Experiment, Aust. Met. Mag. 25(4):175-213.

Zumberge, J. H., ed. 1979. Possible Environmental Effects of Mineral Exploration and Exploitation in Antarctica, SCAR, University Library (Cambridge), 59 pp.

Zwally, H. J., J. C. Cosimo, C. L. Parkinson, W. J. Campbell, F. D. Carsey, P. Gloersen. 1983. Antarctic sea ice, 1973-1976. Satellite passive microwave observations, NASA SP-459, Scientific and Technical Information Branch, National Aeronautics and Space Administration (Washington D.C.) 206 pp.

10.
The Antarctic Treaty as a Scientific Mechanism—The Scientific Committee on Antarctic Research and the Antarctic Treaty System

James H. Zumberge

INTRODUCTION

Both the Antarctic Treaty and the Scientific Committee on Antarctic Research (SCAR) were legacies of the International Geophysical Year (IGY) of 1957-1958. Although the two organizations are independent of each other, they have enjoyed a close working relationship since their creation in the late 1950s. The treaty operates in the world of international diplomacy, while SCAR operates in the world of international science. The treaty functions through the workings of governmental representatives, while SCAR functions through the workings of scientists who act independently of their governments. The treaty is expressed in terms of a precise document signed and ratified by the member nations, while SCAR is expressed in terms of a constitution agreed to by its member delegates. In spite of these differences, the treaty and SCAR have had a mutually beneficial relationship over the years, and it is difficult to conceive of a sound policy for Antarctica without both organizations acting independently, on the one hand, but with interacting roles, on the other.

This chapter deals with the role of SCAR in the Antarctic Treaty System. First, the origin of SCAR is reviewed briefly; second, the structure and procedures of SCAR are examined; third, examples of SCAR's interaction with the treaty system are given; and fourth, an evaluation of the future of SCAR is attempted.

THE ORIGIN AND GROWTH OF SCAR

The Beginnings of SCAR

The IGY began on January 1, 1957, and ended officially on December 31, 1958. Even before the IGY began, however, the U.S. National Committee for the IGY recognized that this 18-month period was too short in terms of the investment in stations and equipment that had been made by the nations planning extensive activities in Antarctica. The U.S. National Committee, therefore, proposed to the Comité Special de l'Année Geophysique (CSAGI) in 1956 that the IGY should be extended for an additional year in order to justify the huge expenditures that had been made for antarctic research.

CSAGI approved this proposal at its fourth meeting in June 1957, and recommended to the executive board of the International Council of Scientific Unions (ICSU) that an ad hoc committee be established under ICSU to examine the merits of further general scientific investigations in Antarctica. That committee met in September 1957 in Stockholm, with representatives from Argentina, Chile, France, Norway, the USSR, the United Kingdom, and the United States, and an observer from Japan and a representative from ICSU. The ad hoc committee identified the need for an organization that could provide coordination for further scientific activities in Antarctica on a more or less continuing basis. The organization that resulted from this deliberation became known as the Special Committee on Antarctic Research (SCAR). SCAR was to be composed of one delegate from each country actively engaged in antarctic research and a representative from each of the following ICSU bodies: International Union of Geography (IUG), International Union of Geodesy and Geophysics (IUGG), International Union of Biological Sciences (IUBS), and Union Radio Scientifique Internationale (URSI).

SCAR was thus born of the desire of scientists to continue the international coordination of research in Antarctica following the IGY.

The Early Years of SCAR

The first meeting of SCAR was held in The Hague in February 1958. Of the 12 nations then engaged in antarctic research, Chile, New Zealand, and South Africa had no representatives at that meeting, even though they were eligible under the terms outlined by the ad hoc committee in 1957.

The main order of business at The Hague was the drafting of a constitution, the election of officers, and the preparation of a budget and making provisions for funding it. The SCAR constitution was ratified by ICSU at its eighth general assembly in October 1958. The first officers of SCAR were also elected at The Hague: president, G. R. Laclavere (France); vice president, K. E. Bullen (United Kingdom); and secretary, V. Schytt (IUG). The SCAR budget was set at $6,000 per year, with each of the 12 members contributing $500 toward this amount. To cover larger budgets in future years, it was decided that members would contribute additional amounts in proportion to the level of antarctic activity as measured by the number of wintering-over personnel.

At the fourth meeting of SCAR, in Cambridge in 1960, delegates from all 12 nations were present for the first time. At the fifth SCAR meeting, it was reported that ICSU wished SCAR to be renamed. The word "special" was replaced by "scientific" so that from then on, SCAR was the acronym for the Scientific Committee on Antarctic Research.

The Growth in SCAR Membership

The initial membership of SCAR consisted of delegates from the 12 nations that participated in antarctic research during the IGY (Argentina, Australia, Belgium, Chile, France, Japan, New Zealand, Norway, South Africa, the USSR, the United Kingdom, and the United States) plus representatives from IUG, IUBS, IUGG, and URSI.

The membership of SCAR remained constant for 20 years until Poland and the Federal Republic of Germany were admitted in 1978, bringing the total representation to 14 countries. Before then, however, representation from other organizations under the ICSU umbrella had increased somewhat.

The fifteenth country to gain representation in SCAR was the German Democratic Republic, whose delegate was seated in 1982. And finally, the sixteenth and seventeenth representatives, from Brazil and India, were voted into SCAR in 1984.

Other applications are pending. The Peoples Republic of China hopes to gain admission by the time XIX SCAR

Table 10-1 Nations with National Antarctic Committees that are Members of SCAR[a]

COUNTRY	YEAR ADMITTED
Argentina	1958
Australia	1958
Belgium	1958
Brazil	1984
Chile	1958
France	1958
Federal Republic of Germany	1978
German Democratic Republic	1982
India	1984
Japan	1958
New Zealand	1958
Norway	1958
Poland	1978
South Africa	1958
USSR	1958
United Kingdom	1958
United States	1958

[a]As of January 1985.

meets in 1986, and Uruguay has reported on its plans for an antarctic research program that could lead to membership in SCAR.

A list of all SCAR member nations with national antarctic committees and the years in which they were admitted to SCAR is given in Table 10-1.

SCAR STRUCTURE AND PROCEDURES

The SCAR Constitution

The SCAR constitution was a rather short and simple document when first formulated at The Hague in 1958. It consisted of a preamble, criteria for membership, and the basic principles to guide SCAR's functioning.

The preamble stated that "SCAR is a Special Committee of ICSU charged with furthering the coordination of scientific activity in Antarctica, with a view to framing a scientific programme of circumpolar scope and significance. In establishing its programme, SCAR will take care to acknowledge the autonomy of other existing international bodies."

The constitution defined the membership as one delegate from "each country actively engaged in antarctic research" plus one delegate from each of the international scientific unions federated in ICSU. Other special committees of ICSU could send observers to SCAR meetings.

The balance of the constitution dealt with the establishment of the SCAR executive (president, vice president, and secretary), authority to establish ad hoc committees, and a procedure for preparing the budget and fixing contributions by members as recommended by the budget committee.

All this information was published in 1966 in the <u>SCAR Manual</u>. That publication has been revised from time to time as necessitated by new developments, new members, and other changes in the structure, organization, and function of SCAR. The various changes are incorporated in a recent SCAR publication entitled "Constitution, Procedures and Structure, 1981." This document contains a revised SCAR constitution in which the relationship between SCAR and the consultative parties of the Antarctic Treaty is spelled out in considerable detail. The following statement, under the section entitled "Guidelines for the Conduct of SCAR Affairs" is particularly pertinent: "SCAR will abstain from involvement in political and juridical matters, including the formulation of management measures for exploitable resources, except where SCAR accepts an invitation to advise on a problem."

The revised constitution also provides for an alternate delegate in addition to the permanent delegate from each national committee adhering to SCAR, but each country is entitled to only one vote.

Other matters of interest in the revised constitution include an expanded SCAR executive consisting of a president, immediate past-president, two vice presidents, and a secretary. It should be noted that a member of the SCAR executive need not be a delegate from one of the countries adhering to SCAR but could be a member representing one of the international unions federated under ICSU.

Another feature of the revised constitution deals with the conditions for national membership in SCAR. The general condition for membership remains the same; that is, only countries actively engaged in antarctic research are eligible. The changes are in the manner in which membership applications are submitted and handled and the addition of a provision for withdrawing the voting rights of any member country that has not been active in the Antarctic or SCAR for a period of four years. Another change has to do with the granting of observer status to countries that are planning to establish scientific research activities in the Antarctic.

The way in which SCAR conducts its business has become more sophisticated since the 1960s, but, basically, SCAR has remained true to the principles and philosophy that have guided its activities since inception.

Procedures of SCAR

SCAR Executive

Continuity of leadership in SCAR is lodged in the five-member executive committee, commonly referred to as the SCAR executive. The executive consists of the president, the immediate past-president, two vice presidents, and the secretary, each of whom holds office for four years. The terms are staggered, however, to allow for continuity of leadership. The executive meets in odd-numbered years, usually at the SCAR headquarters in the Scott Polar Research Institute, Cambridge, the United Kingdom. These meetings are designed to maintain continuity between the biennial meetings of SCAR and normally consist of reviewing matters that were referred to it during the previous SCAR meeting or considering other items that need to be acted upon before the next SCAR meeting. The SCAR executive cannot, however, act on a membership application, since that decision rests only with the SCAR delegates at a regular meeting. The executive does review applications for membership to see that all requirements have been met, after which a recommendation will be forwarded to the delegates for action at the next meeting.

Working Groups

The core of SCAR lies in its nine permanent working groups: Biology, Geodesy and Cartography, Geology, Glaciology, Human Biology and Medicine, Logistics, Meteorology, Solid Earth Geophysics, and Upper Atmosphere Physics. Until recently, a tenth working group, Oceanography, was in existence, but the activities of that group were more or less merged with another ICSU Committee, the Scientific Committee on Oceanic Research.

Members of the working groups are selected from the SCAR member countries through their national committees, but some working groups have scientists from countries that are not members of SCAR if it is believed that their knowledge is useful to the deliberations of the working group.

Working groups meet during SCAR meetings or at times when it is convenient for a group to assemble in connection with some other international meeting. Normally, two to four working groups meet at the time of regular SCAR meetings, and much of the work is done by correspondence among the members. Each working group elects a secretary or chairman from its members, and this individual has the responsibility for calling and conducting meetings and submitting reports to SCAR. A working group may elect to create subcommittees, as is the case with the biology group, which has a subcommittee on conservation.

Groups of Specialists

When matters arise in SCAR that do not fit neatly within the purview of a single working group, groups of specialists are formed. These bodies usually are formed when multidisciplinary problems are involved or when SCAR is requested to provide advice to the Antarctic Treaty governments. All members in groups of specialists need not necessarily be representatives of national committees. When their assignments are completed, groups of specialists are discontinued.

Currently SCAR has five groups of specialists: Antarctic Climate Research, Antarctic Environmental Implications of Possible Mineral Exploration and Exploitation, Antarctic Sea Ice, Seals, and Southern Ocean Ecosystems and Their Living Resources.

It is worth citing an example of how a group of specialists can influence the course of antarctic research. The Group of Specialists on the Living Resources of the Southern Ocean (later merged with SCAR as Working Group 54) designed an extensive program of biological research in the Southern Ocean entitled Biological Investigations of Marine Antarctic Systems and Stocks (BIOMASS). The main objective of BIOMASS was to study the ecosystem of the ocean surrounding the antarctic continent, with special attention to the life history, special distribution, and abundance of krill (<u>Euphausia superba</u>). This small shrimp-like crustacean is one of the main elements of the food web in the Southern Ocean and is a possible candidate for human exploitation.

In its early days, BIOMASS was a child of SCAR, which encouraged the planning and development efforts that later materialized into two major collaborative, multiship investigations of the Southern Ocean. The first of these, the First International BIOMASS Experiment (FIBEX), executed in 1981, involved 13 ships from 10 nations working on synoptic data collection in the Scotia Sea and Drake Passage. The second effort of the BIOMASS plan has the apt acronym SIBEX (Second International BIOMASS Experiment). It is similar in scope to FIBEX but with additional emphasis on the nutritional dynamics of birds, fish, and mammals in parts of the Southern Ocean not covered by FIBEX.

The value of these investigations will be increased once the data collected are analyzed and evaluated in the BIOMASS data center. The results therefrom could represent valuable information for use in managing the living resources of the Southern Ocean under the Convention on the Conservation of Antarctic Marine Living Resources.

The establishment of BIOMASS under the sponsorship of SCAR is an example of how an initiative involving multinational participation can be launched successfully.

While it may not have been impossible for a single nation to put together a program similar to BIOMASS, it seems highly improbable that such an initiative from a single country could have generated the high interest and strong enthusiasm that was encountered by the founders of BIOMASS under SCAR sponsorship.

Publications of SCAR

In addition to the SCAR Bulletin, which is published in January, May, and September of each year, in English in the Polar Record and in Spanish by the Instituto Antarctica Argentina in Buenos Aires, SCAR publishes various reports of a special nature. These reports are ad hoc in nature and do not constitute a series in any sense of the word. An example of one of these is the Report on Possible Environmental Effects of Mineral Exploration and Exploitation in Antarctica published by SCAR in 1979. Because of the importance of this subject, and because the report constituted the response of SCAR to a request from the Antarctic Treaty governments, SCAR decided that it was worth publishing. Other occasional publications by SCAR are issued from time to time.

The main vehicle for communications within the SCAR organization is the SCAR Circular. It provides a means to request information from national committees on various matters that SCAR wishes to address or to convey to national committees information from the SCAR executive or the SCAR secretariat. While the SCAR Circulars are not archival publications in the strict sense of the word, they are serially numbered and contain much information of value and importance to the national committees.

SCAR requires each member nation, through its national committee, to submit an annual report on its ongoing programs of research and other activities in Antarctica. National committees are required to submit to SCAR on June 30 of each year information on research programs of the preceding year, including the current winter season. Additionally, the national report must contain a list of the occupied stations with their latitudes and longitudes, plans for the following year for both summer and winter, and a bibliography on publications related to antarctic research that have been published since the previous report. These national reports are distributed widely by SCAR to all national committees so that there is a continual flow of information circulated to all who are active in antarctic research.

SCAR takes these national reports very seriously and goes so far as to specify the format of the report and the size of the paper on which they are printed. This is the only instance in which SCAR has shown any sign of adopting bureaucratic measures, but given the use to which these reports are put, the uniformity of style and format is entirely justified. To my knowledge, no national committee has ever deviated from the form prescribed.

SCAR Symposia

From time to time since its inception, SCAR has sponsored symposia on a variety of subjects related to Antarctica. Most symposia are sponsored by SCAR in association with some other international organization, and a publication usually results. While SCAR provides some funding for these symposia, additional subventions are required to fund travel of participants and publication of the papers presented. Some of the volumes generated as an outgrowth of SCAR-sponsored symposia represent major additions to the scientific knowledge of the antarctic continent and the Southern Ocean. More than two dozen symposia have been sponsored or cosponsored by SCAR since the first one on Antarctic Meteorology was held in Melbourne in 1959.

SCAR Meetings

In the early years, SCAR met every year, but later the routine of biennial meetings became the norm. Meetings are held in one of the member countries, as invitations are extended from national committees. The host country provides all meeting sites and other amenities to the delegates at no cost to the SCAR treasury. All delegates provide their own travel and other expenses, but SCAR does fund travel expenses for the executive.

It has been the custom of SCAR to alternate its meeting sites between member countries in the Northern and Southern Hemispheres. This informal arrangement may be difficult to follow in the future, given the fact that 11 of the 17 member countries lie north of the equator and six lie south, if Brazil is considered a Southern Hemisphere country. The dates, sites, and designated numbers of all meetings of SCAR through 1984 are given in Table 10-2.

Table 10-2 Dates, Sites, and Designated Numbers of SCAR Meetings, 1958-1984

Meeting Number	City	Country	Dates
I	The Hague	The Netherlands	February 3-5, 1958
II	Moscow	U.S.S.R.	August 4-11, 1958
III	Canberra	Australia	March 2-6, 1959
IV	Cambridge	United Kingdom	August 29-September 2, 1960
V	Wellington	New Zealand	October 9-15, 1961
VI	Boulder	United States	August 20-24, 1962
VII	Capetown	South Africa	September 23-27, 1963
VIII	Paris	France	August 24-28, 1964
IX	Santiago	Chile	September 20-24, 1966
X	Tokyo	Japan	June 10-15, 1968
XI	Oslo	Norway	August 17-22, 1970
XII	Canberra	Australia	August 7-19, 1972
XIII	Jackson Hole	United States	September 2-7, 1974
XIV	Mendoza	Argentina	October 11-23, 1976
XV	Chamonix	France	May 16-27, 1978
XVI	Queenstown	New Zealand	October 14-25, 1980
XVII	Leningrad	U.S.S.R.	July 5-9, 1982
XVIII	Bremerhaven	Federal Republic of Germany	October 1-5, 1984

[a]SCAR refers to its meetings by Roman numerals. For example, the 1968 meeting in Tokyo is designated as X SCAR.

Normally, meetings last for two weeks. The first week is devoted to meetings of working groups and groups of specialists, and the second week is reserved for the plenary sessions and meetings of delegates. This format has worked well and is likely to be followed in the future, even with an expanded membership.

The SCAR Secretariat

Until XI SCAR, in Oslo in 1970, SCAR functioned with a secretary elected from the membership and with no permanent secretariat. At XII SCAR, in Canberra, however, the services of an executive secretary were in place, and a permanent SCAR secretariat had been established at the Scott Polar Research Institute in Cambridge.

George E. Hemmen was the first executive secretary and serves admirably in that capacity to this day. His regular duties are with the Royal Society in London, but he manages about one day a week in Cambridge. He is ably assisted by a secretary, Jane Whiting, who spends full time in Cambridge dealing with correspondence and a variety of tasks connected with the secretariat. SCAR is indeed fortunate to have these two dedicated people looking after the day-to-day affairs of its members.

THE INTERACTION OF SCAR WITH THE ANTARCTIC TREATY SYSTEM

The Antarctic Treaty was signed in 1959 and entered into force on June 23, 1961, after ratification by the governments of the original 12 contracting parties, the same 12 that constituted the initial membership of SCAR. SCAR is not explicitly mentioned in the treaty, but in the minutes of the first Antarctic Treaty consultative meeting, in Canberra in July 1961, SCAR was referred to several times in various recommendations. These included, among others, the following wording in Recommendation I-IV: "(1) that the free exchange of information and views among scientists participating in SCAR, and the Recommendations concerning scientific programmes and cooperation formulated by this body constitute a most valuable contribution to international scientific cooperation in Antarctica; (2) that since these activities of SCAR constitute the kind of activity contemplated in Article III of the treaty, SCAR should be encouraged to continue this advisory work which has so effectively facilitated international cooperation in scientific investigation."

These words leave little doubt about the high regard of the treaty parties for SCAR and its programs. This first indication of the treaty consultative parties' respect for SCAR has continued ever since. Whenever the treaty parties are in need of scientific advice concerning Antarctica, they have come to SCAR.

Requests to SCAR for advice and information are made in a formal way by the treaty parties. These recommendations are designated by a numbering system. For example, for Recommendation VIII-14, the Roman numeral refers to the eighth Antarctic Treaty consultative meeting and the number 14 identifies a specific recommendation. Recommendations to SCAR have emanated from many of the consultative meetings and are too many in number to review in these pages. Generally, these recommendations have ranged over a variety of topics, including a resolution at the first consultative meeting urging the contracting parties to be guided in their conservation policies by the recommendations of SCAR. At the third consultative meeting, in 1964, the treaty consultative parties adopted a series of agreed measures on antarctic conservation based on SCAR recommendations. In the same context, the treaty nations set aside certain parts of the Antarctic as Sites of Special Scientific Interest and other areas designated as Specially Protected Areas on the advice of SCAR.

Other matters on which the treaty nations have called on SCAR for advice and guidance cover such subjects as logistics, telecommunications, living resources of the Southern Ocean, and the exploration and exploitation of mineral resources in Antarctica. The last matter deserves some additional comment because of its general interest not only to the consultative parties, but to other nations around the world.

The question of antarctic mineral resources had never been raised formally in SCAR until 1976. Meeting in Mendoza, Argentina, that year, XIV SCAR addressed a recommendation from the eighth Antarctic Treaty consultative meeting held the previous year in Oslo. Emanating from that meeting was Recommendation VIII-14, which invited SCAR to "make an assessment on the basis of available information of the possible impact on the environment of the treaty Area and other ecosystems dependent on the antarctic environment if mineral exploration and/or exploitation were to occur there."

SCAR was apprehensive about its response, lest it be inferred that by taking on the assignment SCAR was tacitly endorsing a move toward exploitation of mineral resources in Antarctica. Both SCAR and the treaty nations had avoided the minerals issue until it was forced on their agenda by international events beyond their control. The main reason why the minerals question came to the fore at this particular time was the quadrupling of the price of crude oil in 1973-1974 by the Organization of Petroleum Exporting Countries.

To formulate its response to the treaty nations, SCAR organized a group of specialists, which wrote a report in time for the ninth Antarctic Treaty consultative meeting, in London in October 1977. This report was also published by SCAR in 1979. This was the beginning of an ongoing relationship between SCAR and the treaty nations on the minerals issue. SCAR's role has diminished, however, since the consultative parties have met on several occasions to forge a separate accord as part of the Antarctic Treaty System to deal with the question of the extractive industries, should they ever gain a foothold on the continent or in the continental shelves surrounding it. It should be emphasized again that SCAR's role in this matter was confined to factual information and the scientific interpretation of those facts. It must be noted, however, that many individuals connected with SCAR also serve as advisers to their respective governments on Antarctic Treaty matters. In so doing, these individuals are careful to keep their roles in SCAR separate and distinct from their roles in treaty matters.

In summary, it can be said that the consultative parties and SCAR play separate but mutually beneficial roles in the international affairs of Antarctica. The success of this relationship is based on two observations: First, the consultative parties derive their authority from the Antarctic Treaty. Second, the success of SCAR is based not on the authority of the SCAR constitution but rather on the experience and scientific reputations of its members and working groups. Included in these are most of the world's leading experts in antarctic affairs, both scientific and logistic. Collectively, these experts constitute the greatest concentration of talent related to antarctic science and attendant technology ever assembled. For this reason, the consultative parties in the Antarctic Treaty System are likely to continue their dependence on SCAR for scientific and technical information for as long as the treaty and associated agreements and conventions are in force.

A LOOK AT SCAR'S FUTURE

During the first 20 years of its existence, SCAR changed very little. The membership was stable during this period, most of the delegates and working group members were experienced veterans of countless IGY and post-IGY antarctic scientific expeditions, and most of the matters addressed were scientific or logistic. Since the middle to late 1970s, however, SCAR has undergone considerable change in its membership, in its constitution, and in its agenda.

The number of nations with SCAR delegates has increased from 12 to 17, with more to come. The number of delegates representing other ICSU bodies has increased also, although many of these delegates are also delegates from national committees. In addition, the cadre of alternate delegates has swelled the number of individuals in attendance at SCAR meetings to more than twice what could be expected in the early days. Also, women are now active in antarctic science, a development that was unheard of during the IGY.

Many of the IGY veterans who were active in leadership roles in SCAR during its first two decades either have retired or are about to retire. It is therefore inevitable that SCAR will have to replace the old generation of SCAR scientists with a new generation. This may be traumatic for some who resist change wherever it occurs, but the new generation is ready and waiting to assume important roles in the affairs of SCAR.

To accommodate nations that aspire to membership in SCAR but have not yet satisfied SCAR's requirements, SCAR is studying the possibility of creating a new class of members. These "associate members" would consist of nations that are gearing up for ongoing antarctic programs but may be several years away from their implementation. If such countries were accorded the status of associate membership and allowed to participate in SCAR in a limited capacity, their plans for antarctic operations could be enhanced by what they might learn by participating in SCAR. A recommendation is expected on this matter at XIX SCAR in 1986. If such a proposal is adopted, it will require yet another change in the SCAR constitution.

Perhaps the greatest challenge to SCAR lies in its ability to deal with other international bodies and address nonscientific issues without compromising the distinction between science and politics. SCAR has adhered rigidly to this distinction in the past and has been well served by so doing. But the line between science and politics has become more finely drawn, and SCAR must exercise constant vigilance to avoid becoming tangled in policy matters that, while they may relate to scientific activities, are the business of the consultative parties that administer the affairs of the Antarctic Treaty and related agreements.

SCAR must recognize that the scope of its activities will be broadened in future years. It has already responded to a request from the U.N. for information in connection with the U.N. discussion on the question of Antarctica. Along with this broadening of its agenda, SCAR will have to become more responsive to groups and organizations that have developed an interest in antarctic affairs. SCAR can no longer function exclusively as a closed group whose members speak antarctic jargon to one another at SCAR meetings and within the confines of the working groups.

A first step in this direction will be taken when SCAR joins the International Union for the Conservation of Nature and Natural Resources in April 1985 in the sponsorship of a symposium on scientific requirements for antarctic conservation. Other opportunities for joint sponsorship will undoubtedly be forthcoming from other organizations in the future, and SCAR must measure each such request against its basic mission. Moreover, SCAR must avoid being drawn into a position of advocacy, no matter how tempting some of these positions might appear to be.

As an organization, SCAR is an advocate only of the continuation of high-quality scientific research programs in Antarctica in accordance with the words in the preamble of its constitution. So long as SCAR can keep this mission in the forefront of its activities, and so long as its scientists can maintain strong programs of high scientific merit, SCAR will continue to flourish as the only international body dedicated solely to the advancement of knowledge on this unique area of the planet Earth. There is nothing in SCAR's past or present behavior to indicate that it will deviate from the mission that it established for itself at The Hague more than a quarter century ago.

11.

The Role of Science in the Antarctic Treaty System
E. Fred Roots

A distinctive feature of Antarctica is that for the past 70 years most human activities, both on the continent and in the surrounding ocean waters, have been dedicated directly to scientific research. In no other large part of the world is the pursuit of science such a dominant activity. For the past 25 years, science in Antarctica has been carried out under the terms of the Antarctic Treaty, and national and international scientific research programs have been the most conspicuous visible expression of the workings of the treaty. It is therefore pertinent to examine the relation of science to the politics of Antarctica and the role of scientific research in the Antarctic Treaty System today.

BACKGROUND

In 1959, amid the tensions and conflicts of claims, disputes, and the contrasting goodwill and enthusiasm generated by the successful international scientific cooperation of the International Geophysical Year (IGY), the nations that had IGY programs in Antarctica were invited to a meeting in Washington. They were asked to conclude a treaty designed to preserve the antarctic continent as an international laboratory for scientific research and to ensure that it be used only for peaceful purposes. Thus, from the beginnings of discussion leading toward the Antarctic Treaty, science, international cooperation, and maintenance of peace were inextricably linked.

However, the recognition of the value of science in the polar regions and the essential need for international cooperation to achieve worthwhile scientific results in these areas did not originate with the IGY; the IGY was a direct descendant of the International Polar Year (IPY) of 1882-1883. The IPY had marked an enormous step forward, not only in the gathering of systematic knowledge through the organized cooperation of many nations but also in the realization and acceptance of the concept that careful study of natural processes in the polar regions would be of direct benefit and relevance to nations in all parts of the world.

Although like all major international scientific events it would not have come about had not the state of science been ready for it and had not many people in many countries endorsed the idea, it grew from the inspiration, conviction, and persuasiveness of one man, Karl Weyprecht of Austria. Weyprecht was a physicist and naval officer, interested in the aurora and magnetism, and he had a passion for polar exploration. During the German North Polar Expedition of 1872-1874, of which he was coleader, he became convinced that geographic exploration for its own sake or for national glory should be replaced by international science carried out according to a cooperative plan. On his return he campaigned vigorously in academies of science and prestigious scientific institutions throughout Europe for a new approach to the study of polar regions. He stated that traditional polar exploration had been nothing more than an international steeplechase to reach the poles, where "immense sums were being spent and much hardship endured for the privilege of placing names in different languages on ice-covered promontories, but where the increase in human knowledge played a very secondary role".

Weyprecht claimed that the polar regions offered opportunities unparalleled anywhere on the planet for scientific studies of the Earth's physical and natural processes. He drew up a set of principles for research in the polar regions that would integrate polar science with the science of the rest of the planet. Some of those principles, enunciated 110 years ago, are very pertinent in the light of present discussion on Antarctica:

> The earth should be studied as a planet. National territories, and the Pole itself, have no more and no less significance than any other point on the planet, according to the opportunity they offer for the phenomena to be observed;

Science is not a territory for national possession;

Small nations must be able to take part in polar research;

Scientific knowledge of lasting value can result from coordinated and cooperative studies undertaken according to an agreed plan, with the results of the observations freely shared without discrimination.

The ideas of Weyprecht and his friends led to the creation of the 11-nation International Polar Commission, which drew up a coordinated and synchronized program of observation and study in meteorological and geophysical subjects in the north and south polar regions. This program became the IPY, in which 11 nations sent 14 separate but simultaneous expeditions to the Arctic and the sub-Antarctic. The IPY marked a new approach to polar investigation: instead of going where no one had been before, the sites for study were selected in advance for their expected ability to provide particularly useful observations of meteorology, magnetism, and aurora. About 20 special observing stations were also set up in subpolar regions, which cooperated in taking simultaneous observations with 39 permanent observatories already established in 25 countries around the world. Research in many other fields, from anthropology to biology and oceanography, was carried out by the expeditions as afforded by opportunity and location. In southern latitudes, the IPY was represented by a German expedition to South Georgia and a French expedition to Cape Horn.

An important feature of the IPY was that all data from the basic program were standardized and reduced to a common format as far as practicable. They were published promptly by the International Polar Commission.

Starting with the idea of international cooperation to achieve effective study of the polar regions, it became truly the first coordinated multidisciplinary study of the whole planet. Its direct contribution to new scientific knowledge was impressive; but what is perhaps more important in the context of current discussions on Antarctica is the influence that this one, coordinated activity, focusing on science in the polar regions, had on international cooperation and in bringing about a change in the character of science. From being an exclusive or elitist pursuit, often jealously guarded for reasons of national or institutional prestige, science developed into an open activity, in which everyone qualified could take part and in which the results belonged to the whole world and the quality of the science was judged by criticisms of other knowledgeable scientists, not by patrons or clients.

The IPY also marked the beginning of the practice of governments to sponsor scientific research in a number of fields using public funds, even if the research was not carried out in the sponsoring country and the results would not be the sole property of the sponsor. Sponsorship was justified on the grounds that basic knowledge of natural phenomena in polar regions, particularly in subjects relevant to a country's own issues and research, was useful in itself and that pursuit of such knowledge was a good investment for the country concerned.

The conviction that openly shared scientific knowledge of the polar regions, obtained through international cooperation, was a good investment for the sponsoring country independent of any economic return from the region was confirmed through the second IPY in 1932-1933 and the IGY in 1957-1958 and became a philosophical cornerstone of the Antarctic Treaty. The IGY, as Dr. Rutford has pointed out in his presentation, brought systematic, international, multidisciplinary science to the antarctic continent.

Articles II and III of the Antarctic Treaty make it clear that among the principal purposes of the treaty are the guarantee of freedom to conduct scientific research, the encouragement of international cooperation in research, and the open exchange of scientific information and research results. But science and the world have changed in the 25 years since the treaty was first drawn up. In light of these changes, how has the treaty served scientific activities in Antarctica? Is open sponsorship of scientific activities in Antarctica, whose results are available to all, a good political and economic investment for the sponsoring countries? How well, in the present context, do scientific activities serve as the principal vehicle to carry out the objectives of the treaty; or do the present scientific programs make more difficult the achievement of other objectives of the treaty, such as resolution of international conflict and maintenance of nonmilitary status? To answer these questions, it may be useful to examine frankly the need for and importance of antarctic science in world affairs, the expectations that countries or scientists may have of antarctic research, and the use made of the results of antarctic research within the spectrum of interests and concerns of various countries.

THE POLITICAL ROLE OF SCIENCE IN ANTARCTICA

If science, as the open and organized pursuit of knowledge, is a cornerstone of the Antarctic Treaty, for what reasons do member countries endorse antarctic science? Do the adhering countries value the scientific knowledge sufficiently to justify their adherence to the treaty and their investment in antarctic programs, or is the cooperation in science a front or cover for some other commercial or strategic interest or perhaps merely a way of maintaining some involvement in the region in case some other opportunity may become apparent in the future?

It is clear that the involvement of 30 nations in antarctic activities under the treaty over a period of 25 years cannot be purely altruistic or a response to the enthusiasm or persuasiveness of a few individuals. Nor can it be simply a gesture of endorsing international goodwill or the national worship of pure science as an end in itself. Over this period of time, with all the political and economic vicissitudes that occur in different countries, the reason for sustained scientific activity by many nations in Antarctica has to be national self-interest. And, to justify continuing political as well as scientific attention, the national benefits must not only have outweighed the costs but also have prevailed against other competing uses for resources or expertise.

Scientific studies and their results in the form of new knowledge are the main products that emerge from the political attention and economic resources devoted to Antarctica under the treaty. It is thus the science itself that must justify the political and national interest and investment. And if, as was pointed out by Weyprecht 100 years ago and amply confirmed by subsequent activities in both polar regions, science is not a territory for national possession, and the knowledge of most lasting value is obtained through international cooperation and open sharing of information, the contribution of science in Antarctica to the political self-interest of the individual countries involved must be through nonexclusive, cooperative actions and information that is available to all. How can states maximize their self-interest by participating in an activity in which others are equal beneficiaries and by accepting an obligation to share the results of their own efforts freely with everyone else?

There appear to be three ways in which open, freely shared science can serve the political interests of the member states in Antarctica: (1) by delivering a product, in the form of publicly available scientific knowledge or information, that contributes in some identified way to national issues, contributes to national wealth, or helps overcome recognized national problems; (2) by serving to increase domestic scientific and technical expertise and capacity; (3) by providing a vehicle for international contact and influence.

DIFFERENT APPROACHES TO SCIENCE IN ANTARCTICA

Over the years, the political and public justification for scientific activities in Antarctica must be in terms of its contribution to the economy; to social, cultural, and domestic development; or to foreign policy or strategic objectives. It is therefore useful to examine the antarctic science program, as summarized in Chapter 9 by William Budd, to see how the results of the program contribute to those objectives. As was pointed out by W. Ostreng in connection with science in the Arctic regions, the contribution of science to national objectives tends to be viewed, by the public and the politicians, in three ways:

(1) In a functional sense, which views science as a tool or a means to achieve some previously agreed objective, such as scientific prospecting to find mineral deposits, to make money;

(2) In a pragmatic or opportunistic sense, in which the objective is to increase scientific knowledge or capacity in areas of national interest and then to use that knowledge to solve problems or develop opportunities, for example, studies of climate change, which could lead to knowledge of immense economic or social importance but in ways as yet unknown;

(3) In an ideological or academic sense, in which every opportunity to increase genuine knowledge or understanding should be taken because all

knowledge is ultimately of value and the practical
utility of even a portion of the new knowledge
will be sufficient to justify the total
investment.[1]

All three points of view are valid as ways at looking
at the contribution of science to national objectives.
All science is, of course, a gamble, in that the results
of research cannot be predicted in advance and the practical economic or political payoff, if any, may be very
protracted in its effect or may benefit someone other
than the sponsor. Experience has shown that science
undertaken in a functional sense, while it can contribute
most directly to identified goals, tends to be marginal
in its net payoff, for when it is successful it tends to
be increased to the point of nonsuccess. If you find one
gold mine, you or others are tempted to use up all the
profit looking for the next one. Academic research, in
the best sense, usually requires smaller investment and
has a larger net payoff, although the individual beneficiaries are rarely identifiable in advance. The
scientific work done to date in Antarctica has shown few
positive results from the so-called "functional" approach
to science except, and this is still unproven, with
regard to marine living resources. But the payoff from
science in a pragmatic or academic sense has already been
great. It has sustained antarctic activities in terms of
national self-interest since at least the IGY, and the
future looks even more exciting.

WHAT RESULTS CAN ANTARCTIC SCIENCE DELIVER?

Seen from a nonscientific national or international
interest viewpoint, the results of scientific research in
Antarctica may provide a number of services:

(1) *Prediction of environmental and geophysical conditions and changes.* Obvious examples are weather
observing and forecasting activities and, on a longer
scale, the studies that lead to better estimates of what
might happen to the climate because of natural or human-
induced influences. Other areas in which antarctic
science is vital to predictions with important socio-
economic effects lie in the fields of oceanic circulation,
geomagnetism, ionospheric variations, earth tides, and
geodesy. In each of these examples, it should be noted

that it is not what is happening within the antarctic regions themselves that is of prime importance to interested nations, although what is happening there may be of outstanding scientific importance. It is that the information from Antarctica is important, or in some cases absolutely vital, to improved scientific predictions about phenomena that affect the home country and other parts of the world.

(2) **Monitoring of environmental, biological, or geophysical conditions.** Many of the scientific programs in Antarctica take advantage of the unique location and conditions of Antarctica in order to make continuing observations of a kind that could not be made anywhere else on the planet, to detect and keep track of regional or global changes in natural or human-induced conditions or characteristics. Such monitoring may range from timely information on short-term effects due to human activities, such as fallout of radioactive particles in the atmosphere or the spread of persistent chemical pesticide residues around the world, to more subtle but potentially enormously important changes in the thickness and stability of the large ice sheets or the position of the antarctic convergence in response to climate changes. (The antarctic convergence defines the northern boundary of the Southern Ocean, where cold, southern surface waters are pushed below warmer, northern waters moving south. It fluctuates between 47°S and 63°S latitude)

Many features of Antarctica make it ideal and unique as a monitor of global conditions and changes. Some of these features are

- Its remoteness from significant human population and centers of industry and large-scale conversion of energy;
- The circumstance that much of the air descending onto the interior of Antarctica has traveled at higher levels from lower latitudes and thus has a long "residence time" in the atmosphere, while that near the surface over the Antarctic Ocean is thoroughly mixed in one of the most turbulent zonal circulation belts in the world;
- The thick cover of material of uniform composition and physical properties (ice) that blankets most of the continent and provides continent-scale laboratory cleanliness and measurable near-homogenous background to observations ranging from reception of low-frequency radio waves to the

sampling of microbes and the history of atmospheric chemistry for the past 30,000 years;
- The circumstance that Antarctica covers the geographical South Pole and the south magnetic and geomagnetic poles, so that variations and events connected with rotation of the Earth and the nearly vertical magnetic-field lines can be observed directly. This is the only place from which observations of these phenomena can be made, because many of the counterpart locations in north polar regions are on the Arctic Ocean, where fixed observing stations are not possible.

In many subject areas, Antarctica thus serves as a unique world sentinel of global environmental and geophysical changes. Its value to the world is much greater than the value of the information about Antarctica itself; measurements from Antarctica provide information about what is happening in the rest of the world that cannot be obtained from anywhere else in the world.
(3) <u>Knowledge and understanding of basic geological, geophysical, biological, and oceanographic processes and of human adaptation</u>. As science in Antarctica has matured from exploratory reconnaissance and careful description of what is there, the questions of how it got that way and how its natural processes operate have become more sharply focused. Today, "process" studies, rather than descriptive survey, lie behind much of the most important and challenging research in antarctic regions. The list of disciplinary and multidisciplinary projects to explore the frontiers of human knowledge that can be carried out only in Antarctica is limited only by the imagination and originality of scientists themselves. The pertinence of such studies to world science lies in that, in many ways, Antarctica is positioned quite literally at "one end of the Earth"; it provides one pole or extreme natural example for many living and nonliving natural processes.

In addition to the natural conditions that make it singularly useful as a global monitor site (see above), other characteristics of Antarctica make careful research in and around the continent of particular value to world science and its application. Only a few examples will show the scientific significance of this unique region:

- The area has a unique geological history. Antarctica was part of the ancient supercontinent Gondwanaland, which included the other Southern

Hemisphere continents and India. After the breakup of
Gondwanaland, Antarctica remained physically isolated
and remarkably free from later geological convulsions
until it collided with the "Pacific rim" in
comparatively recent geological times. Study of
antarctic geology is providing distinctive information
of direct value to the geological interpretation of
the other fragments of Gondwanaland; and some of the
basic geological processes well displayed in antarctic
rocks, despite the absence of known mineral deposits,
are of value in understanding the processes of
mineralization and aid in the search for mineral
deposits elsewhere in the world.
- The low angle of solar radiation and its marked
seasonal variation throughout the year have led to a
distinct climatic regime, which in turn has led to
distinctive oceanographic, geomorphological, and
biological processes and responses. For example, the
annual formation and disappearance of sea ice around
Antarctica means that an enormous area undergoes an
annual change from open water to ice cover and back
again. This change causes a drastic change in surface
reflectivity and in exchanges of heat and gases between
water and atmosphere. It is by far the largest annual
cyclic variation of surface conditions anywhere on the
planet, with all sorts of biological, physical, and
planetary thermal implications that are, at this
stage, poorly understood.
- Antarctic ecosystems have adapted to the lowest
levels of net intertrophic flows of biological energy
of any ecosystems known. The processes of this adaptation
and evolution are seen in genetic, morphological,
and behavioral characteristics whose understanding
promises to add much to our fundamental understanding
of the processes of life itself and thus to the relationships
between the biological and the nonbiological
worlds, to the stimuli and processes of biological
evolution, and to the responses of organisms and
populations to physical stress and energy availability.

These, and many other examples that become apparent in
any review of antarctic research, confirm after 25 years
the wisdom inherent in the invitational letter to the
1959 Washington Conference, which promoted the value of
an international agreement "to preserve the antarctic
continent as an international laboratory for scientific
research."

(4) _Information for management, apportionment, exploitation, and conservation of antarctic resources._ Scientific study is also an obvious component both of the identification and exploitation of living and nonliving resources in the modern context and of their management and conservation. In Antarctica, the use of science in connection with resource exploitation and management applies most directly to marine living resources, where international programs such as the Biological Investigations of Marine Antarctic Species and Stocks provide the essential scientific background to treaty activities under the Convention on the Conservation of Antarctic Marine Living Resources. The scientific work carried out to this end has elements of the research noted above related to prediction, monitoring, and understanding of natural processes in Antarctica. In addition, sophisticated science is involved in the development of means to acquire an adequate data base for management; for development of technologies for detection, harvesting, and processing of the resources; and for obtaining information pertinent to enforcement of provisions for sustained management and environmental protection.

With respect to mineral resources, there is generally a distinct difference in approach and method between geological and geophysical studies carried out for purely scientific purposes, and prospecting or minerals exploration carried out to locate mineral deposits. Nevertheless, almost all good geological and solid-earth geophysical information is useful to assessment of the potential for mineral deposits. The geological information obtained to date in Antarctica has not revealed areas of high promise for the discovery of metallic mineral deposits on the continent or of hydrocarbons in the offshore areas. Combined with the estimates of the technological difficulties, the expense of finding, developing, and producing minerals, as well as the distance and difficulty of transport to a major market, has made Antarctica a poor prospect indeed, compared with almost any other place on the planet, to search for mineral or hydrocarbon resources or to hope to extract them profitably if found. This is the case even before the severe problems of producing minerals or oil while preserving the integrity of the vulnerable antarctic environment are taken into account. (For further discussion of potential minerals development in Antarctica, see Chapters 17-20.)

At the same time, as noted above, the study of basic geological and environmental processes in Antarctica is contributing to improved understanding of the formation and location of mineral deposits. This can be applied to the search for and development of minerals in other parts of the world. Science in Antarctica is therefore relevant to minerals exploitation because it facilitates a realistic but, on the basis of present knowledge, not encouraging assessment of commercial minerals potentials there. Worldwide, antarctic science also enhances the prospects for profitable minerals development in other parts of the world.

The possibility of development of the ice resources of Antarctica as a practical source of fresh water represents a special case of the application of science to mineral resources development and export. There is no problem of adequacy of supply. But glaciological, oceanographic, meteorological, and engineering sciences will be heavily involved in studies of technical feasibility, economic feasibility, reliability, and environmental effects if current concepts are to be developed to a serious design and economic evaluation stage.

(5) <u>Services to Conservation, Awareness, and Communication</u>. Apart from, and also in addition to, the roles that science in Antarctica can play in providing knowledge that is pertinent to problems of the world environment and management and optimum use of resources, scientific study and knowledge of Antarctica play an important cultural role in helping to achieve conservation of the region's unique features, characteristics, and ecosystems.

Antarctica is in large part bleak, forbidding in human terms, and less supportive of life than any other extensive region on the surface of the planet. It is also one of the most beautiful parts of the world. Despite its short human history, it has generated powerful myths and holds an impressive grip on the minds and emotions of several societies. Penguins, whales, icebergs, and even blizzards have an appeal far beyond any rational appraisal of their place in nature. Evidence of this can be seen every year in the high national and international public interest attached to amateur sporting and quasi-scientific exploration ventures to Antarctica, in the popularity and high quality of antarctic picture books in nearly every language, and in the growing industry (and the growing environmental problems) of tourism to the region.

Scientific study of antarctic biology, glaciology, geomorphology, and oceanography plays an important role in shaping the worldwide popular interest in Antarctica. It also has a heavy responsibility, under the treaty, to ensure that public interest is channeled in constructive directions. The establishment of designated protected areas [Sites of Special Scientific Interest (SSSIs)] reflects an initial and timely response to the recognition that some areas are particularly vulnerable to disturbance, either inadvertently or in the course of narrow minded scientific investigation. The agreement to designate SSSIs under the Antarctic Treaty System is not only a positive act of conservation, however local, and a recognition that some places and parts of the antarctic web of life are particularly vulnerable to disturbance; it is also an international statement that science in Antarctica is of international importance and that special scientific places must be protected, even against disturbance by scientists themselves.

As Antarctica becomes more "popular," its special features will on the one hand be likely to become more highly valued by people in other lands who are concerned about the environment and conservation, as noted by the special chapter on Antarctica in the World Conservation Strategy produced by the International Union for the Conservation of Nature and Natural Resources, the United Nations Environment Program (UNEP), and the World Wildlife Fund International. At the same time, the emotional and cultural appeals of the region will lead to increased visitation, tourism, and exploitation of its scenery and wildlife in ways that individually appear to be benign or insignificant but that collectively can be very destructive. Scientific study is essential for understanding the effect of human intrusion on the antarctic ecosystem and environment, in order to preserve priceless antarctic values and at the same time to contribute to the cultural and educational benefits that Antarctica can bring to the world at large. In the long run, this popular service by science may be one of the most important proofs of the effectiveness of the Antarctic Treaty System.

THE SETTING OF SCIENTIFIC PRIORITIES IN ANTARCTICA

If science is important in helping to serve the political needs of the Antarctic Treaty, and the treaty has a basic function to preserve the opportunity to pursue science, how are the priorities for science in Antarctica arrived at? Clearly all interesting or desirable subjects cannot be pursued, even by the most wealthy and ambitious research programs. Each country is responsible for the content and emphasis of its own studies; yet it is apparent that too many studies, each focusing on the same fashionable scientific problem of the moment, are not good for the science or for the working of the treaty.

It is in this area that antarctic science and the treaty are singularly fortunate in the existence and activities of the Scientific Committee on Antarctic Research (SCAR). SCAR, which was in existence before the treaty was drawn up, has served as an international coordinator and focus of scientific interest in Antarctica. Through its various panels and groups of experts, its symposia, and the annual reports of its national committees, SCAR serves as an international peer review group and custodian of the quality and relevance of science in Antarctica. Being a nongovernmental agency and nonpolitical, it has been remarkably free from political pressures. Yet its organization into a national committee from each consultative party state has enabled it to translate general scientific priorities into recommended national programs that are compatible with the respective priorities and capabilities of participating countries. One of the fortunate aspects of the Antarctic Treaty situation is that a nonpolitical, international, professional scientific body of the International Council of Scientific Unions, SCAR, has made it possible for science to play a strong political role within the treaty without becoming politicized or compromising its scientific integrity.

Article III of the treaty not only calls for exchange of plans for scientific programs to permit maximum economy and efficiency of operations, and for the exchange of personnel, observations, and scientific results, but also gives every encouragement to the establishment of cooperative working relations with other organizations having a scientific interest in Antarctica. This means, inevitably, that not only is science in Antarctica linked inexorably with science in the rest of the world but also that the priorities for science in Antarctica will be, to an increasing extent, determined by the priorities for science in the world as a whole.

THE FUTURE

Throughout the life of the Antarctic Treaty, scientific activities, and the openly shared scientific knowledge resulting from those activities, have become ever more firmly entrenched as the central reasons for the interests of many nations in Antarctica. The support of scientific activities has been found, over 25 years, to be a good investment by many nations as viewed in their own national and international interests. Antarctic science has proven that its principal value is as a contribution to world knowledge, providing information from Antarctica that is vital to environmental development and resource problems in other parts of the world. A secondary but important aspect of antarctic science is its essential contribution to environmental protection and the management of living resources in the antarctic region. To date, scientific activities in Antarctica have given very little or no encouragement to prospects for commercially exploitable mineral or hydrocarbon resources in Antarctica, with the possible exception of development of ice as a freshwater resource, but they have provided information on which any potential mineral deposits can be evaluated and the technical, economic, and environmental assessments carried out.

It can be expected that present trends regarding science in Antarctica will continue. Antarctic science will likely continue to grow more sophisticated and even more closely tied to world science as a whole. At the same time it will also focus on the "local" goals of protection of the environment, conservation of distinctive ecosystems, and management of resources within the antarctic region itself. There are a number of major world problems of enormous social and economic importance for the future, such as climate change, the distortion of biogeochemical cycles, and the accelerating disappearance of living species on a worldwide basis, to whose solutions science can make real contributions. But these solutions increasingly call for specialized and compatible information from all parts of the planet. Antarctica is a distinctive and very specialized part of the planet. What happens in Antarctica, and the physical and biological processes that can be studied more clearly in Antarctica than in any other planet, have an importance to the rest of the world that is only now being realized. Therefore, it can be expected that research in Antarctica will to a degree even greater than now become an essential component of major worldwide science programs.

For example, polar atmospheric and thermal transfer processes, and the climatic history recorded in the major ice sheets, are an important part of the World Climate Research Program. The need for research in Antarctica has been identified as a part of the world conservation strategy. Antarctic stations are becoming integrated into the Global Environmental Monitoring System of UNEP.

The International Geosphere-Biosphere Program (IGBP) now being developed under the International Council of Scientific Unions and in many ways a direct descendant of the IPY and the IGY is raising many questions about the relationships between physical phenomena and biological processes. Some of these relationships can best be studied in the most extreme of all terrestrial environments, Antarctica. It can be expected that in the next few years an increasing amount of important antarctic research will be related to the IGBP.

Thus, future science in Antarctica, conducted through the international cooperation and open exchange made possible by the Antarctic Treaty, can be expected to serve the purposes of the treaty. At the same time, it will be increasingly integrated with scientific research in other parts of the world, and its results will be judged in terms of its contribution to the major scientific, environmental, social, and economic problems of populated areas. Science in Antarctica is coming of age and affirming the statement made by Weyprecht 110 years ago that "The polar regions offer opportunities unparalleled anywhere on the planet for the study of the Earth's physical and natural processes."

REFERENCE

[1] Ostreng, W. 1985. Report of the Nordic Conference on Arctic Research, N. Aalesund, Svalbard, Norway, August 1-8, 1984. (In Norwegian.) Published by University of Trondheim (Trondheim, Norway), pp. 19-26.

12.

Panel Discussion on Antarctic Science

The panel consisted of Robert B. Thomson (moderator), Lewis M. Branscomb, and Omar bin Abdul Rahman.

SUMMARY

The papers presented in this session provided detailed and informative background on the development of scientific research in the Antarctic. Their subjects ranged from the earliest days of south polar exploration, through the exciting days of the "heroic age" at the turn of this century, to the period of the 1957-1958 International Geophysical Year (IGY), which established the international cooperation in the Antarctic that has been successfully maintained to this day.

All the speakers drew attention to the fact that the Scientific Committee on Antarctic Research (SCAR) and the Antarctic Treaty had originated with the IGY. They described SCAR as a nongovernmental organization composed of scientists from all countries active in Antarctica and intent on promoting continuation of the IGY's international scientific cooperation. The Antarctic Treaty was characterized as "operating in the world of international diplomacy," and the treaty nations as the U.N.'s "antarctic rangers."

REMARKS BY LEWIS M. BRANSCOMB

Branscomb identified three major phases in antarctic research and linked them to technological improvements. The initial phase covered the early explorers' meticulous recording of observations and collection of materials and provided an important base of information for those who followed. The second period began after World War II and culminated in the IGY of 1957-1958. During this phase, improved technology for logistics support (especially icebreakers, tractors, aircraft, and radio communications) permitted better trained professional scientists to take part in antarctic science. Nevertheless, field work conditions still limited the sophistication of instrumentation and support facilities that could be employed and thus continued to restrict scientists to largely descriptive work.

The third, and current, mature phase, according to Branscomb, has been emerging throughout the years of the Antarctic Treaty. It draws on the full panoply of sophisticated modern scientific tools permitted by improvements in antarctic field support capabilities. Branscomb noted that solar-powered satellite telemetry allows Imre Friedmann's laboratory in Florida to receive measurements of the environmental conditions in the Ross Valley sandstones, within which the endolithic organisms he is studying live. (Dr. Friedmann's lecture was one of three presented later to participants in the workshop. See below.) Other examples of modern technology include bringing the insights of molecular biology to bear on studies of highly complex ecologies in Antarctica, utilization of specialized telescopes that have been developed specifically to study longwave solar oscillations observable only at the South Pole Station, and inexpensive construction of a 26-mile-long antenna about a mile above the ground plane at Siple Station, located at the magnetic conjugate point from a sister station in Quebec.

In Branscomb's view, these examples illustrate four aspects of current antarctic science:

(1) That highly sophisticated field work can now proceed hand in hand with base laboratory science, assisted by antarctic base capabilities that supplement support available in home laboratories. This adds significantly to the depth of knowledge that can be obtained from any given study.

(2) That much of the most exciting antarctic science relies on imaginative synergy of the unique physical circumstances found in Antarctica, such as the ice sheet (which can be used as a concentrator of meteorites), the continuous availability of an overhead sun, an atmosphere uniquely lacking in water vapor, the lack of diurnal terrestrial tides, and the absence of pollution and other consequences of human modification of nature.

(3) That the application of modern scientific tools in Antarctica can reveal the history of the planet in many of its most important aspects, going back hundreds of thousands of years and, in geology and paleontology, millions of years. This history is recorded in ice cores and in the evolution of living species in Antarctica; it is waiting to be analyzed to provide insights about the origin, survival, and evolution of life. (Living species are relatively small in number in Antarctica, but extraordinary in their levels of adaptation to the unique environment.)

(4) That antarctic science deals more and more with the interdependence of the geological, biological, and climatological history of the continent, drawing on interdisciplinary studies and the increasingly complete mapping and classification of all antarctic regions. This requires increased cooperation among scientists working in the different regions and thus among active participants in the Antarctic Treaty. Since the enhancement of science programs depends on modern logistics and communications, this would also suggest that increased cooperation among respective national programs will pay dividends in the future to all participants.

These trends led Branscomb to conclude that antarctic research had become a carefully structured activity, integrated into the world knowledge base and the mainstream of world science. To illustrate how diverse its applications are, he remarked that the study of the glycopeptides in antarctic cod--that prevent the cod from freezing--could well find its first application in the use of synthetically produced compounds to make smoother ice cream.

With respect to resources in Antarctica, Branscomb commented that while exploration of biological and geological resources may reveal resources of value there, the greatest likelihood of practical reward from such studies is in the generation of useful knowledge that will facilitate the search for valuable resources in much more accessible parts of the world. In his view, the peoples of Third World countries will gain more benefits from antarctic science, at least in the near future, if they focus their attention on the knowledge benefits rather than on potential raw materials exploitation.

Finally, he advocated that countries with well-established support facilities and active programs of scientific investigation in Antarctica be encouraged to invite the participation of interested and qualified scientists from Third World countries. In this way they could establish firsthand relationships with the world community engaged in these studies.

REMARKS BY OMAR BIN ABDUL RAHMAN

Rahman confessed that as a neophyte to antarctic science he has been greatly impressed by the amount of excellent scientific work carried out in Antarctica. He identified the various factors motivating those conducting scientific research there as follows:

For countries close to Antarctica, their proximity requires that they learn as much about this "back yard" as they know about their front yard because it has direct bearing on their everyday life.

The wealthy, developed countries that can afford to devote substantial economic resources to scientific research are involved in Antarctica for the same reasons that they are involved elsewhere: their scientists are pushing back the frontiers of knowledge. They have other reasons as well, including that of "presence," which could mean many different things.

Countries in equatorial regions are directly affected by the cold continent with respect to their weather, oceanic currents, and the nutrients in adjacent seas. He cited India as an example of a developing country that feels the need to be involved in antarctic science because of the direct influences of Antarctica on the Indian subcontinent.

For small developing countries such as Malaysia, there are more important priorities closer to home; Antarctica is remote or at least is not of pressing interest. This does not mean that their concern for what goes on in Antarctica is less sincere or less genuine. Malaysia's scientists, according to Rahman, few in number though they may be, view themselves as part of the worldwide community of scientists and are interested in what goes on in Antarctica. He noted that what happens in the name of science in Antarctica can have global ramifications and that Malaysians are inhabitants of that globe.

Moreover, Rahman pointed out that research is not always carried out for the sake of acquiring knowledge;

it may be exploitative in nature or merely a front, as suggested in Dr. Roots' presentation (Chapter 11). He noted that it has also been said that in antarctica scientists are proxies of the big powers and carry out work of interest to the powers concerned.

While the speaker acknowledged that the international nature of antarctic science is generally accepted, he stressed that what was being questioned was the administration of Antarctica, the political/legal aspects of Antarctica. The recent response to this questioning from those in control had been initially patronizing, if not downright condescending. He characterized the reactions of those powers controlling the well-funded, high-technology programs as reminiscent of the old "go play with your colored beads; leave big magic well alone."

Rahman concluded by commending the exemplary international cooperation in Antarctica and expressing his wish that this same singlemindedness of purpose could be applied to solving such urgent world problems as the control of desertification in Africa.

SUMMARY OF THE DISCUSSION

Comments on the presentations emphasized the global significance of much of antarctic research and the consequent importance of continuing and expanding international cooperation in this research. This cooperation was deemed especially valuable in studying subjects for which a great deal of work remains to be done, such as the understanding of relationships among the atmosphere, oceans, and ice.

Motives behind the accession of new countries to the Antarctic Treaty were also discussed. There were those who believed that at least in some cases accession was driven by an interest in antarctica's resource potential rather than in scientific research. Other speakers tended to dispel this idea, demonstrating a genuine interest in participating in antarctic research based on its international importance. They stated that past actions should not be construed to indicate that these countries are not interested in antarctic science; they now wish to become better informed and realize that if they do not begin to show interest now, it will be more difficult for them to learn in the future.

There was widespread agreement that the scientific achievements during and after the IGY could be described as immense and that resulting publications have provided a wealth of scientific information. The point was made that this information, covering the whole spectrum of antarctic science, has not been restricted to the antarctic community, as believed by some; a vast amount of published material is available to the international community, in addition to all the data regularly provided to established world data centers. Antarctic research is, in effect, already internationalized, and its results are the "common heritage."

Nevertheless, in the view of some participants, the discussion underscored that while many attendees appeared generally well informed on Antarctic Treaty matters, they lacked knowledge of the role and work of SCAR and had little appreciation of the importance of antarctic research to improved understanding of global phenomena and problems. They noted that there appears to be a need for wide distribution of an informative publication on the real value of science in Antarctica.

One speaker noted that while scientific activities in Antarctica are likely to go on forever, commercial activities, with the likely exception of tourism, will probably occur only for a finite period on the order of 10 to 20 years. Resource activities will be important for their short-term effects, but they will be far less significant than science to the future of Antarctica. Over the long term, Antarctica is an ideal platform from which to monitor the rest of the globe and to continue the wide range of research identified by the speakers as so necessary and beneficial to humankind. Even in the short term, in discussing various uses of the Antarctic, science was believed to be the most important and immediate activity of economic importance to the international community.

There was some discussion of the types of activity that have occurred over the years in the north polar regions. Despite considerable resource development there, some participants maintained that science continues to be one of the main industries in the north as well.

It was noted that research is vital to the establishment of effective regimes for the conservation and management of resources as well as to the identification of potentially exploitable resources, such as minerals and icebergs.

One participant suggested that recent developments in high technology can be of great assistance to many antarctic programs, particularly those requiring continuous monitoring. For instance, he believed that the increased use of polar-orbiting satellites presents excellent opportunities for contributions to antarctic science programs.

The lateness of the hour and the proposed early start the following morning limited the discussion period. The chairman closed the meeting, expressing hope that additional time might be provided the following day to permit further discussions on the subject of antarctic science. Unfortunately, time did not allow for a further formal session to be scheduled, but many references were made to science during sessions devoted to related topics.

In addition, the U.S. National Science Foundation arranged for workshop participants to attend three informative lectures by U.S. Antarctic Research Program scientists at the Beardmore Camp. Anna C. Palmisano spoke on the ecology of sea ice microbial communities in McMurdo Sound, Imre Friedmann on endolithic microorganisms within antarctic rocks ("la dolce vita," as he characterized it), and William Cassidy on his work with meteorites discovered in Antarctica. Each scientist addressed the wider implications of his/her specific research project.

THE ANTARCTIC ENVIRONMENT:
MANAGEMENT AND CONSERVATION OF RESOURCES

A. CONSERVATION AND ENVIRONMENT

13. The Antarctic Treaty System as an Environmental Mechanism—An Approach to Environmental Issues
   John A. Heap and Martin W. Holdgate

14. Panel Discussion on Conservation and Environment

B. LIVING RESOURCES

15. The Antarctic Treaty System as a Resource Management Mechanism—Living Resources
   John A. Gulland

16. Panel Discussion on Living Resources

C. NONLIVING RESOURCES

17. Arctic Offshore Technology and Its Relevance to the Antarctic
   K. R. Croasdale

18. Discussion on Technology and Economics of Minerals Development in Polar Areas

19. The Antarctic Treaty System as a Resource Management Mechanism—Nonliving Resources
   Christopher D. Beeby

20. Panel Discussion on Nonliving Resources

13.

The Antarctic Treaty System as an Environmental Mechanism—
An Approach to Environmental Issues

John A. Heap and *Martin W. Holdgate*

INTRODUCTION

The Antarctic Treaty System is a management tool. It regulates human activities, by international consensus, over the whole area of land and ice shelf south of 60°S latitude. It does so with certain defined objectives in view, laid down in the treaty and interpreted in consultative meetings. Additional conventions have extended the regulation of certain human activities to a wider area of the Southern Ocean. From an environmental standpoint, the primary element in the treaty system is the requirement that the unique features of the Antarctic environment be safeguarded and made available to people of all nations for scientific research and their peaceful enjoyment. The ultimate objective of the treaty as an environmental mechanism is the harmonization of utilitarian, conservation, and aesthetic values. In this chapter we examine how far the treaty system has achieved this objective.

CHARACTERISTICS OF THE ANTARCTIC ENVIRONMENT

Publications about the Antarctic stress its distinctive and extreme features. Superlatives jostle one another. It is the coldest, highest, iciest, and most remote continent on Earth. It is surrounded by a continuous belt of the world's stormiest seas. It is fringed by the world's greatest expanse of floating ice and the world's largest icebergs. The two percent of land that is not covered in perennial ice supports the most impoverished plant and animal life of any continent, although the surrounding seas support comparatively productive marine ecosystems and immense populations of seabirds and marine mammals.[1,2,3]

These generalizations point to the unique characteristics of Antarctica, but they are of little point for our present analysis. We need to start by recognizing the contrast in the Antarctic environment between two broad types of subsystem:

(1) Small, but numerous, terrestrial areas where human activity can have a considerable impact, even if it is itself on a relatively modest scale;
(2) Large, broadly uniform marine and terrestrial areas capable of absorbing substantial human activity with little or no impact.

These subsystems require somewhat fuller description. The two percent of ice-free land exists as a series of mountainous rock outcrops, coastal strips, and islands. In these areas, and especially around the coasts of the Antarctic Peninsula and its off-lying island groups, there are areas of primitive soil that support a surprising complexity of moss, hepatic and lichen vegetation, and associated soil animals. In these regions there are a few patches of the two kinds of higher plant native to Antarctica (the grass Deschampsia antarctica and the cushion plant Colobanthus crassifolius), and in a few areas, higher insects (small wingless midges) are also to be found. In some places the slow growth of the mosses has built up frozen peat to a depth of 2 m. These ice-free coastal areas are also the breeding ground for very large populations of seabirds, and they contain many small lakes, ice-free in summer and themselves supporting communities of plants and invertebrate animal life. In addition, this is a region where seals haul out both to breed and to molt, fertilizing the soil and lakes with their excreta but damaging vegetation and soil by their wallowing. These areas thus exhibit a substantial degree of ecological interaction, which gives them high scientific interest. They stand out in any objective classification of Antarctic habitats. But they are also extremely fragile.

The soils and moss mats have developed since ice retreat many centuries ago. They are still evolving under the influence of percolating moisture and the deposition of nutrients by birds and seals, by spray from the sea, and by the action of plants and the soil fauna and microflora. Although they are naturally disturbed through the regular physical alternation of freezing and thawing, they do not quickly regain their natural patterns if they are disturbed by people or vehicles. The moss mats are very slow growing (rates of approximately 1 mm a year) and bear the scars of human pressure (for example, footprints) for years or even decades. The bird populations, although apparently tolerant of human intrusion, are actually affected in a subtle but nonetheless significant way. The metabolic rates of incubating penguins, for example, can be raised by the mere presence of an observer to such an extent that the birds' food reserves are insufficient to sustain them for their proper spell of incubation before being relieved by the other partner: eggs are therefore abandoned and breeding success reduced. Disturbance of these vulnerable coastal areas, which contain much of the attractiveness of the Antarctic to the tourist and much of its value to the scientist, therefore tends to have a cumulative impact, which is far from obvious to the casual observer. The same probably applies to the shallow seas, which support a rich marine fauna below the limits of ice scour, although less is known about both the diversity and the resilience of these ecosystems in the face of disturbance from small boats and ships and from incidental pollution from vessels and shore stations.

In contrast, the great expanses of the Antarctic ice cap support virtually no life except snow algae and are resilient in the face of human traverse and pressure. Snow obliterates the marks of man and vehicles, or wind scours them away. Likewise, the immense expanses of ocean and floating pack ice around the Antarctic, driven and mixed before wind and current, have a great capacity to disperse pollution and are most unlikely to bear any detectable impact from localized human activities, even up to the scale of a substantial oil spillage. The main extensive human activity in these areas--fishing for krill (<u>Euphausia superba</u>), now running at approximately 300,000 tons per annum[4]--is still tiny compared with the estimated productivity of this species. Nonetheless, it is worth recalling that the major human impacts on Antarctica in the past have come through the disastrous overexploitation of two kinds of animal that depend on those extensive seas for their food--fur seals and whales. Too relaxed an attitude to the resilience of these oceans and their life forms is accordingly unwise. Krill occupy a central place in the Antarctic marine food web, account for about half of the total biomass of animal plankton,

and sustain many species of whale, seal, and bird. All
exploitative activities have small beginnings, and for
this reason alone, biologists are right to be concerned
that the present small catches of krill may be the
harbinger of much more to come.

HUMAN IMPACTS ON THE ENVIRONMENT OF ANTARCTICA

Antarctica is a classic example of a frontier environment. It lies, and almost certainly always will lie, beyond the bounds of permanent human settlement. People have made incursions into it, usually for short periods, in pursuit of resources or information that they can carry away. This was the approach of the initial sealing incursions between 1780 and 1830, which rapidly brought the fur seals of South Georgia, the other sub-Antarctic islands, and the South Shetland and South Orkney island groups to near extinction.[5] The pressures were intensely competitive, and because the resource was open to all comers, with no sovereign interest in regulating its exploitation, it was quickly depleted and its habitat abandoned.[6] A very similar approach was adopted by the twentieth century whaling industry, although here regulation was attempted once it was clear that unchecked open access and competitive exploitation threatened to destroy the interests of all the exploiters. The regulatory efforts have nonetheless brought only partial success, and many resource biologists would say that this is because there is no unquestioned authority able to impose a solution. In contrast, the incursions of scientists into Antarctica have been better organized and regulated, with international discussion of programs, exchange of data and observers, and the formulation of agreed cooperative programs through the Scientific Committee on Antarctic Research (SCAR).[7] But the commercial element has been absent from these activities.

Interest in Antarctica over past centuries, and especially since 1900, has arisen largely because of the continent's distinctive environmental features. Exploration of geographical, geological, glaciological, and biological attributes has gained ground and involved an increasing number of scientists. The research has provided insights into the working of species and ecosystems in an extreme habitat and has contributed to understanding of how the Earth as a whole functions as a geophysical and biogeochemical system. We know that the Antarctic has a substantial impact on the climate of a wide zone of the Earth, and its marine circulation patterns interlink with those over the world ocean northward to the equator, and in some cases beyond.[5]

Today scientists go to the Antarctic because its environment offers insights and opportunities for study not to be found anywhere else on Earth. A primary objective for the Antarctic Treaty System (ATS) is to keep this environment free from damage and open to research. Increasingly, however, the Antarctic oceans are also being seen as the source of important food resources for humanity, with a potential crop of krill estimated by some in the tens of millions of tons per annum, with a further potential yield if conservation measures eventually allow the resumption of whaling, and a possible additional resource of seabed hydrocarbons. Land mineral resources have also been the subject of increasing speculation. The ATS now needs to prove its effectiveness as a framework for the management of these commercial activities, which could easily hamper both the scientific uses of the region and the enjoyment of its unique wilderness qualities by an increasing number of visitors.

One way of viewing the present challenge to the ATS is to ask whether it can ensure the implementation in Antarctica of the broad objectives of the World Conservation Strategy (WCS).[8] This analysis recognizes the importance of development of the world environment for human welfare but stresses that it is in the interest of all people for this development to be managed so that it provides for the sustainable use of the renewable resource base. Conservation of ecological systems and their genetic diversity is important from the human standpoint because these systems form a crucial part of the human life-support system.

In the remainder of this chapter we analyze how far the ATS is implementing the objectives of the WCS in the region and ask whether the mechanism needs to be developed or adapted to ensure achievement of this objective in the future.

THE EVALUATION OF ENVIRONMENTAL GOALS

The WCS and other analyses have led to the formulation of certain broad goals of environmental policy, which are applicable to all regions. These can be summarized in nine points:

(1) There shall be a conscious plan for managing and developing the environment of a region;
(2) The long-term productivity of ecological systems and other renewable resources shall be sustained under that plan, and the use of these resources shall be controlled by competent authorities;
(3) Damage from chemical contamination and energy releases, which could threaten sustainable use, shall be held, by effective regulatory action, within limits formally established by the proper authorities;
(4) Representative samples of the range of ecological diversity of the region shall be set aside as reserves, with conservation as the priority for their use;
(5) Outstanding aesthetic qualities of the environment shall be safeguarded, with the establishment of "national parks" in key areas;
(6) The likely impact of any activity liable to change the environment shall be evaluated in advance, and a regulatory system established to prevent activities deemed likely to cause unacceptable damage;
(7) There shall be proper scientific study as a basis for environmental assessment and management;
(8) The state of the environment, the productivity of its systems, the degree of pollution, and the operation of activities permitted within conservation and management plans shall be monitored and periodic reports prepared and published; and
(9) There shall be a consultative process, in which interested parties may participate, to adjust activities that threaten established environmental goals or appear liable to create unforeseen hazards, and this process shall include effective procedures for the resolution of disputes.

ENVIRONMENTAL CONSERVATION WITHIN THE ANTARCTIC TREATY SYSTEM

Two quite distinct approaches to environmental questions, in a broad sense, are evident in the ATS. The first attempts to define principles that should govern the protection of the Antarctic environment from the damaging impact of present or future activities. The second identifies and guards against particular activities that could have a deleterious effect on the Antarctic environment. The first approach is general and essentially nonspecific; the second is activity specific. Both are precautionary in their approach.

The overall approach is defined in several general statements, especially those set out in Recommendations VIII-13 and IX-5 adopted at Antarctic Treaty consultative meetings. The key elements in these recommendations are the following:

Extract from Recommendation VIII-13

The Representatives [of the consultative parties] RECOMMEND to their Governments:

1. In exercising their responsibility for the wise use and protection of the Antarctic environment they shall have regard to the following:
   (a) that in considering measures for the wise use and protection of the Antarctic environment they shall act in accordance with their responsibility for ensuring that such measures are consistent with the interests of all mankind;
   (b) that no act or activity having an inherent tendency to modify the environment over wide areas within the Antarctic Treaty Area should be undertaken unless appropriate steps have been taken to foresee the probable modifications and to exercise appropriate controls with respect to the harmful environmental effects such uses of the Antarctic Treaty Area may have;
   (c) that in cooperation with SCAR and other relevant agencies they continue, within the capabilities of their Antarctic scientific programme, to monitor changes in the environment, irrespective of their cause, and to exercise their responsibility for informing the world community of any significant changes caused by man's activities outside the Antarctic Treaty Area....

Extract from Recommendation IX-5

[The consultative parties] **DETERMINED** to protect the Antarctic environment from harmful interference;

HAVING PARTICULAR REGARD to the conservation principles developed by the Scientific Committee on Antarctic Research (SCAR) of the International Council of Scientific Unions;
RECALLING their obligation to exert appropriate efforts, consistent with the Charter of the U.N., to the end that no one engages in any activity in Antarctica contrary to the principles or purposes of the Antarctic Treaty;
DECLARE as follows:
1. The consultative parties recognize their prime responsibility for the protection of the Antarctic environment from all forms of harmful human interference;
2. They will ensure in planning future activities that the question of environmental effects and of the possible impact of such activities on the relevant ecosystems are duly considered;
3. They will refrain from activities having an inherent tendency to modify the Antarctic environment unless appropriate steps have been taken to foresee the probable modifications and to exercise appropriate controls with respect to harmful environmental effects;
4. They will continue to monitor the Antarctic environment and to exercise their responsibility for informing the world community of any significant changes in the Antarctic Treaty Area caused by man's activities.

The general principles in Recommendations VIII-13 and IX-5 are in full accord with the WCS and give a general direction to actions under the treaty to protect the environment. In particular, they emphasize the need to act in the Antarctic in the interests of _all_ humankind [Recommendation VIII-13, 1(a)]; to plan activities to avoid significant and avoidable environmental damage [Recommendations VIII-13, 1(b) and IX-5, 1, 2, and 3]; and to maintain continuing scientific and administrative surveillance and monitoring [Recommendations VIII-13, 1(c) and IX-5, 4]. They define a broad strategy for sound conservation of environmental resources, as required under the first of the nine objectives listed above.

The specific approach has led to actions under the treaty in the following eight areas, with a further theme under active discussion.

(1) _The Agreed Measures for Conservation of Antarctic Fauna and Flora._ These measures, concluded in 1964 at the Third Antarctic Treaty consultative meeting, were an exercise in forethought aimed at preventing any repetition of the near extermination of species that took place in the nineteenth century. These Agreed Measures, which have been characterized as a "minitreaty," or a treaty within a treaty, prohibit the citizens of any party to the treaty from killing, capturing, or molesting without a permit any mammal or bird native to Antarctica. They also establish the basis on which Specially Protected Areas shall be established and the rules that shall operate regarding access to them and define the concept of Specially Protected Species. Subsequently two Specially Protected Species and seventeen Specially Protected Areas have been designated, and it has been agreed that the statistics of animals killed or captured under permit will be published.

(2) _The Convention for the Conservation of Antarctic Seals._ This convention was also an exercise in forethought. It had its beginnings in 1964 when an exploratory sealing voyage examined the potential harvest of crabeater seals in the South Atlantic pack ice. It was concluded in 1972, after a number of drafts had been examined by both SCAR and the consultative parties. The convention provides for closed areas, closed seasons, and for what would now, in the light of the Law of the Sea Convention and customary international law, be termed total allowable catches (TACs) of seals. These TACs were based on a ten percent take of what have turned out to be very conservative estimates of the total populations of pelagic seals. Since the convention was concluded there has been no commercial sealing in the Antarctic, but it is not clear how far this is an effect of the convention rather than of logistics difficulties, costs, and consumer resistance to seal products.

(3) _The Convention on the Conservation of Antarctic Marine Living Resources._ The convention (CCAMLR) was concluded in 1980 and entered into force in 1982. This is an ambitious instrument that sets out to conserve the marine living resources of the Antarctic and Southern Ocean area south of the Antarctic convergence, including birds, in accordance with principles of ecosystem conservation. Following the steady decline of Antarctic whaling through the 1950s and 1960s, fishing for finfish and experiments to see if krill could be located, caught, processed, and marketed began in 1969-1970. Initial

catches of finfish were good, but because of the slow rate of growth of these species in the cold Antarctic waters, catches on the four main grounds tailed off rapidly. Catches of krill rose fairly rapidly to 300,000 tons in 1983-1984, but the economics of this fishery remain doubtful because of processing and marketing problems. At present the forecasts of being able to double the world's marine resource catch from krill, made in the late 1960s, seem wildly far off the mark. There are major ecological doubts about how far it is prudent to harvest krill in view of its central role in the Antarctic marine ecosystem and the possibility that krill depletion will impair the recovery of whale populations. The questions facing CCAMLR in fulfillment of its own objectives are whether it can so regulate fishing of the depleted finfish stocks that there will be a return to higher catches on a sustainable basis and whether it can so regulate krill exploitation that the health of the Antarctic marine ecosystem as a whole is sustained.

(4) Recommendations to foresee and guard against the impacts of tourism. From a study carried out in Britain and the Antarctic, it seems that the interests of tourists are in Antarctic stations, wildlife, and scenery in that order. The impact tends therefore to concentrate on the stations. But many of these stations are located in areas of particularly diverse and vulnerable coastal environment, and long-term scientific studies are often in progress in their neighborhood. The recommendations provide for a government to say that it will not accept visits to its stations from tour ships and to regulate the activity of tourists when visiting stations. A reporting system has been established to monitor where tourists land elsewhere than at stations, and a statement of principles and practices of the Antarctic Treaty has been compiled for the information of all visitors including tourists (Recommendation X-8). It must be admitted that this statement is not very compelling reading, but some governmental expeditions have produced more interesting pamphlets about Antarctic wildlife with some simpler dos and don'ts for tourists.[10]

(5) Recommendations on the protection of historic sites. Some expeditions, notably from New Zealand, have restored and provided wardens for the huts of the "heroic age" explorers, which are most frequently visited.

(6) Recommendations on the preparation by SCAR of a code of conduct for waste disposal in Antarctica. The code, which it is hoped will be applied to all expedi-

tions, is now being revised in view of improvements in waste disposal techniques.

(7) Recommendations on the establishment of Sites of Special Scientific Interest (SSSIs) where there is a likelihood of inadvertent interference with scientific studies. Proposals are first evaluated by SCAR and are then made applicable to all expeditions for a limited term of years by a recommendation. It is a failure of the treaty consultative partners that they have so far not found a way of resolving a perceived conflict of interest between designating marine SSSIs and freedom of navigation.

(8) Environmental impact assessment. It was the recommendation of the last consultative meeting that all research activities and supporting logistics activities that are likely to have a significant impact on the Antarctic environment should be subject to such assessment.

The same specific approach has also been evident in the discussions, at successive consultative meetings, on how to regulate possible minerals exploration and exploitation. These discussions continue and are described in other chapters of these proceedings. While they are centered on the administrative machinery that would regulate the development of economically useful minerals, including hydrocarbons, should these be discovered (and they have not been so far), the strongest underlying theme has been how to ensure that such activities do not damage the environment.

The general principles in Recommendations VIII-13 and IX-5 have been put into practice by these various specific measures, which address a number of the environmental policy objectives set out above.

The CCAMLR and the Convention for the Conservation of Antarctic Seals (Seals Convention)--and especially the former--seek to maintain the productivity of renewable living resources [objective (2)], and the first of these conventions is unusual in the stress it lays on the need to adjust harvesting so as not to impair the balance of the supporting components of the ecosystem. The recommendations on waste disposal fit within objective (3), while the Agreed Measures, recommendations on SSSIs, and recommendations on tourism together cover much of aims (4) and (5), which are concerned with the conservation of wildlife, scientific interest, and natural beauty. But there are no areas meeting the internationally accepted definition of national parks in Antarctica. The protected areas are small, and their selection was not originally based on a deliberate plan to safeguard a series of representative habitats and ecosystems. Objective (6) is directly covered by the proposals for environmental assessment, which would certainly be applied to any minerals-related activity. Finally, scientific study and monitoring are built into all parts of the ATS and into the program of SCAR. Open publication and broad international discussion of scientific findings are implemented through a well-established and effective consultative process.

THE ANTARCTIC TREATY SYSTEM AS A MECHANISM FOR ENVIRONMENTAL CONSERVATION

The Antarctic has commonly been looked on as one of a progression of "last frontiers"--the American "West," Antarctica, space. Except for the slaughter of the seals in the 1820s and of the whales in the first half of this century--comparable to the slaughter of the North American bison--human presence in Antarctica has been organized in a way that the pushing back of other "frontiers" was not. Most of the activity in the Antarctic since World War II has been carried out by governments. Under the Antarctic Treaty, they set out to regulate their own activity, bringing to bear on it an awareness of the earlier and darker history of Antarctic exploitation and of the newer concepts of conservation, ecosystems, and the environment. So often, elsewhere in the world, these concepts have been brought into play for purposes of regulation only after damage has been done. In the Antarctic, by contrast, the treaty powers have set out both general and specific rules before the activities that they have sought to regulate are far advanced. That is an encouraging start. The same environmental consciousness continues. As one of us put it at one stage in the negotiation of the minerals regime: "Here are the claimants and nonclaimants going at each other about minerals, and what they are working out is not a minerals exploitation regime but an environmental protection regime."

The main virtue of the activity-specific measures for environmental protection is their precautionary approach. They have addressed possible impacts of specific activities before the activity itself develops. In this respect, they observe what is becoming an increasingly prevalent feature of environmental policy. It could be said that the measures on sealing and tourism have not been tested, but the fact that regulations were in place before activities began or had reached any considerable scale must have had an influence on anyone wondering whether to invest money in such activities. The treaty system flashes an amber light signifying a clear intent to regulate. This warning light appears to have failed only with respect to finfish exploitation, but this came about because of the sudden arrival in the region of distant-water fishing fleets evicted by the rapid extension of 200-mile fishing zones elsewhere in the world.

Using examples like this, some critics have argued that the success of the ATS is more illusory than real.[6,9] It is contended that the various measures have been agreed on without undue difficulty because nobody has a strong interest in breaching their provisions. Many of these provisions have not been tested by real pressure. It can also be argued that none of the conventions and recommendations gets to the root of the most proven difficulty in protecting such resources: the restriction of open, competitive access. Although regulatory mechanisms that could include TACs are provided for under the CCAMLR and the Seals Convention, their observance depends on voluntary restraint. This has notoriously had only partial success in conserving open access stocks elsewhere, notably those of oceanic whales. Virtually all regional fisheries conventions have encountered comparable difficulties. Is it any more reasonable to expect a system founded on voluntary self-restraint, by a whole series of governments with differing attitudes to the ownership of the resources, to work effectively in Antarctica if strong economic incentives to exploit limited resources arise and if that exploitation is assigned to industries to whom a competitive approach is second nature?

What is needed is an approach that, where there is doubt about the effects of exploitation, consistently gives the benefit of the doubt to the resource rather than to the exploiter. This approach has not yet been adopted anywhere on an international basis despite widespread recognition of its validity. For fisheries, the alternative process of extending coastal state jurisdiction has been followed in most areas. The ATS has certain features that should, in contrast, allow a conservationist approach to be followed at the international level. The first is the proven ability of the system to lay down regulatory ground rules before large-scale investment has been committed. The second is the common wish of the parties not to have to fall back on the use of territorial jurisdiction, although this remains a possible alternative regulatory basis if all else fails.

It is clear that there are gaps in the series of measures adopted under the treaty. The coverage of Specially Protected Areas is not fully representative of the diversity of habitats and ecosystems in Antarctica--although these areas do cover many samples of particularly vulnerable, small coastal localities. The absence of large designated areas with the equivalent of national park status is seen by some as a serious gap--although we would argue that the Antarctic as a whole enjoys a degree of protection and an absence of threat unique in the world, so that this gap is more apparent than real. But we do accept that further action to extend the series of designated sites in relation to an objective scientific classification of the range of variation in Antarctic habitats is desirable. Such action was initiated by treaty Recommendation VII-2 in 1972, and SCAR is now preparing proposals. It is also important that such areas, once designated, be respected, something that has not happened in all respects so far.

Another gap relates to pollution prevention. Recommendation VIII-11 on waste disposal deals with only one possible source of chemical contamination of the region. So far, there are no specific agreements to reduce the risk of oil pollution from ships (other than the various International Maritime Organization (IMO) conventions) or to control the use of substances such as pesticides that could cause persistent low-level contamination of living organisms and so reduce their value as monitors of pollution dispersed from outside. Radioactive contamination has been contained by agreements--including the voluntary decommissioning by the United States in 1972 of the small nuclear reactor erected at McMurdo in 1962, and the removal of its wastes and other material from the area. But this whole subject of contamination and pollution is another topic for continuing discussion.

There is also the problem of competing use. The treaty system currently provides no guidance on how the values of, for example, scientific research, shore-based minerals development, fishery potential, and the conservation of wildlife and aesthetic qualities are to be weighed against one another in circumstances where there

is perceived competition between them. These conflicts
need to be resolved individually and locally: it is no
answer to give one use of Antarctica absolute and universal priority over others. There needs, therefore, to be
machinery for reasoned judgment among alternative uses of
the environment.

To catalog these and other gaps is not, however, to
establish failure on the part of the treaty system.
Rather, it is to demonstrate that the present measures of
environmental protection require both extension and
consolidation--and that the mechanism tested over the
past 20 years provides a good basis for both. SCAR
provides an authoritative source of scientific judgment
about Antarctic ecosystems and their likely response to
impact. The consultative process has proved its ability
to create agreed recommendations and conventions. Extension of the various provisions to cover some of the possible gaps mentioned above, and to deal with the impacts
of possible minerals exploitation, should therefore be
possible. Greater challenges lie in the need to demonstrate that these agreements can be implemented effectively in the face of conflicts of interest and economic
pressures. This will require the wholehearted commitment
of all the consultative parties and of any other governments undertaking Antarctic activities.

The WCS called on individual states to prepare national
conservation strategies applying the broad concepts of
the strategy to their own national circumstances. Such
an approach is not appropriate in Antarctica--where something much more significant lies within our grasp. We
suggest that the general and specific actions set out
above are the ingredients of a __continental__ conservation
strategy embracing the philosophy of Recommendations
VIII-13 and IX-5 of the treaty and the nine aims noted
above. It would not be difficult to present the specific
actions taken to date, together with the broad framework
of an agreed policy to prevent environmental damage from
minerals-related activities, in such a form. Such a
consolidation would demonstrate in a unique way how a
group of nations has been able to work together to care
for the environment of an entire continent and would
demonstrate the achievements of the treaty system.

We believe that this would do much to explain a consistent program that has lasted for many years. We suggest that what has been achieved under the ATS is remarkable and gives grounds for optimism. In many ways that
system has been ahead of its time in pioneering the

preventative approach to environmental impact that is now widely accepted, for example, in the United Nations Environment Program. The central need now is for the various elements of the treaty system as an environmental mechanism to be bound together as a coherent whole and endorsed and applied by all national groups that pursue activities within the southernmost zones of the Earth.

NOTES

1. Holdgate, M. W. ed. 1970. Antarctic Ecology, Academic Press (London), 998 pp.
2. Laws, R. M. ed. 1984. Antarctic Ecology, Academic Press (London).
3. Holdgate, M. W. 1977. Terrestrial ecosystem in the Antarctic. Phil. Trans. R. Soc. London Ser. B 279:525.
4. Knox, G. A. 1983. The living resources of the Southern Ocean: A scientific overview. In F. Orrego Vicuna, ed. Antarctic Resources Policy, Cambridge University Press (Cambridge).
5. Holdgate, M. W. 1984. The use and abuse of polar environmental resources. Polar Rec. 22(136):25-48.
6. International Union for the Conservation of Nature and Natural Resources. 1984. Conservation and development of Antarctic Ecosystems, paper submitted to the U.N. Political Affairs Division. International Union for the Conservation of Nature and Natural Resources (Gland, Switzerland), p. 36.
7. See, for example, SCAR Manual (Cambridge). 1972. Scott Polar Research Institute, p. 128; SCAR Bulletins, published regularly as annexes to Polar Record.
8. International Union for the Conservation of Nature and Natural Resources, U.N. Environment Program, World Wildlife Fund International. 1980. World Conservation Strategy. International Union for the Conservation of Nature and Natural Resources (Gland, Switzerland), p. 46.
9. Mitchell, B., J. Tinker. 1980. Antarctica and Its Resources, Earthscan International Institute for Environment and Development (London), p. 98.
10. British Antarctic Survey. 1984. A Visitor's Introduction to the Antarctic and Its Environment, British Antarctic Survey, Natural Environment Research Council (Cambridge).

14.

Panel Discussion on Conservation and Environment

The panel consisted of Martin Holdgate (moderator), Sachiko Kuwabara, Kenton Miller, and W. Timothy Hushen.

REMARKS BY SACHIKO KUWABARA

Kuwabara stated that the protection and preservation of the unique environmental value of Antarctica from harmful impacts of human activities merit the broadest kind of international cooperation. The significance of Antarctica in understanding and maintaining the ecological balance of planet Earth makes it a matter of interest to all nations, both within and outside the Antarctic Treaty System (ATS). Moreover, as with scientific research, international cooperation to safeguard the Antarctic environment could, and should, in her view, be promoted without having to resolve the legal status of Antarctica.

Kuwabara outlined three steps to accomplish this objective. The first step is to utilize existing systems of cooperation in environmental protection, first and foremost the ATS, which has demonstrated its ability to cope successfully with emerging environmental concerns in Antarctica and has provided a viable framework for further collaboration in this field. Kuwabara believes, however, that states party to the Antarctic Treaty should provide more opportunities to interested nonparty states and to competent international organizations to contribute to environmental management policies for Antarctica. She suggested that the ATS should increase the flow of information on measures relevant to the protection of the Antarctic environment to these entities and that their participation as observers in Antarctic meetings should be promoted.

Second, coordination among existing international organization programs and forums relevant to environmental protection in Antarctica should be increased. On the one hand, exchange of information on a continuing basis would facilitate wider appreciation on the part of the international community at large of the work carried out under the ATS. On the other hand, the ATS could benefit from drawing on internationally agreed guidelines, principles, and standards developed by these programs. For example, the results of the monitoring programs carried out under the ATS would be a valuable input into the Global Environmental Monitoring System (GEMS) of the United Nations Environment Program (UNEP). Conversely, UNEP environmental management guidelines and programs could offer useful contributions to the work of the ATS in such areas as environmental impact assessment, control of marine pollution from offshore mining and drilling, hazardous waste management, and the protection and environmentally sound development of regional seas and their coastal areas.

Third, at some time in the future it might be necessary to improve or develop mechanisms to fill in the gaps in international cooperation for the protection of the Antarctic environment. Kuwabara believes that some of these gaps have already been recognized, such as the limited scope of measures relating to pollution prevention and the lack of guidance regarding environmentally sound development of Antarctica and its resources. In addition, it might be necessary to strengthen existing mechanisms for consultation and conflict resolution with respect to environmental concerns. Equally important, collective procedures should be developed to review potential environmental impacts of proposed activities in Antarctica and to ensure that damaging activities do not take place or continue. In this context, Kuwabara noted that the discussions among the consultative parties concerning the application under the ATS of environmental impact assessment procedures are of great interest to the international environmental community.

REMARKS BY KENTON R. MILLER

Miller noted that the twelfth consultative meeting had considered the question of inviting relevant international organizations to participate in future treaty meetings with the status of observer. As the director general of the International Union for the Conservation of Nature and Natural Resources (IUCN), he expressed IUCN's interest in supporting the efforts of the ATS and noted his organization's qualifications for observer status at consultative meetings and with the Scientific Committee on Antarctic Research (SCAR).

In summary, he believes that IUCN has interests and expertise in Antarctica and can make serious scientific contributions; represents a wide range of interests embracing scientific, technical, aesthetic, and moral considerations; has experience in theoretical and practical questions of natural resources planning and management as well as in promoting public awareness; and maintains linkages with a broad world community. He noted in particular IUCN's expertise in protected areas.

In assessing the ATS, Miller commended its operation as a preventive mechanism with respect to environmental damage but questioned the lack of an environmental management review mechanism under the ATS comparable with peer review within the scientific community. In this context, he stressed the need for the ATS to create a sense of confidence in the effective implementation of its objectives and responsibilities. He supported the proposal for a continental conservation strategy for Antarctica as a positive step, believing that this would offer a major opportunity to deal with mechanisms to evaluate and manage Antarctica as well as to build confidence in that process.

More specifically with respect to the role of IUCN, Miller noted that the General Assembly of IUCN has given mandates and directives to the organization on Antarctica, indicating IUCN's interest in the area, and that IUCN has already made contributions that demonstrate its expertise in the subject.

He cited the following examples of IUCN's activities in this regard:

(1) Provision of guidance to IUCN on its policy and program by IUCN Antarctic Experts Advisory Committee, consisting of recognized scientists and lawyers;
(2) IUCN representation at the meetings of the Convention on the Conservation of Antarctic Marine Living Resources (CCAMLR) and contributions to the work of CCAMLR committees;
(3) Drafting, promotion, and monitoring of major international treaties and conventions dealing with natural resources management and conservation by IUCN's Environmental Law Center in Bonn;
(4) Development of a data base on the status of species and genetic resources, protected areas, and trade in endangered species by IUCN Conservation Monitoring Center in the United Kingdom;

(5) Joint sponsorship with SCAR in April 1985 workshop held in Bonn on the scientific requirements for Antarctic conservation, in which IUCN scientific and technical commissions, made up of networks of experts in law, ecology, protected areas management, education, and species and environmental planning, collaborated;
(6) Development and promotion of the World Conservation Strategy (WCS), in close collaboration with UNEP and the World Wildlife Fund International, and the forthcoming peer review of the application of the WCS at a major conference in Ottawa in 1986; and
(7) Benefit to IUCN's council from the close and regular participation by the International Council of Scientific Unions, the U.N. Food and Agriculture Organization (FAO), the U.N. Educational, Scientific and Cultural Organization, and other organizations with major interests and mandates on the conservation of living natural resources.

Finally, Miller noted that a unique and relevant feature of IUCN is its global membership of 57 states, 125 government agencies, and 339 nongovernment organizations, including important universities, research facilities, and citizens conservation groups.

REMARKS BY W. TIMOTHY HUSHEN

Hushen discussed the role of SCAR in Antarctic conservation. He noted that SCAR, as requested by the Antarctic Treaty's twelfth consultative meeting, is completing a document containing advice on (1) the categories of research and logistics activity in Antarctica that might reasonably be expected to have significant impacts on the environment in Antarctica, and (2) procedures for assessing and monitoring these impacts.

SCAR has prepared a publication on conservation areas in the Antarctic. At the XVIII SCAR meeting, in Bremerhaven, Federal Republic of Germany, in September/October 1984, SCAR developed proposals for additional Sites of Special Scientific Interest, including marine sites. These documents and recommendations will be considered by the thirteenth Antarctic Treaty consultative meeting, in October 1985.

Hushen indicated that SCAR has also been discussing the possible development of a new type of protected area in Antarctica. Such areas will be larger and will include plans for multiple uses such as tourism and scientific research.

DISCUSSION

There was unanimous agreement on the importance of the Antarctic environment as a component of planetary systems. Study of the Antarctic environment provides broader insight into global climate, atmospheric geophysics, and the structure and history of the southern continents. It is also valuable as a baseline for monitoring climatic change (from the long record in ice cores) and changes caused by humans (for example, the accumulation of carbon dioxide in the atmosphere, the possible reduction in stratospheric ozone, and variations in aerosol and particulate deposits). It provides opportunities for research on unique ecological systems, including those within translucent rocks, those of primitive soil and vegetation, and those attached to floating sea ice. It also allows for examination of the adaptation of organisms to extreme conditions. Finally, the Antarctic environment is a legitimate object of concern because of its great natural beauty and relative immunity from human disturbance. There was no dissent from the conclusion that the ATS has provided an invaluable framework for the development of scientific research on these and other environmental features.

It was agreed that the conservation of the Antarctic environment is of high priority. Moreover, although much of this environment appears robust in the face of human interference, this apparent resilience should not be taken for granted. The approximately two percent of land not covered by permanent ice and snow includes many habitats vulnerable to human pressure, and it is these areas that are the most likely to attract such pressures. Ice-free coastal lowlands, for example, support the most advanced Antarctic vegetation and the largest seabird and seal colonies; but they are also the most attractive sites for scientific stations and their logistics support facilities. Suggestions to reduce potential environmental impacts by sharing program facilities received some attention.

It was agreed further that the ATS has proved itself as an evolving series of agreements and institutions within which measures to protect the environment have been developed. The preventive nature of these measures, which have almost all been drawn up in advance of the threats that they set out to regulate, is an important characteristic and one that fully accords with the modern philosophy of environmental protection. In this sense the ATS has been a true pioneer, and the conservation agreements achieved under it are more comprehensive than can be found in any other area of similar size. It was noted that these measures in Antarctica go a long way toward meeting the objectives of the WCS.

There is now a good case for consolidating what has been achieved and defining what else needs to be done. At the twelfth Antarctic Treaty consultative meeting it was recognized that there is a need to consider whether further coordination is necessary of the various elements of environmental protection contained in the ATS. At its 16th General Assembly (Madrid, 1984) IUCN recommended "that a comprehensive review be carried out under the Antarctic Treaty system of the existing environmental and conservation Conventions and measures, with a view to determining whether any new Conventions or measures are needed for the environmental protection of the Antarctic environment and the Southern Ocean," and it was noted that the director general of IUCN has a mandate to contribute to this work. It was felt that IUCN, in cooperation with SCAR and especially following the SCAR/IUCN Symposium (Bonn, April 1985), is well placed to cooperate with the ATS in what was seen as the need to prepare a conservation strategy for the Antarctic and the Southern Ocean. It was believed to be important that in the preparation of such a strategy the participants in the ATS should broaden the base of their work and should make use of the expertise and experience available to them in other international organizations, such as IUCN and UNEP.

The panel participants recognized that additional action might be needed in a number of areas, including

(1) **Monitoring and assessment.** It was noted that SCAR, in answer to a request from Antarctic Treaty governments, is developing advice on procedures for evaluating and monitoring impacts of science and logistics activities on the Antarctic environment and that the resulting document is to be submitted to the thirteenth Antarctic Treaty consultative meeting.

It was noted, however, that there is a shortage of published data on, and assessments of, the state of the Antarctic environment, although research in these areas is ongoing. One of the problems is that the scale of Antarctica makes overall monitoring difficult. One participant suggested that a secretariat could be established to help determine what studies should be undertaken.

(2) **Information**. It was also suggested that insufficient information about the state of the Antarctic environment was being published or otherwise made available.

(3) **Coverage of protected areas**. There is a need to ensure that effective, long-term protection is afforded to a truly representative, and adequately extensive, series of Antarctic habitats. Recent experience in the Nordic countries was cited as a precedent for managing a series of selected areas in a coordinated way to meet the needs of science and tourism. It was noted that SCAR is currently publishing **Conservation Areas in the Antarctic** (March, 1985) and is also considering the case for a new category of conservation area.

(4) **Inspection and enforcement**. It was suggested that the existing arrangements for inspection under the treaty might be used to check that protected areas are being respected and that other environmental measures--for example, on waste disposal and impact assessment--are being implemented properly. As the number of activities in Antarctica increases, the inspection system should be expanded and perfected, since it represents a unique adaptation to enforcement in the situation of jurisdictional ambiguity existing in Antarctica. Joint inspections by one or more countries might increase the frequency of inspection and allow more countries to conduct them by reducing costs to any single country.

The consultative parties should not assume that the measures that they have adopted are operating effectively without such checks. Moreover, there is a need to **demonstrate** the effectiveness of the Antarctic management system.

In response to a question about whether the inspection provisions currently apply to environmental issues, it was pointed out that they apply to all measures adopted pursuant to the Antarctic Treaty, including those having to do with the environment. It was also noted that there should be a link between such inspections and monitoring and assessment arrangements and that the CCAMLR has an inspection and enforcement system of its own.

Thought might also need to be given to the nature of enforcement with respect to individual offenders--a wider issue within the treaty system. One participant questioned whether an international approach to enforcement might be pursued, while another doubted whether enforcement could be effectively carried out in the absence of a permanent structure under the Antarctic Treaty.

(5) <u>Development of a capacity to handle economic pressures</u>. It was noted that some conservation measures under the treaty system, such as the Convention for the Conservation of Antarctic Seals or CCAMLR, have not been truly tested; the effectiveness of the latter will be judged by its capacity to set and enforce catch limits (for finfish as well as krill) and other regulations with sufficient flexibility to permit progressive adjustment as scientific knowledge grows. (This subject is considered further in Chapters 15 and 16.)

(6) <u>Development of an effective regime to prevent environmental damage from minerals exploration or exploitation</u>. It was clear that many doubted the likelihood or the desirability of minerals development in the Antarctic. Granted the possibility of such development, however, it was agreed that a strongly protective regime should be drawn up as a further step in the development of a preventive approach to environmental management. (This subject is considered in more detail in Chapters 17-20.)

As part of the adaptive evolution of the ATS in the environmental field, many participants supported the case for stronger links with international organizations with relevant expertise. The unique role of SCAR was generally appreciated. Climatic data from Antarctica were being drawn on by the World Meteorological Organization in the World Climate Research Program. The IUCN can contribute substantial understanding of wildlife and habitat conservation principles and practices. UNEP can offer ideas on environmental impact assessment, marine pollution control, and hazardous waste management and itself needs to draw on Antarctic data in the GEMS and in state-of-the-environment assessments. The UNEP regional seas program might provide assistance to consultative meetings or CCAMLR discussions. The workshop heard suggestions that consultative meetings might benefit from the presence of UNEP and IUCN as observers, as sources of relevant expert ideas and suggestions.

The institutional machinery for developing Antarctic conservation may need to be developed further. Most participants accepted, on pragmatic grounds, that the sensible course is to build on the existing institutions of the treaty system, and it was noted that these have evolved and are still doing so. It was suggested that some kind of full-time Antarctic environmental protection agency staffed by trained professionals might be justified, to carry out monitoring, assessment, and inspections; manage protected areas; produce publications; and act in support of the treaty consultative meetings. Such a group might be backed by a special fund administered by nongovernmental agencies. The case for these and other mechanisms might usefully be considered in the process of preparing the proposed conservation strategy for the Antarctic.

It was, however, stressed that the ATS depends on consensus among the independent, sovereign consultative and contracting parties. While the treaty system could be viewed as a management tool, it has not fully internationalized or unified the continent in an administrative sense. Moreover, its responsibilities are broader than environmental protection alone.

Outside organizations should work by persuasion, based in turn on the quality of their ideas, and need to recognize that the ATS is a unique attempt to provide a framework for the management of geopolitical tensions that, if they were not so managed, would represent a far greater threat to the Antarctic environment and wildlife conservation than any activity in Antarctica being conducted at present or foreseen in the future.

15.

The Antarctic Treaty System as a Resource Management Mechanism— Living Resources

John A. Gulland

INTRODUCTION

In the Antarctic the living resources offer a complete contrast between marine and terrestrial systems. The Southern Ocean is rich with life--among which krill, whales, seals, and penguins are the best known. Historic visits to the Antarctic have been largely those of sealers, whalers, and fishermen engaged in harvesting these resources. The management problem is thus one of ensuring that this harvesting is carried out in a rational manner, with due regard to future interests in the resources. The Antarctic land mass is cold and barren and extremely hostile to life. What life there is, is probably vulnerable to disturbance through other human activities, such as research, tourism, and possibly in the future, extraction of minerals. The management problem is one of diminishing or minimizing such disturbance.

The management problems of sea and land are therefore best discussed separately, with the exception of seal and penguin requirements for a firm base (land, or ice shelves) on which to breed. These essentially marine animals can be vulnerable to damage to, or disturbance at, these breeding sites, and mechanisms to prevent this are best discussed together with other aspects of terrestrial management.

MARINE RESOURCES

Background

While the Antarctic Treaty applies to the area south of 60°S latitude, in considering the marine resources of the region it is better to look at the whole area south of the Antarctic convergence. The exact position of the convergence is variable but on average corresponds closely to the boundary of the area of responsibility of the Commission for the Conservation of Antarctic Marine Living Resources established by the 1980 Convention on the Conservation of Antarctic Marine Living Resources (CCAMLR). South of the convergence, the current systems ensure a good supply of nutrients to the surface, and in the summer, primary production in parts of the region in the form of microscopic plants (phytoplankton) is among the highest in the world, with the exception of a few spectacular areas such as the upwelling zones off Peru and California.

Further favorable factors from humanity's point of view are that the food chain is generally short, with two steps from phytoplankton to baleen whales via krill, and that most animals are long lived. This means that a relatively high proportion of the original primary production appears in a harvestable form, and the standing stock represents the accumulated production of several years (several decades in the case of the whales) and is therefore high.

The waters of the Antarctic and sub-Antarctic, despite being cold, rough, and far from civilization, have attracted sealers, whalers, and fishermen for the past two centuries. The use of these resources shows that unmanaged exploitation can be disastrous. By the end of the nineteenth century the fur and elephant seals and the right whales of the sub-Antarctic had been brought close to extinction; by the middle of this century the larger baleen whales (blue, fin, and humpback) had also been greatly reduced. Partly through the measures introduced by the International Whaling Commission (IWC), the depletion of the baleen whales has been slightly less extreme than that of right whales or fur seals, but of the baleen whales only the minke whale is now present in numbers similar to those at the time when human visitors first came to the Southern Ocean.

Exploitation is now focused on krill and demersal (bottom-living) fish. Several fish stocks seem already, as reported at the third meeting of the CCAMLR in September 1984, to be greatly reduced from their original level. Krill are so far probably little affected. Current catches, at approximately 500,000 tons per year, are still far less than potential yield. Estimates of this potential range from tens of millions of tons upward--

comparable to the present total yield of all types of fish from the oceans of the world.

The declines or collapses of seals, whales, and fish stocks are not unique to the Southern Ocean. Virtually every commercially attractive, exploited fish stock has been allowed to decline below its most productive level. These declines have not often turned into catastrophic collapses largely because of two factors: the high productive capacity of most fish makes them less vulnerable than mammals to sustained over exploitation, and most fishing fleets can move to alternative resources when stocks and catch rates in one area begin to decline. The events in the Southern Ocean should therefore not be ascribed to some unusual degree of greed or shortsightedness on the part of the harvesters; they are the predictable results of unmanaged exploitation of a common-property, open-access resource.

From the point of view of managing the marine resources of the Antarctic, three points can be made:

(1) The resources are rich and have made, and could continue to make, a significant contribution to world food supplies.
(2) Several important elements (fur and elephant seals and whales) of the marine ecosystem have already been greatly altered, and this has probably had effects on many other elements; that is, in the ocean we are not dealing, as is the case for the Antarctic land mass, with an undisturbed system.
(3) If the resources are to be maintained in an optimum condition, management will be essential.

The balance of this chapter will discuss the mechanisms required to provide the necessary management actions [including deciding on what is meant by "optimum condition" in (3) above] and then, with special reference to the purposes of the workshop on the Antarctic Treaty, examine the past, present, and potential future role of the Antarctic Treaty System in ensuring that these mechanisms are established and operate correctly.

Mechanisms for Management

The basic requirements for successful fishery management have been discussed on a number of occasions, notably by the U.N. Food and Agriculture Organization's Advisory

Committee on Marine Resources Research and its working parties. The need is not merely to have a mechanism to introduce management measures. Before such action and any decision on specific measures can be taken, it is necessary to have discussion and agreement on the objectives that the measures should achieve as well as adequate technical and scientific information on the immediate and long-term results of alternative management actions (including the possibility of doing nothing). The actions must then be followed by steps to ensure that the agreed measures are actually implemented and to review their success and, where necessary, revise them.

At each stage, it is important that all those actually or potentially interested in the resource participate. In an open-access situation, such as the Antarctic, unless there is virtually unanimous agreement, few measures are likely to be effective. Any participant that does not abide by these, for example, by limiting total catch volume, or by avoiding catching small or immature animals, will gain nearly all the benefits of the management actions of others, but there will be little net conservation effect. Bitter experience has also shown that management measures, as well as being unanimous, should also be introduced as early as possible, before the industry builds up excess capacity; the aim should be to have to do little more than put the brakes on development as the optimum harvesting rate is approached rather than to have to cut back on overcapacity, with all the economic and social problems that this is likely to bring about. Unfortunately, the first criterion works against this: all the participants are likely to agree on the need to do something only after the need has become incontrovertible, with the collapse or severe depletion of the resource.

INTERNATIONAL WHALING COMMISSION

How do events in the Antarctic fit in with these requirements? The fur seal and right whale stocks collapsed before there was any attempt at management, so few lessons can be learned from these experiences. Much more can be learned from the IWC: although the commission has suffered from several serious flaws, it must not be assumed that everything that the IWC has done is wrong. Indeed, when it was first established in 1946, the IWC was considered to be in the forefront of what would later become the environmental movement.

The first flaw in the IWC to become apparent was the lack of good, quantitative information on what was happening to the stocks and what, in quantitative terms, would be the effect of different management measures (in particular, the effect of a reduction in the quota). Only when fishery scientists, who had long had a more quantitative approach, were brought in as members of the IWC's Committee of Three (later Four) Scientists, to advise the commission in the early 1960s, did the commission have adequate information on which to base its policies. The blame for this—if it is fair to talk about blame with the benefit of many years of later experience—lies further back, with those who determined the level and direction of research in the Southern Ocean. It was particularly unfortunate that, after the government of the United Kingdom in 1925 had established the Discovery Committee to carry out research in the Southern Ocean in support of the whaling industry, no one ensured that quantitative studies of the dynamics of whale stocks and their reactions to exploitation were actually carried out. Scientific benefits, for example, in terms of knowledge of krill stocks, are still being reaped from this program.

Other flaws that became apparent after the Committee of Four had made its report, and when it was clear that great reductions in quota were necessary, concerned the objectives of the commission and participation in its deliberations. From 1964 onward, once the commission was receiving clear scientific information on what needed doing, it had, until the mid-1970s, great difficulty and encountered considerable delays in acting on it. This was because the objectives of the whaling convention implied, or could be interpreted as implying, that considerable weight should be given to the economic interests of the whaling industry. This problem was compounded by the fact that the membership of the commission was largely confined to countries with active whaling industries, whose immediate economic interests were a major factor in determining national policies.

This situation is now reversed, with many of the IWC member countries having no direct connection with whaling and the policies of some former whaling countries determined largely by active environmental lobbies. The result is that the decisions of the commission tend to be as strongly weighted against any harvesting of whales as they previously were in favor of a volume of catches in excess of what the resource could sustain. There is, for example, as little scientific justification (assuming that the objective of the commission is to ensure rational use of the resource) for the currently recommended moratorium on all minke whale catching as there was for the continuation of a total quota of 15,000-16,000 blue whale units around 1960.

Events in the utilization and management of other Antarctic resources (krill, fish, and the seals of the ice shelf) have not proceeded so far as to facilitate a historical judgment of the management arrangements. Special mention should, however, be made of the 1972 Convention for the Conservation of Antarctic Seals (Seals Convention). This convention was drafted at the initiative of the Antarctic Treaty powers at a time when there was no commercial harvesting of seals, although there was a possibility that such sealing might start. Since it recognized that commercial sealing was an intrinsically legitimate activity, if done in a responsible manner, it has sometimes been regarded as a pro-sealing convention. This is very far from being true. Those responsible for identifying the need for a convention were well aware of the dangers of uncontrolled harvesting and the need to set up proper controls early. For about the first time in history, the necessary framework to institute control was set up in advance of any commercial activity. The Seals Convention gives the interested countries rights and responsibilities to manage any sealing, and, in an annex, spells out specific controls (annual catch limits and a pattern of closed areas), which should ensure that any harvesting is well within the productive capacity of the stocks, pending more precise scientific analysis.

The 1980 convention establishing CCAMLR was set up while the krill fishery was still growing, while catches were well below the likely level of the sustainable yield and before there was any evidence that the volume of fishing was affecting the stock. The taking of demersal fish around several of the Antarctic and sub-Antarctic islands had reduced some of the stocks well below their unfished level, but there are some biological reasons for being less seriously concerned about such reductions in fish stocks than about similar reductions in mammals. The real concern that led to the establishment of CCAMLR was less for krill or fish in themselves than for the impact that a large-scale krill fishery might have on those species that feed on krill. Krill play a central part in the Southern Ocean ecosystem, being the most abundant herbivore and the major food of many of the larger animals, including baleen whales, penguins, and crabeater seals. There is a fear that if krill abundance were reduced by fishing, this could endanger the recovery of the depleted stocks of baleen whales.

This probable interaction between krill fishing and the dynamics of whale populations illustrates two points that are becoming increasingly apparent in present-day resource management: that the objectives of management are complex and that, in order to achieve whatever objective is decided on, quite detailed scientific research is likely to be needed.

If the Southern Ocean were to be managed purely in order to maximize the supply of food from the region, it is probable that harvesting should be concentrated on krill (assuming that the technological and economic problems of catching, processing, and marketing of huge quantities of krill can be solved), allowing whale and seal stocks to decline. Against this, there are those who feel that no activities should be allowed that would prevent a rapid recovery of whale stocks to their original unexploited level. A more balanced and a more generally acceptable policy might be for a krill harvest at rather less than the maximum possible, sufficiently small to offer no threat to the survival of the whale populations, though probably not so small as to permit any harvesting of whales.

Whatever objective is chosen, it is clear that those setting quotas or other appropriate measures for krill will need to know what the long-term effects are of different patterns of krill harvest on stocks of krill, whales, and other consumers. It is probable that the answer will depend not only on the gross magnitude of krill harvest but also on where and what sizes of krill are caught and on how closely these correspond to the location and sizes of krill eaten by whales. It may also depend on the detailed population dynamics and other aspects of the biology (feeding behavior, etc.) of both krill and whales. This sort of research cannot be done overnight, and effective management of the Southern Ocean as a complete ecosystem will have to be based on good, long-term multidisciplinary research. CCAMLR has available to it a good scientific basis for starting its work, as a result of the research that has already been carried out by the Scientific Committee on Antarctic Research (SCAR) and especially by its Biological Investigations of Marine Antarctic Systems and Stocks (BIOMASS) program.

THE ROLE OF THE ANTARCTIC TREATY

In the narrowest sense, the Antarctic Treaty has had very little direct impact on the management of marine resources. Article VI, which states that nothing in the treaty should prejudice the rights of any state with regard to the high seas, might seem to preclude such impact, except on seals or penguins, which can be harvested when they come ashore. In this connection, one of the earliest actions under the Agreed Measures for the Conservation of Antarctic Fauna and Flora was to give special protection status to fur seals and to the Ross seal.

It is also true that, while the responsibility under Article IX(1)(f) of the treaty for the "preservation and conservation of living resources in Antarctica" makes no distinction between terrestrial and aquatic resources, the initial focus of those concerned with the treaty was on terrestrial animals and plants and those aquatic animals that came to land, or the ice shelves, to breed. For example, the Agreed Measures explicitly discuss mammals (excluding whales), birds, and plants but make no mention of fish, krill, or other marine animals.

Such a narrow interpretation would ignore the very significant contribution that the Antarctic Treaty mechanism has made to managing marine resources in bringing into existence CCAMLR and the Seals Convention and in helping determine the content of the two conventions and the way in which CCAMLR is likely to operate.

Though the conferences at which these conventions were finally signed took place outside the formal framework of the Antarctic Treaty, these final acts were the results of lengthy discussions and negotiations, many of which took place, formally or informally, during Antarctic Treaty consultative meetings.

As remarked above, these conventions are notable for the fact that they were established wholly or largely before there was large-scale harvesting of the resources concerned and well before such harvesting began to have obviously harmful effects. Such public forethought is highly exceptional in the history of the utilization of national resources and owes much to the forethought and initiative of a few individuals in a number of the treaty countries.

Such private forethought would, however, have been of little use if the Antarctic Treaty did not give it a framework in which it could be effective. Despite the initial lack of focus on marine resources, the treaty did give its signatory countries a joint interest and a responsibility for all the resources south of 60°S latitude. Because of the nature of the marine resources, for which the 60°S latitude has no special significance, this joint interest tended to extend northward to the natural boundary at the Antarctic convergence.

Governments do not like taking action unless it is very clear that failing to take action will cause much more trouble than doing something. If the Antarctic Treaty had not existed, providing the framework, and, to a large extent, the political justification for the initial discussion, it is fairly certain that there would now be no sealing convention. It is also unlikely that negotiations over a more general "ecosystem management" convention corresponding to CCAMLR would have got much beyond the initial confrontations between those interested in conservation in the narrow sense of discouraging any exploitation and those interested in immediate exploitation with little concern for long-term interests.

The potential for confrontation is well illustrated by the recent history of the IWC. Its problems were briefly mentioned earlier. They illustrate that a management body cannot work well unless there is a reasonable balance among different interests. In the IWC the balance has swung from undue dominance by short-term economic interests to dominance by conservation interests (and to a large extent the more extreme conservation views).

Unfortunately, the IWC conflicts have spilled over into wider discussions on the Antarctic marine ecosystem as a whole taking place under the CCAMLR, since the countries and some of the individual scientists who take part in these two forums are much the same. The conflicts in the IWC have reached the stage at which each side distrusts the objectives and scientific integrity of the other. In particular, those countries still whaling believe, not wholly without justification, that some of those attending meetings of the IWC are interested only in stopping all whaling, without much consideration of how this is done or whether it would be in accordance with the spirit and intention of the original convention.

The whaling countries are largely the same countries as those interested in harvesting krill and fish in the Antarctic. They certainly would not have been prepared to consider a new commission that seemed to be nothing more than a repeat of the IWC, and probably it was only because of the existence of the Antarctic Treaty mechanism, and in response to a treaty initiative, that those countries were prepared to consider actively a new conservation mechanism for Antarctic marine resources.

The Antarctic Treaty states in Article IX(2) that while any state that is a member of the U.N. may accede to the treaty, participation in consultations is dependent on serious interest in the Antarctic, as demonstrated by substantial scientific research activity. The same principle is expressed in the CCAMLR, in which a qualifying interest for membership on the commission may be in terms of either research or harvesting activities [Article VII(g)]. While there are some good reasons for believing that any body with long-term responsibilities for conservation should in principle be open to as wide a membership as possible, including states whose current interest in the resources is more potential than real, there are equally good reasons for believing that the extremely wide membership of, for example, the IWC is not always conducive to finding a constructive solution. It can be a very difficult problem when membership of an organization, whether of nations or of individuals, is open to all, without the qualification of a serious interest in the objective of the organization. While the founders of CCAMLR would probably have found a solution to this potential difficulty, the way was made much easier for them by the precedent of the Antarctic Treaty and the existence of the consultative powers as a "club" of countries with a proven and responsible interest in the Antarctic.

In summary, therefore, the contributions of the Antarctic Treaty System to current actions to manage Antarctic marine resources have been very significant. Without the treaty, the different interests would never have got together to agree on a convention, and the treaty provided a model of the form of membership that should ensure a workable commission. Conservation interests may well believe that the commission is working slowly, and the measures so far taken and contemplated (minimum mesh sizes to be used when trawling for fish and minimum fish sizes that can be landed) are more cosmetic than real. However, the commission is moving toward more effective measures and toward the collection of data that would enable these measures to be soundly based. This is clearly preferable to the only likely alternatives, which would be an absence of any commission, or a commission dominated by conservation interests but ignored by the fishing nations.

TERRESTRIAL ACTIVITIES

There is little interest in harvesting living resources on land. The exception is the possible harvesting of essentially marine animals when they come ashore to breed. This is a special aspect of the general question of the rational use of these resources, already discussed. Otherwise the conservation of living resources on land involves no serious conflict, and the task is one of ensuring that any activity carried out does not accidentally damage the resource.

This task the Antarctic Treaty, through its Agreed Measures for the Conservation of Antarctic Fauna and Flora, has done well.

Three broad types of control are used: those applying to all human activities and additional protections for certain species (Ross seal and fur seal) and for certain areas. The choice of appropriate measures, and the general agreement to be bound by them, has been helped by the fact that scientific activities, combined with tourism on a limited scale, have so far been the only activities on the Antarctic land mass. In both cases, those involved can be expected to have an above-average interest in and knowledge of the resources as well as an interest in their conservation.

The general measures include the prohibition, unless a permit has been required, of the "killing, wounding, capturing, or molesting of any mammal or bird" (Article VI), the need to "take appropriate measures to minimize harmful interference" (Article VII), and the limitation under Article IX of the introduction of species that might upset the ecological balance. (In effect, only sledge dogs and laboratory animals can be introduced.)

At the present scale of human activities, these general provisions are probably sufficient. They accept that nearly any activity must have some immediate impact on the ecosystem but that all ecosystems have some natural resilience; that is, if the impact is sufficiently small, it will be only temporary. Thus, when permits are given to kill animals as food, the numbers taken should be capable of being replaced in the following breeding season. There is, however, some belief that the Antarctic terrestrial ecosystem, at the limits of conditions under which life can exist, is less resilient than most others, including the Antarctic marine system, in which, for instance, the fur seal population at South Georgia has shown dramatic powers of recovery. There is little quantitative knowledge of the degree of resilience of the Antarctic terrestrial and inshore life, that is, how much disturbance different types of environments can withstand without permanent damage. Such knowledge, and the incorporation of this knowledge into quantitative regulations—such as the amount of waste that can be discharged into coastal waters and the number of visitors that can be allowed to a given site—may become more important if Antarctic activities increase. They would seem to be essential if industrial activities such as oil extraction ever became a practical possibility. In that case, the Agreed Measures could still be seen as important first steps.

Controls for certain areas applied initially to the creation of Specially Protected Areas (SPAs), in which many activities, including driving vehicles, were prohibited. Some 17 SPAs have been designated and are given very effective protection. However, the strictness of the measures and especially the resultant difficulty in carrying out research in these areas have probably lessened the enthusiasm for designating SPAs. There has been in addition some feeling that the SPAs do not cover wide enough areas to achieve all the purposes that such areas could serve. These purposes were identified at the seventh consultative meeting, in 1972, at which it was concluded that SPAs should include inter alia representative examples of the major Antarctic land and freshwater systems and undisturbed areas to be used for comparison with disturbed areas. The same meeting also made provision for the creation of Sites of Special Scientific Interest (SSSIs), which are governed by less rigorous restrictions and are of a temporary rather than a permanent nature, although their designations may be extended.

While there have been problems in achieving full and rigorous observance of all the Agreed Measures, most recently in the case of the proposed French airstrip in Adelie Land, the measures have in general been obeyed and have worked well. Even the exceptions seem to have involved no serious threat to the conservation of resources; they are important mainly in showing that no agreement will necessarily be obeyed unless there is some monitoring of compliance and that the Antarctic Treaty is no exception to this.

THE ROLE OF SCAR

No discussion of conservation in the Antarctic would be complete without a mention of the special role of SCAR. SCAR is formally an organ of the International Council of Scientific Unions, which comprises national academies of science, rather than governments. However, the qualifications for national membership in SCAR are the same as those for consultative status under the Antarctic Treaty: substantial research interest in the Antarctic. Thus, the list of members of the two groups is very similar. This list forms what is often considered the Antarctic club, with the treaty being the political and SCAR the scientific arm. The achievements (or failings) of SCAR can be taken as reflecting in some way the merits (or demerits) of the treaty system.

The immediate contribution of SCAR has been in providing, or helping to provide, the scientific base for the three main substantive actions for conservation: the Agreed Measures under the Antarctic Treaty, the Seals Convention, and the CCAMLR. SCAR also carries out, until such time as the majority of the contracting parties to the Seals Convention establish their own scientific advisory committee, the functions of compilation and analysis of data relating to sealing and advising the contracting parties on these matters.

In the long run, the most important function of SCAR is to provide the coordinating mechanism for the long-term fundamental studies that are essential to effective scientific management of any natural system. CCAMLR has its own scientific committee, and the boundaries between the two scientific groups are not at present clear. The CCAMLR committee may be expected to concentrate on the more immediate practical aspects, such as analysis of commercial catch and effort data, while SCAR concentrates more on less immediate issues, such as the physiology of krill. The latter may (or may not) turn out in the long run to be of direct practical significance to answering management questions.

In the context of CCAMLR's ecosystem approach, the studies carried out by the BIOMASS program, coordinated by the SCAR Group of Specialists on Living Resources of the Southern Ocean, deserve special mention. These have looked especially at the structure of the Antarctic marine ecosystem and at the central role played in this system by krill. The BIOMASS program accomplished a number of studies that could have been carried out under the auspices of the CCAMLR institutions. Nevertheless, there are various data requirements under the CCAMLR that could not be met by BIOMASS.

Other SCAR groups, especially the Working Group on Biology in relation to the Agreed Measures, and the Group of Specialists on Seals in relation to the Seals Convention, have played, and continue to play, important roles in conserving Antarctic living resources, both marine and terrestrial.

16.

Panel Discussion on Living Resources

The panel consisted of Peter D. Oelofsen (moderator), Alexandre Kiss, James Barnes, and Yoon Kyung Oh.

SUMMARY

Mr. Gulland's presentation on the management of Antarctic marine living resources provided concrete examples of two subjects raised in Chapters 13 and 14 on the environmental management of Antarctica:

(1) That a type of continent-wide approach, as suggested by Heap and Holdgate, is already being initiated with respect to marine living resources management based on the ecosystem standard adopted in the Convention on the Conservation of Antarctic Marine Living Resources (CCAMLR); and
(2) That the Convention for the Conservation of Antarctic Seals and CCAMLR illustrate the preventive approach toward conservation, attempting to foresee and address problems that might arise once commercial exploitation of these species develops.

It also pointed out how the contracting parties to the CCAMLR have sought to avoid mistakes made by the International Whaling Commission (IWC). First, they are attempting to acquire the knowledge necessary to establish a solid foundation for management decisions before over exploitation occurs. Second, to get away from the IWC problem of a fluctuating balance between proexploitation and proconservation members, CCAMLR operates according to consensus decision-making procedures, seeking agreed objectives for conservation measures.

In general discussion, the CCAMLR received strong support both as an effective mechanism for the conservation of Antarctic marine living resources and, on political grounds, because it has dealt effectively with opposing positions on national claims. Some participants cautioned against distancing fishery management objectives under the CCAMLR from the political priorities of the Antarctic Treaty System (ATS).

As a resource management regime, it was noted that the continent-wide approach would have to be capable of coping flexibly with local and regional exploitation pressures. Questions raised by the panelists and the audience focused primarily on the implementation of the CCAMLR, still in its early stages. Various suggestions were put forward to help guarantee that its objectives were fully achieved. These addressed the following points:

(1) Progress in adopting conservation measures and acquiring the data required to underpin them, including the possibility of allowing the krill fishery to develop as an experimental fishery and the role of SCAR's Biological Investigations of Marine Antarctic Systems and Stocks (BIOMASS) program;
(2) A strategy for monitoring living resources;
(3) The establishment of a CCAMLR inspection system;
(4) Policies with respect to international organization observers;
(5) Measures to avoid adverse effects of marine debris on living species; and
(6) The possibility of creating whale habitat sanctuaries.

REMARKS BY ALEXANDRE KISS

On the whole, Kiss was optimistic that the ATS has and will develop legal regimes adequate to protect Antarctic ecosystems for the benefit of the whole of mankind. He endorsed the continent-wide approach of the ATS, stressing the need to treat issues as part of one interrelated system.

Nevertheless, Kiss called attention to a certain paradox, as he saw it, between developments under the Antarctic Treaty and those under CCAMLR. Despite the fact that the Antarctic Treaty contained a fairly narrow basis for drafting regulations to protect the environment [Article IX(1)(f)], the consultative parties have turned their attention to this subject beginning at their first meeting. CCAMLR, on the other hand, provided a detailed mandate for enactment of conservation regulations; yet three meetings have produced fairly limited results. The first meeting dealt with important administrative rules, but it was not until the third meeting, in September 1984, that the first two conservation measures were adopted. These covered minimum mesh size regulations for nets throughout the convention area and a prohibition on fishing other than for scientific research purposes in waters within 12 nautical miles of South Georgia.

Kiss reported that a third proposal to limit fish size did not obtain consensus and that no practical measures had yet been adopted on how to avoid entanglement of marine mammals, birds, and other nontarget species in marine debris, such as lost or discarded fishing gear. This problem, along with monitoring, will be considered at the fourth CCAMLR meeting, in September 1985. Moreover, few steps have been taken toward implementing the inspection procedures provided for in Articles IX(1)(g) and XXIV of the CCAMLR. In Kiss' view, these achievements fall short of the far-reaching aims of the CCAMLR.

On the subject of data requirements, Kiss was critical of the fact that the CCAMLR scientists have to rely to some extent on data provided by the commercial fishermen. He believed that states have an important responsibility to collect and transmit scientific data to the commission because these data are indispensable for the preparation of regulations.

Kiss warned against the danger of consensus decision-making procedures, which could allow a minority of voting members to paralyze the decision-making mechanism, but he noted that the caution and thorough discussion required to obtain consensus also favor sound agreement. At the same time, he expressed confidence that the CCAMLR Scientific Committee, composed of scientific experts, will help the commission to accomplish its tasks.

REMARKS BY JAMES BARNES

Barnes agreed that the CCAMLR was a potentially valuable instrument, but he identified potential problems with it as well. He, too, was afraid that consensus decision-making procedures could allow fishing states that are members of the commission to block measures that restricted their fishing. In his view, this is already a problem in the Southern Ocean finfishery. He also wondered whether the commission would exercise the will to set national quotas. Although national quotas are not actually listed among the types of measures that may be adopted by the commission, it was his understanding that the commission could adopt them if it wished.

With respect to the inadequacy of data available for rational decision making under the CCAMLR, Barnes believed that more baseline data, as well as a better understanding of the dynamics of the marine ecosystem, are needed. He reminded the group that a number of nongovernmental organizations have proposed that the Southern Ocean fishery be managed as a giant scientific experiment, with various closed areas, closed seasons, and use of the "indicator species" concept. He advocated more substantial international support for the BIOMASS program as another approach to helping to meet data needs.

Barnes shared Kiss' regret that the inspection system called for in CCAMLR has not received much attention to date in commission meetings.

On the subject of observer participation, he noted that the Antarctic and Southern Ocean Coalition had requested observer status under CCAMLR in 1978. He stressed that conservation organizations wish to be involved in CCAMLR meetings, in keeping with normal international organization practice, in which representatives from conservation groups may express views and circulate documents. In his opinion, policies with respect to observers will be important in helping to gain widespread acknowledgment of the legitimacy of the ATS.

Finally, regarding whale habitat sanctuaries, Barnes noted that several species of whale are at one and five percent of their historical stock sizes and that Article II(2)(b) of the CCAMLR calls for measures to rebuild depleted stocks. He advocated giving attention to creating "marine protected areas" in order to preserve the feeding grounds of the great whales, and he was pleased that SCAR, at the thirteenth biennial meeting of the Antarctic Treaty consultative parties in October 1985, will recommend approval of marine protected areas, even if none of these is a habitat sanctuary. (Chapters 13 and 14.)

REMARKS BY YOON KYUNG OH

Oh expressed the Republic of Korea's support for the contributions of the ATS in facilitating cooperation in many fields of science among the countries active in Antarctica. He acknowledged the system's role in establishing a zone of peace, prohibiting military activities, protecting the pristine environment, and conserving living resources in the area. He said that the Korean government has sent ships to Antarctica four times since 1978 to conduct fishing activities and oceanographic surveys and that it plans to continue its surveys in the years ahead. He added that reports on these activities have been distributed to the parties to the Antarctic Treaty and to relevant international organizations.

Oh noted his country's willingness to accede to the CCAMLR and to work together with the contracting parties to the Antarctic Treaty in promoting international understanding and peaceful use of the Antarctic continent. He urged that the ATS be open to all new participants without reservation. (The Republic of Korea, which is not a member of the U.N., is in the anomalous position of requiring the consent of the consultative parties to accede to the 1959 Antarctic Treaty. See the discussions in Chapters 7 and 27.)

DISCUSSION

The discussion on progress toward adopting specific conservation measures for marine living resources turned on the two points made by Gulland in his presentation: the need for agreement among those concerned on the objectives of such measures and the adequacy of scientific and technical data on which to base them.

There was general agreement that the process of formulating objectives and methods is an ongoing one that is important with respect to information requirements as well as to specific conservation regulations. The second and third meetings of the CCAMLR commission and scientific committee had concentrated on how to establish data reporting and analysis systems. The scientific committee had agreed that these topics deserved highest priority and that the formulation of objectives in this area would be extremely important, because effective implementation of the CCAMLR depends on capabilities to detect trends, changes, and potential changes in species populations and the surrounding environment.

Debate over CCAMLR accomplishments to date and future prospects took note of the fact that the CCAMLR institutions had adopted some conservation measures at their September 1984 meeting, as noted by Kiss, and that the scientific committee, taking note of the assessment of fish stocks in the South Atlantic had agreed to recommend that states desist from directed fishing for a specific fish species (Notothenia rossii) in the entire convention area. On the other hand, lack of adequate data had prevented the scientific committee, despite intersessional meetings, from making certain additional recommendations regarding assessment.

Some participants argued that the slow pace in implementation of CCAMLR would discredit the convention and that the commission would be in no better position at its September 1985 meeting to make judgments on conservation measures than it had been in 1984. They questioned whether lack of political will, rather than lack of data, is the problem, and stressed that at some point commission members will have to proceed to adopt additional, substantive measures. One participant questioned why no interim conservation measures had been adopted and stated that there should be some discussion of the need for a moratorium on the harvesting of certain species.

Others countered that this assessment showed little appreciation of ongoing efforts to establish an effective assessment capability. They noted that stock assessment efforts are being conducted primarily with a view toward the 1985 meeting and that it would be a bad precedent if determinations were made in the absence of soundly based scientific judgment. (The 1984 meeting established ad hoc working groups on fish stock assessment and ecosystem monitoring. Workshops are to be held before the 1985 CCAMLR meeting.)

Gulland endorsed this view, pointing out that the CCAMLR secretariat and the nations party to the treaty are working to collect more detailed data on fish stocks. He questioned whether any arrangement other than the CCAMLR would have been more successful in obtaining data, getting agreement on conservation measures, and pressing countries to adopt these measures.

Elaborating on experiences under existing fishery agreements as a measure of progress in the implementation of CCAMLR, Gulland stated that a number of fishery commissions have developed good data collection and assessment techniques. In fact, since the introduction of 200-mile fishery zones, coastal nations have found it difficult to carry out these tasks by themselves. Moreover, most commissions commenced work with a good storehouse of data. CCAMLR, on the other hand, started with virtually no data, so its progress to date is quite good.

On the question of how much information is required to enable decision makers to be confident of a reasonably sound basis for judgment, Gulland alluded to the amounts of data needed for increasingly complex questions, such as, "Does a ten-million-ton krill recovery have a serious effect on whale populations?" He believed that the relevant question is rather whether the incoming data was up to the standard required by scientists; the answer for CCAMLR at this time would be that they are not.

BIOMASS

The Biological Investigations of Marine Antarctic Systems and Stocks (BIOMASS) program was commended for its data collection contributions to information needs under CCAMLR, and it was urged that the program be continued and expanded to include more cooperative international efforts. (BIOMASS will officially conclude in 1986 unless its life is extended.) National support for the program has varied in the past. The BIOMASS data are now being computerized and will be lodged in the British Antarctic Survey and be available for analysis.

There was some discussion of Gulland's point that the respective roles of the CCAMLR scientific committee and SCAR are not clearly delineated. Some participants expressed doubt that the first and second BIOMASS expeditions, FIBEX and SIBEX, would have been undertaken by the scientific committee. The scientists pointed out that their work is driven by curiosity, although they recognized that at some point its results may have practical application as well.

EXPERIMENTAL FISHERY

The discussion of developing Southern Ocean fishing as an experimental fishery contrasted the situation of Antarctic krill with that of finfish in the Southern Ocean. With respect to finfish, CCAMLR is in the same position as virtually every other convention whose objective is to conserve finfish species; that is, the negotiations take place after the stock is already seriously depleted (although the Southern Ocean stocks might not have been so depleted had there not been a rush to establish coastal state 200-mile zones of fishery jurisdiction elsewhere).

With respect to krill, the managers are in a position to make use of their time. Nonetheless, they could find themselves in the same situation as have other fishery commissions if they spend too much time deciding whether sufficient data exist and do not act until it is too late; there is time, but not an unlimited amount of time, to collect basic data and to design research programs.

There are also good biological reasons to be more concerned with krill, because of its place in the Southern Ocean ecosystem. Moreover, the CCAMLR expressly adopted the ecosystem standard in its approach to conservation.

There is little doubt that krill pose certain difficulties for determining when conservation measures should be adopted. The classic means of data gathering and of studying species distribution and abundance for management purposes--catch per unit of effort and the monitoring of key indicator species--might not be applicable to krill. Specific interactions between krill and other Southern Ocean species would cause some difficulty in identifying sound management practices: krill is relatively short lived, and its position in the ecosystem is difficult to establish. The criterion of catch per unit of effort might not be particularly useful when applied to krill, because these figures would be distorted by krill swarm density as well as by the processing ability of the harvester. Acoustic surveys will be helpful only in small areas.

Gulland suggested that a better method might be to see what distances a harvester would have to sail before encountering a new swarm; that is, to rely on catch per unit of fishing time rather than on catch per unit of effort. The breeding success of nonharvested, dependent species such as the albatross could also be utilized as an index of krill abundance, although albatross breeding success reflects other factors as well.

For these reasons, the idea of a krill fisheries experiment was viewed as an attractive one, although no one was certain of the best way to get the information required to manage the krill fishery most effectively. It was suggested that the SCAR/IUCN symposium (Bonn, April 1985) might be helpful in answering this question. In addition, it was noted that the meeting of the ad hoc group on ecosystem monitoring established at the 1984 CCAMLR meeting (Seattle, May 1985) would look at what sorts of observations of penguins and other creatures would be required.

Returning to the experience of existing fishery commissions, Gulland noted that the situations in which they had not worked well were those cases in which the stocks were already depleted and fishermen were being forced to reduce their catches. Another participant noted that most of these agreements had covered a single species. He wondered whether, with respect to single species--there might be some similarities with the current situation in Antarctica; that is, vested economic interests would be affected.

Gulland agreed with Barnes that in general the lack of accord on allocation of national quotas is a weakness, but he pointed out that many existing commissions do not themselves allocate quotas, leaving this to member countries to agree on. What the CCAMLR commission might do is to determine a maximum allowable catch and leave it to the interested member countries to decide directly among themselves how to allocate this catch. Thus, he did not believe that the absence of national quotas is a fundamental weakness in the CCAMLR.

Another speaker reaffirmed Barnes' comment that the fact that the CCAMLR has not expressly provided for allocation of national quotas on the list of permissible conservation measures in no way detracts from the commission's full power to adopt any kind of measure, including national quotas, to attain its objectives. He explained that the impression that the commission does not have such competence derives from a misinterpretation of the report of the consultative meeting at which it was decided to initiate the negotiation of the CCAMLR. That report indicates that the agreement will not include national quotas on other forms of economic regulation. What this means is that the CCAMLR will not itself set forth entitlements to resources among its parties (e.g., as between fishing states or states with coastal claims in Antarctica). This is in keeping with the nature of CCAMLR, the purpose of which is to establish the legal obligations and mechanisms necessary to conserve Antarctic marine living resources. It was not intended--nor is it reflected in the text of the convention--that the commission will be limited in taking measures including national allocations--that it believes necessary to achieve the CCAMLR objective. The list of possible conservation measures in the convention, which does not specifically mention national quotas, is illustrative, as indicated by the final point in the list: "the taking of such other conservation measures as the commission considers necessary for the fulfillment of the objective of this Convention" [Article IX(2)(i)].

Finally, according to Gulland, operating by consensus has not posed major problems for other fishery commissions.

INSPECTION AND OTHER CCAMLR MEASURES

Several participants shared the panelists' view that the commission should demonstrate that it is making progress on establishing a system of observation and inspection, as called for in Articles IX(1)(g) and XXIV. They pointed out that a lack of scientific data does not affect this issue.

Gulland made a distinction between inspection and enforcement in commenting on the effectiveness of existing fishery commissions. While most commissions have established effective inspection systems, it is difficult to find any international agreement that is effectively enforced. Unless fishermen agree that conservation regulations protect their long-term interests reasonably well, the regulations are unlikely to work.

Measures to ensure that marine mammals, birds, and other nontarget species are not adversely affected by lost or discarded fishing gear or other marine debris was cited as another substantive area in which the commission could act without waiting for additional data. On the other hand, it was noted that the adoption of a strategy for management and monitoring of living resources and the study of relationships with neighboring ecosystems, while important topics for the commission to consider, were in part dependent on acquiring data and outlining strategies to do so.

17.

Arctic Offshore Technology and Its Relevance to the Antarctic
K. R. Croasdale

INTRODUCTION

In considering the issue of potential antarctic oil and gas resources, especially offshore, it is perhaps relevant to look to the Arctic for an analog of what might be possible. This chapter will provide a briefing on the technology being used and/or developed for Arctic offshore oil and gas operations. For our purposes, the Arctic will be defined as northern offshore areas subject to major ice coverage. Therefore, the Canadian east coast, with its iceberg problems, is also included. My major focus will be on operations in Canada, where most oil and gas activity has taken place. Other nations bordering on the Arctic, however, also have interests in Arctic off-shore resources and are developing technology similar to that which I will describe.

Where appropriate, reference will be made to the similarities and contrasts between the Arctic and the Antarctic. The chapter will conclude with some specific comments on the possible adaptation of Arctic offshore technology to the Antarctic.

GEOGRAPHY AND OIL AND GAS RESOURCES

The Arctic consists mainly of a large ocean surrounded by land that is divided here and there by straits, channels, and small seas. The Arctic Ocean can be considered a polar Mediterranean--an inland sea with very little communication, except through the Fram Strait, with the other oceans of the world.

It is a hostile region, but because of the populated, surrounding land masses, human presence around the

fringes of the Arctic Ocean has existed for at least
40,000 years. Only during the past three centuries,
however, have Europeans endeavored to explore and exploit
the region. Early ventures used wooden sailing ships and
were conducted either for whaling or for exploration,
especially in the search for a northwest passage to Asia.
Arctic technology used before about 1900 was not really
adequate; many expeditions spent years at a time trapped
by ice in the Arctic, and many of the explorers perished.

Today, the major incentive for Arctic operations is
the exploitation of minerals, primarily oil and gas.
During the past 15 years or so, the Canadian oil industry,
encouraged by government incentives and the prospect of
large discoveries, has been very active in exploring
Canada's Arctic and offshore areas. Although no production has yet occurred from Canada's offshore Arctic,
several promising discoveries have been made:

- gas and oil in the Beaufort;
- gas and some oil in the High Arctic (Arctic Islands);
- gas off Labrador; and
- oil off Newfoundland.

Furthermore, there is the promise of large oil and gas
reserves yet undiscovered. Canadian government sources
predict mean potential recoverable reserves of oil for
the Canadian Beaufort of about 8.5 billion barrels; for
the Arctic Islands, about 4.3 billion barrels; and for
the Canadian east coast, about 12 billion barrels. On
the Alaskan continental shelf, a mean value for potential
recoverable reserves is about $18 billion. This, plus
the desirability of North American self-sufficiency in
energy, is the incentive that has led the major oil
companies and governments actively to pursue the
exploitation of Arctic oil and gas (despite its high cost
relative to oil and gas operations in more temperate
regions). Of course, the high cost of Arctic oil and gas
operations is mainly a function of the climate, the
remoteness, and especially the presence of offshore ice.

THE ARCTIC OFFSHORE ENVIRONMENT

The Arctic Ocean is covered with drifting ice. Only in
the more southerly coastal zones does the ice clear every
summer. The major drift patterns of the Arctic ice are

shown in Figure 17-1. The Pacific or Beaufort gyral dominates the ice drift in the North American sector of the Arctic Ocean, rotating clockwise, completing one revolution in about 10 years at a rate of several kilometers per day. It consists mostly of multiyear ice with an average thickness of about 5 m. Around the edges of the Arctic Ocean, the ice melts and reforms every year, creating first-year ice with a maximum thickness of about 2 m.

Close to shore, out to about 20 m water depth and between islands, the ice becomes landfast each winter. The approximate demarcations among the various ice zones in the winter in the Beaufort Sea are shown in Figure 17-2; this is typical of most of the coastal regions of the Arctic Ocean.

As depicted in the cross section (Figure 17-3), pressure ridges, both first year and multiyear, can occur in all the ice zones. The probability of seeing multiyear ice in the near-shore zones is quite low; in fact, it usually invades only during those summers when the permanent polar pack is driven south by onshore winds. This occurs on average about once every five years. Pressure ridges, especially multiyear ridges, are the ice features that create the greatest difficulty for offshore operations. The movement of multiyear floes and ridges against offshore platforms causes high lateral loads, and these also create the most difficult obstacles for icebreaking vessels. Multiyear ridges up to about 25 m thick can occur.

The pressure ridges that ground in the near-shore zones cause gouging of the seafloor. Numerous gouges several meters deep have been observed. Any seafloor facilities, such as pipelines and wellheads, have to be protected against this ice action. The present approach is to put such facilities into trenches, so that they are below the depth of the deepest gouge likely during their lifetimes.

The open water in the coastal zones does allow floating operations to be conducted during the short summer season. In the southern Beaufort Sea, the period of open water averages about 100 days. This period gets shorter with increasing distance offshore.

Off Canada's east coast, annual pack ice occurs south to about 45°N latitude. The open-water period off Newfoundland can range from about 200 to 365 days per year. Further north in the Davis Strait, it reduces to about 100 days.

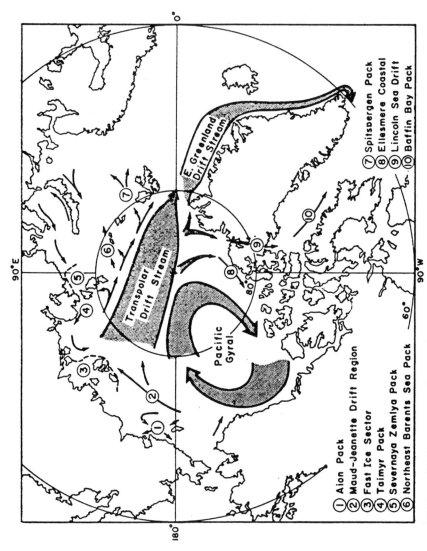

FIGURE 17-1 The major drift patterns of the arctic ice.

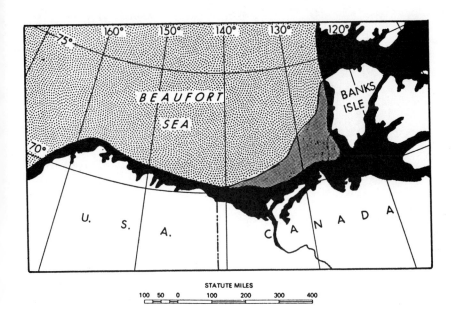

ARCTIC PACK ICE

 MAINLY OLD ICE 3-4 METERS THICK MOVING SLOWLY YEAR ROUND.

SOLID AND UNMOVING

 OLD OR FIRST YEAR ICE OR A MIXTURE OF BOTH COMPLETELY COVERING THE WATERWAY.

MORE THAN 7/8ths

 MAINLY FIRST YEAR ICE 1-2 METERS THICK IN INTERMITTENT RESTRICTED MOTION.

FIGURE 17-2 Typical winter ice conditions in the Beaufort Sea area (February) (Department of Energy, Mines, and Resources, 1970) (Reprinted with permission).

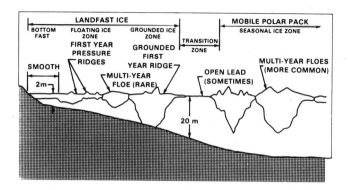

FIGURE 17-3 Near-shore winter ice conditions in the Beaufort Sea.

In addition to pack ice, there are many icebergs off Canada's east coast. The major source of icebergs in the eastern Arctic is the Greenland ice cap, which annually calves about 240 km^3 of glacier ice into the surrounding seas. This results in about 20 to 34 thousand icebergs per year, although on average only about 500 per year will reach the Grand Banks area off Newfoundland. These icebergs are much smaller than those in the Antarctic.

There are no icebergs in the Beaufort Sea, but large pieces of ice shelf ice can occur, which are called ice islands. Ice islands are calved from a relic Pleistocene ice shelf that still exists along the north coast of Ellesmere Island. When calved, the thickness of an ice island can be greater than 50 m, and ice islands can be as much as 10 km in extent. However, large ice islands are so rare (perhaps only one still exists in the Arctic Ocean) that offshore platforms justifiably need not be designed for them. The probability of a collision is extremely low, and ice islands could be detected well ahead of a collision so that people could be evacuated and oil wells sealed off below the ocean floor. (A similar approach is used for hurricanes in the Gulf of Mexico.)

The water depths of the continental shelves surrounding the Arctic Ocean are shallow relative to the Antarctic. In the Arctic the shelf break starts at about 100 m depth. Most offshore operations to date have been conducted within this water depth (although wells off the East Coast of Canada have been drilled in water out to about 1,000 m).

TECHNOLOGY FOR ARCTIC OFFSHORE PETROLEUM OPERATIONS

General

Potential hydrocarbons have to be drilled into in order to confirm their presence. So one of the first tasks for the oil industry in the Beaufort Sea was to consider various methods of offshore drilling in ice covered waters. Figure 17-4 was drawn about 10 years ago. It shows the various concepts being considered at that time.

These ideas range from building artificial islands in the very shallow waters (which can be used year-round) to floating drilling in the deeper waters during the ice-free periods. Also shown are gravity-founded mobile drilling units, which could be ballasted down onto the seafloor, and drilling off the landfast ice during the winter. Active systems using ice cutters in order to remain on location in moving ice were also proposed.

Some of the systems shown in Figure 17-4 have been put to use during the past decade, and these will now be described in the context of main geographic areas.

The Beaufort Sea

Of all the concepts shown, in the Beaufort Sea the artificial island has been one of the most successful, especially in water depths out to about 20 m. Esso Canada built the first artificial island in 1972. A typical island in summer conditions is shown in Figure 17-5. To date, well over 20 islands have been built and

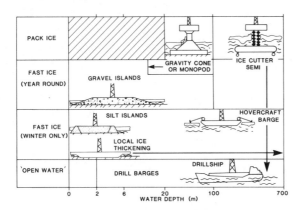

FIGURE 17-4 Exploratory drilling concepts for ice infested waters (Croasdale, 1983) (Reprinted with permission).

FIGURE 17-5 Beaufort Sea Artificial Drilling Island (summer conditions)

drilled from in the Beaufort (Figure 17-6). This map also shows locations drilled using ships, which I will discuss later. Islands constructed in the summer by dredging have been a favored technique for Canadian Beaufort exploration in the shallow water. In contrast, most islands constructed off Alaska have used land sources of gravel. Islands have proved very adequate in resisting the ice, but their cost effectiveness decreases with water depth. In water depths beyond about 20 m the fill requirements tend to be prohibitive, especially for islands with very shallow slopes. Also, islands are susceptible to wave erosion during the summer months, particularly during construction. For these reasons, retained islands were conceived both to reduce fill volumes and to provide immediate wave protection at the waterline.

One of the first designs for a caisson-retained island was Esso Canada's. This consists of a ring of steel caissons retaining an interior sand fill. The caissons are placed on a berm, which can be adjusted in size according to water depth. The Esso retained island has drilled two exploration wells to date in water depths out to 26 m.

A different design of caisson-retained island was used at Tarsiut in about 22 m of water. Here, four concrete caissons were placed on a berm and filled with sand to provide the retaining structure for an interior sand fill. The Tarsiut island also utilized steep dredged slopes; one in 5 compared with one in 15 on most Esso islands. The Tarsiut island set a record for drilling at the very edge of the landfast ice in the winter of 1981/1982 and is shown in Figure 17-7. The island proved well able to resist the ice forces but had a rather limited surface area. Another disadvantage of the two retained islands shown so far is the need to place a land rig on them after construction. In a short Arctic summer this can be a problem, and icebreaker support is often required.

The Gulf Canada Molliqpak was conceived to overcome this problem by designing it with a rig permanently mounted on the caisson structure. The caisson has a hollow core, which is sand filled to provide sliding resistance. The Molliqpak was constructed in Japan and is now on location in the Canadian Beaufort, drilling its first well in 26 m of water.

A similar philosophy of a ready-mounted rig was adopted with the design of the dome single steel drilling caisson

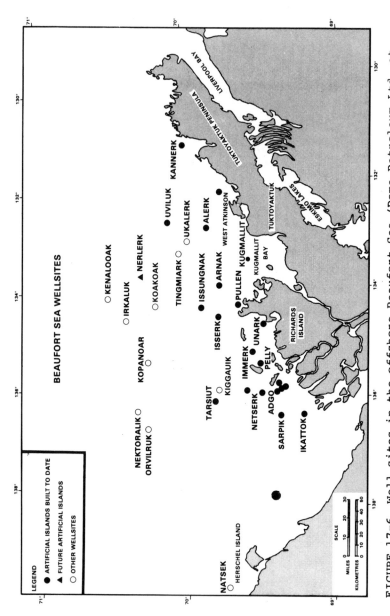

FIGURE 17-6 Well sites in the offshore Beaufort Sea (Dome Petroleum Ltd. et al., 1982) (Reprinted with permission).

FIGURE 17-7 The Tarsiut Drilling Island in the Beaufort Sea (Courtesy of Dome Petroleum Ltd.).

(SSDC). This concept was designed for rapid deployment by water ballasting down onto a prebuilt underwater berm (at -9 m below seabed).

The main structure of the SSDC is actually the converted forebody of a very large crude [oil] carrier specially strengthened to resist the ice loads. It was designed and brought into the Beaufort ice in less than nine months. The SSDC also holds the record for the greatest water depth drilled by fixed platforms in the Arctic. This was at Uviluk in 31 m of water, well outside the landfast ice.

Meanwhile, the U.S. operators have been busy in the Alaskan Beaufort. Several artificial islands have also been built there in recent years for exploratory drilling. This year the Global Marine concrete island drilling system (CIDS) is drilling its first well for Exxon in 15 m of water. The CIDS is a self-contained caisson system that relies on water ballast and can, therefore, be rapidly deployed. Also of significance is that at its present location no foundation preparation was done; it was put down directly onto the seafloor. It is also worth noting that at this location Exxon will be building a spray-ice barrier to provide additional sliding resistance against late winter ice. Spray-ice barriers and rubble-protected systems may enable exploration concepts to be used in the future.

In the even deeper Beaufort waters, Dome Petroleum has pioneered the use of drill ships, which were brought into the Beaufort in 1976. The drill ships are, of course, ice strengthened, and when the ice is managed with icebreakers, drilling can continue into freeze-up. However, it is not usually possible to drill much beyond late November or when the ice is about 0.3 m thick. A late-season drilling situation is depicted in Figure 17-8.

The summer drill ship season is short, hardly long enough to drill and test one well. For most of the year the drill ships remain unused in a winter harbor.

To extend the floating drilling season, the Gulf subsidiary Beaudrill designed and built a round drill ship, the Kulluk (Figure 17-9). Kulluk is designed to stay on location with 1.3 m of ice moving against it. It is held on location with twelve 9 cm mooring lines. Ice management is provided by two Class 4 icebreakers and two ice-class supply ships. So far Kulluk has drilled two discovery wells and apparently has a performance record that exceeds expectations. Incidentally, Kulluk was

FIGURE 17-8 Late summer drilling in the Beaufort Sea (Courtesy of Dome Petroleum Ltd.).

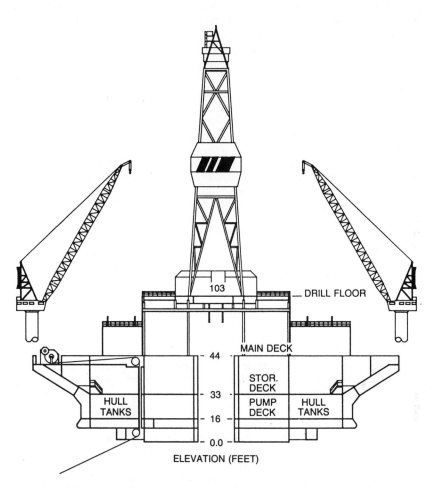

FIGURE 17-9 Typical cross section of <u>Kulluk</u>, showing mooring wire routing (Frankovich, 1984) (Reprinted with permission).

recently approved by the regulatory agency--the Canada Oil and Gas Lands Administration--as a relief well-drilling system for the fixed-caisson systems described earlier. For floating drilling, regulations require that no drilling into hydrocarbon (or potential hydrocarbon) zones take place during the last month of the season, in order that time be available to drill a relief well in the event of a blowout.

During the decade of Beaufort Sea drilling there have been no blowouts and (as far as I can judge) no damage to the environment. However, because of the high logistics costs, the special equipment, and the short season of use, it is not untypical for an exploration well to cost $100 million.

No commercial fields have yet been found, but many encouraging finds suggest that oil and gas production will occur sometime in the next 10 years.

The kinds of production concepts that have been considered for ice-covered waters are shown in Figure 17-10. In the near-shore Beaufort Sea it is expected that artificial islands and/or bottom-founded steel or concrete structures will be used. Fixed platforms in ice-covered regions are probably feasible out to 100 m water depth. Beyond that, some kind of subsea production technology, which is currently being developed for other deep-water offshore areas, will likely be needed.

If oil production does occur from the Beaufort Sea, the oil will be transported to market by either pipeline or tanker; both alternatives are being studied. On the U.S. side, the existing Alaskan pipeline will likely be used for future offshore production. On the Canadian side, a new pipeline down the Mackenzie Valley would be required. However, as the reader may be aware, a large-diameter gas pipeline down the Mackenzie Valley was ruled out in 1977. At that time, an independent review commission advised the Canadian government that construction of a large-diameter line would have an adverse effect on local communities. Since then, however, a small 30 cm oil line has been built to Normal Wells (halfway down the valley), and by analogy it could be extended to the Beaufort by the early 1990s.

The alternative, icebreaking tankers, has of course been under consideration since the late 1960s when the <u>Manhattan</u> performed trials in the Arctic in 1969. Since that time, there have been significant advances in icebreaker design, motivated by Beaufort Sea exploration. Some of the small icebreakers designed by Beaufort operators do incorporate features that could be used on large icebreaking tankers. It is claimed by their proponents that icebreaking tankers would be much safer than conventional tankers. This is because, to combat the ice, they have to be much stronger and incorporate significant redundancy. Thus, an icebreaking tanker

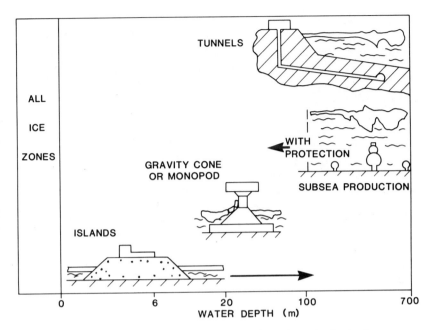

FIGURE 17-10 Production concepts for ice infested waters (Croasdale, 1983) (Reprinted with permission).

would have double hulls, twin screws, and twin rudders; it would also have a much greater power-to-displacement ratio than conventional tankers.

The other commodity of interest is, of course, natural gas. From the Beaufort we would see pipelines as the favored system, either linking with a gas pipeline from Alaska or being part of a polar gas pipeline bringing gas from the High Arctic. For the gas from the High Arctic, liquefied natural gas (LNG) tankers have been considered. These would ply the Northwest Passage and take gas to eastern markets or even to Europe. LNG tankers have also been designed and would contain many of the features discussed for oil tankers. Currently, the Arctic Pilot Project, proposing to ship LNG from the High Arctic, has been shelved, primarily because of lack of markets for gas rather than concern about technical feasibility.

It should be noted that the distances that the Arctic tankers would have to travel to reach either Europe or the eastern United States are about the same as the distances from the Antarctic to such areas as South

America, Australia, or New Zealand. Furthermore, they are less than the distance from the Middle East to the United States.

The Canadian Arctic Islands (High Arctic)

The ice between the islands of the Canadian archipelago is quite different from that in the Arctic Ocean, primarily because it moves less. Over much of the area it becomes landfast every winter. This, combined with the deeper water (typically 200 to 300 m), has enabled exploration drilling to be conducted using ice as the platform for the drilling rig (Figure 17-11).

This method requires ice thickening by flooding, to support the rig and avoid creep failure of the ice under sustained loading. Such techniques are now routine, and several oil and gas discoveries have been made using this very cost effective method. Small ice movements can be tolerated by the marine riser system (as used in floating drilling). Larger movements can be accommodated by moving the rig, but these movements have to be less than about 10 m during the period of drilling the well (say, 30 to 50 days).

Production methods from this area will need to be based on subsea production technology. A trial subsea completion for a gas well with pipeline to shore has already been implemented.

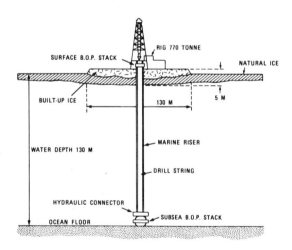

FIGURE 17-11 Drilling off the landfast ice in the Canadian Arctic (Croasdale, 1977) (Reprinted wih permission).

The Canadian East Coast (Labrador and Newfoundland)

Off Canada's east coast, exploratory drilling has been under way for about 15 years. All the drilling has been performed using either drill ships or semisubmersibles during the pack ice free periods. However, during the ice-free periods, icebergs can still occur, and techniques for dealing with them have been developed.

Canadian operators off Labrador and Newfoundland have pioneered drilling among icebergs. The approach is essentially one of avoidance. Alert zones are established around the vessel; as icebergs enter these alert zones, various operational responses are implemented. In the zone farthest from the vessel, typically several miles, the iceberg is tracked using radar. If the iceberg continues to move in the direction of the vessel, then certain actions may be taken in the drilling operation that would enable the well to be shut in. Also, towing vessels will be dispatched to tow the iceberg away from the drill ship. If for some reason this is unsuccessful, in the next alert zone the well is shut in. If the iceberg continues to move toward the ship, then the drill ship disconnects from the seafloor wellhead (and blowout preventer) and moves out of the way of the iceberg. All vessels used in iceberg areas do not use mooring lines but are dynamically positioned (using thrusters). Therefore, the final move off can be accomplished in minutes. After the iceberg threat is passed, the drill ship moves back to the wellhead and continues drilling.

To give some idea of the need for disconnects, it is useful to quote some statistics for the Labrador Sea, where drilling started in 1971. In the first season, one drill ship was used, and the season length was 67 days. During this period, 167 icebergs were tracked, 10 were towed, and there was one disconnect. From 1971 to 1982 there were 21 drill-rig seasons. About 2,600 icebergs were tracked, more than 600 were towed, and there was a total of 13 disconnects.

It is envisaged that in certain areas of the antarctic offshore, methods similar to those developed for drilling off Canada's east coast could be used.

Production from this area will utilize either fixed or floating platforms. It is considered technically feasible to build fixed platforms out to about 200 m; at this distance the platforms can still withstand iceberg impacts. Floating systems, combined with subsea wellheads, are a more likely alternative in deeper waters. Any subsea wellheads and pipelines would of course be placed in pits and trenches for protection from grounding icebergs.

CONCLUSIONS

It should be stressed that, even from the point of view of technology, the Arctic cannot be considered a direct analog for the Antarctic. There are some obvious differences. For instance, water depths are generally greater in the Antarctic, icebergs are larger, and the distances to support infrastructure and markets are greater. On the other hand, pack ice conditions are probably less severe in the Antarctic than in the Arctic.

Nevertheless, despite these differences, some of the methods developed for the Arctic could be used for exploration. These include

(1) Drilling off the landfast ice and ice shelves; and
(2) Floating drilling in the summer months combined with iceberg management, including quick disconnects from a protected subsea wellhead.

The costs of exploration in the Antarctic are expected to be as great as or greater than the Arctic, where an exploration well costing $100 million is not uncommon.

As far as production from the Antarctic is concerned, only in the narrow zone of shallow water (say, less than 50 m deep) could fixed platforms be used. In deeper water, subsea production systems combined with seasonal floating service platforms might be envisaged. These technologies are under development but have not yet been widely used.

So far I have said little about research and lead times. It is important to recognize that in frontier areas, many years of research and data gathering are often needed before commencement of operations. For example, in the Canadian Beaufort, offshore leases were taken by the oil companies in the mid-1960s. Environmental data gathering and research commenced in the late 1960s. Exploratory drilling started offshore in 1973. Production has not yet taken place and is unlikely to occur before 1990.

By analogy, even if a decision were made now to start looking for antarctic hydrocarbons, and even if feasible technology became available, and even if the economics were favorable, production and cash flow would probably be 20 to 30 years away. Furthermore, before then, vast investments would be required, and extensive research and data-gathering programs would have to be implemented.

REFERENCES

Croasdale, K. R. 1977. Ice Engineering for offshore petroleum exploration in Canada. In Proceedings of Fourth International Conference on Port and Ocean Engineering under Arctic Conditions, Memorial University of Newfoundland, Vol. 1:30.

Croasdale, K. R. 1983. The present state and future development of arctic offshore structures. Proceedings of seventh POAC Conference, Technical Research Centre of Finland, Helsinki, Vol. 4:514.

Department of Energy, Mines, and Resources. 1970. The Pilot of Arctic Canada (Ottawa, Canada), Vol. 1, 2nd ed., p. 89.

Dome Petroleum Limited, ESSO Resources Canada Limited, and Gulf Canada Resources Limited. 1982. Beaufort Sea Production Environmental Impact Statement, Vol. 2:3.10.

Frankovich, E. W. 1984. Kulluk extends the arctic offshore drilling season. In Proceedings of the Arctic Offshore Technology Conference. (Calgary), p. 17.

18.

Discussion on Technology and Economics of Minerals Development in Polar Areas

REMARKS BY GEOFFREY F. LARMINIE

Larminie elaborated on the projected costs of minerals exploration and recovery in Antarctica and termed antarctic logistics "murderous." He presented the following example: North of 62°N latitude in the Norwegian sector of the North Sea, it costs about $20 million to drill an exploration well. Two to three million dollars of that cost is for actual drilling, and the rest is accounted for by logistics activities. In Antarctica, the ratio of logistics to drilling costs would be an order of magnitude greater. He suggested that it could cost about $300 million just to drill a <u>land</u> well on an island adjacent to Antarctica (for example, on James Ross Island).

DISCUSSION

In the discussion of offshore technologies, the experts reiterated that, in contrast to existing exploration technologies, which if costs could be accommodated make it nearly possible to conduct exploratory drilling in Antarctica, production systems suitable for potential hydrocarbon exploitation in Antarctica are still in the development phase; complete subsea production systems will be required in Antarctica because water depths there are too great to permit floating operations.

Croasdale and Larminie stressed again the effect that the high costs of working in Antarctica will have on the rate of technology development and the development of a cost-effective production system for Antarctica. Moreover, before any development activities could commence, as Croasdale mentioned (Chapter 17), research on environmental impacts and safety will be required, and the effectiveness of the technology will have to be demonstrated.

The economics of antarctic offshore development will also be affected by distance from large consumer markets; the southern lands nearest Antarctica are not comparable to the high-population/high-oil-consuming areas of North America and Europe.

With respect to the question whether gas, rather than oil, might exist in offshore Antarctica, geological science in Antarctica is rudimentary, and geochemical data will be required to determine whether gas exists there.

With respect to environmental considerations, an example of Canadian-Danish cooperation in the transport of liquefied natural gas (LNG) from Arctic Canada was presented--the Arctic Pilot Project. When Canada was faced with a choice between carrying LNG by means of a pipeline or a very large crude (oil) carrier (VLCC) through the Northwest Passage and south between Canada and Greenland, the Greenlanders became concerned about possible effects of VLCC transport on their environment. The Canadians favored pilot tests, but the Danes were concerned that even a test project could have long-term detrimental effects prejudicial to the wellbeing of the Greenlanders.

The Arctic Pilot Project was set up as a working group of technical experts from the two countries to study potential impacts of VLCC transport. It focused particularly on noise that might result from VLCC propellers and icebreakers and on how the noise might affect the use of the water column for communication by whales and seals. There was also concern that if permanent sea lanes were designated for the VLCCs, this could cut off local Eskimo traffic. On August 26, 1983, Canada and Denmark signed an agreement that, inter alia, provides for consultations with respect to tanker routing if such transport takes place.

During these deliberations, the question arose of access by foreigners to the ongoing environmental impact assessment process in Canada, and the Greenlanders were ultimately admitted to the Canadian process. [As was noted by Croasdale (Chapter 17), the Arctic Pilot Project has been postponed because of low market demand for LNG.]

With respect to the effects on marine mammals of noise from offshore exploration and production activities, one participant pointed out that research in the Arctic Beau-

fort Sea area indicates that the frequency ranges used
for transmitting by marine mammals are different from
those where noise resulting from offshore oil operations
travels. He stated that there is a high level of ambient
noise generally in the Arctic because of such natural
processes as the creation of pressure ridges; this has
necessitated the development of special submarine-
detection equipment, as conventional noise detection
procedures are virtually useless. In his view, high-
frequency sounds produced by gas turbines in power
stations located on the coast could have far more effect
on marine mammals than the sounds associated with conventional offshore exploration and production activities.

19.

The Antarctic Treaty System as a Resource Management Mechanism—Nonliving Resources

Christopher D. Beeby

Antarctica generates powerful myths. Antarctic minerals and the negotiations under way concerning them have not been immune from the myth-making process. In its developed form, to be found, notably but not exclusively, in unresearched press articles, the mythology is that Antarctica is a vast reservoir of wealth, a cornucopia of riches for which a race is on, regardless of the risks to a unique environment. The discussions initiated by the Antarctic Treaty consultative parties about a regime for antarctic minerals are designed to allow that race to be won.

Exactly why such a popular view should have developed is not entirely clear to me. It may have something to do, as one commentator has suggested, with the inevitability of speculation about the last wilderness. Publication some years ago of figures, of a wholly theoretical kind, suggesting that X barrels of oil and Y barrels of gas will be found in Antarctica, may also have been a contributing factor. In all fairness, there should probably be added to this catalog the fact that when the consultative parties began to talk in 1981, in earnest and as a matter of urgency, about a regime for antarctic minerals, they did not do as good a job as they might have in explaining to the rest of the world why they were tackling this issue.

I do not intend to investigate every aspect of the myth that I have outlined. Others, to whose views I shall refer, are much better equipped than I am to inform us in detail of the current state of knowledge about minerals in Antarctica. What I can attempt, as somebody who works for the government that first drew attention to a potential problem about minerals in Antarctica, is to say something about the origins and purpose of the current minerals regime negotiations. Since I have been fairly closely involved in those negotiations, I shall also assay an account of their progress to date.

The common perception of those countries that are parties to the Antarctic Treaty--a perception that I think is now more widely shared following two debates in the U.N. about Antarctica--is that the major achievement of the treaty was to put a lid on the large potential for tension, rivalry, dispute and even conflict that was inherent in the preexisting situation. Those who initiated the Antarctic Treaty in 1959 had to confront a massive dispute about sovereignty in Antarctica. It is hard to imagine a setup that had a larger potential for governments to display their worst side. Here were seven countries that, over a rather long period, had claimed to exercise sovereignty in Antarctica, and the claims of three of them overlapped in the Antarctic Peninsula. As if this were not enough, five additional countries neither made claims themselves, nor recognized the claims of others (although two of them had asserted that their activities had established a basis of claim).

Early on in the 1959 conference that produced the Antarctic Treaty, it became quite clear that there was no possibility of resolving the claims conflict either through the claimant states' or the nonclaimant states' giving away their basic positions. But a means was found--and enshrined in Article IV of the treaty--to put disputes about sovereignty to one side. The treaty founded on that article was, then, a very significant measure in favor of stability and security in Antarctica. It represented, and it still represents, that Antarctica will remain peaceful and stable--will not become, in the words of the treaty, "the scene or object of international discord."

The examination, within the Antarctic Treaty System, of the question of antarctic minerals had its origins there. It was a concern to preserve the stabilizing regime of the Antarctic Treaty (coupled with a worry about the impact on the environment if mining ever occurred on the continent) that led New Zealand, as far back as 1970 at the consultative meeting held in that year, to raise the question of antarctic minerals and to urge its treaty partners to consider the need for a comprehensive regime governing them.

The treaty does not deal with resources of any kind, whether living or mineral. The New Zealand fear was that, if at some stage important resources were found, an unregulated scramble could develop in relation, in particular, to minerals. A scramble of that kind could obviously have the most damaging effect on a very special environment. But it could also bring back to center stage the conflicting positions about sovereignty that the treaty had put to one side. This is so because, in the absence of agreed rules, states that claim sovereignty and states that neither make claims to sovereignty nor recognize the claims of others would give totally conflicting answers to a whole series of obvious questions arising if a minerals deposit of economic significance were ever found in Antarctica: "Who has title to the deposit? Who should license its exploitation? Who would regulate it? Who would get the taxes and royalties?" The answers to questions of this kind are entirely dependent on the view taken as to the location of sovereignty, if any, over the area in which the deposit is located.

As Arthur Watts has pointed out (Chapter 6), Article IV does not resolve the issue of sovereignty in Antarctica or remove entirely the possibility of serious disagreement arising from conflicting positions on sovereignty. The effective operation of the whole Antarctic Treaty System depends on restraint, on a willingness on the part of the parties to the treaty not to press their views about sovereignty to a logical conclusion. In the absence of agreed rules, discovery of significant mineral deposits in Antarctica would place maximum pressure on the willingness of all to continue to exercise that restraint.

What motivated New Zealand in 1970 was a belief that the unregulated exploitation of antarctic minerals, if it ever occurred, would present a threat to the Antarctic Treaty. In the worst case, the result could be a breakdown of the treaty, the loss of the disarmament regime that it contains, and, more generally, of the stabilizing effect that it has had on the entire area south of 60°S latitude.

The origins of discussions about antarctic minerals are therefore essentially political rather than economic in nature.

That this continues to be so is confirmed by what the experts tell us 14 years later about the mineral resources of Antarctica. The mythology, they say, is just that: there is no evidence at all that antarctic minerals represent an economic treasure house ready for the taking in the foreseeable future. Despite many years of research, a limited amount is known about antarctic minerals. A great deal more work has to be done to make other than speculative assessments. There are minerals onshore, but none of proven economic significance. What is more, 97 or 98 percent of the continent is covered by a mile or more of ice; nobody is going to have an economic interest in penetrating that for many years. Offshore, the prospects look a little more interesting. There may be oil and gas, but whether there is and, if so, where and in what quantities, is simply not known. There seems little doubt that it will be well into the next century before the extraction of oil or gas, if they exist, becomes an attractive economic possibility. That prediction is not merely a result of the current oil-supply situation. It is also a function of the extremely adverse conditions in Antarctica, logistical and marketing problems, and the huge costs that would be involved at any time in mining in the region. We are told, again by the experts, that a giant or perhaps a super-giant oil field would need to be found in Antarctica before there would be an economic justification for proceeding to exploit it. There is no evidence that such a field or fields exist.

Summing up the studies that had been undertaken in 1982, James Zumberge, who is one of the experts whom we are fortunate to have as a contributor to this workshop, said: "All who have dealt with this matter have come to the same conclusion: no mineral deposits likely to be of economic value in the foreseeable future are known in Antarctica. This statement is not to say that Antarctica has no mineral resources, but rather, if they exist, they have no economic significance today or in the near-term future."[1] (Zumberge goes on to point out that of course economic circumstances can change rapidly.) Another expert, Franze Tessensohn, writing a year later, said, after surveying the geophysical work that has been undertaken with respect to offshore areas frequently regarded as likely to be of economic interest before onshore areas, said: "If one tries to draw a conclusion based on all these investigations it can be stated that several interesting sedimentary basins exist on the antarctic shelves, but that we do not know much about their extent, about [the] nature and age of the sedimentary infill, about the thermal history they have undergone, and about possible source and reservoir rocks. So I think we are still much in the phase of guessing."[2]

I return now to the history of the discussions about antarctic minerals within the Antarctic Treaty System. After the topic had been broached at the sixth consultative meeting, in Tokyo, the 1970s saw a progressively more intensive discussion about minerals, both inside and outside the regular meetings of the consultative parties. The help of the Scientific Committee on Antarctic Research was asked for; other groups of experts were set to work. A landmark was reached at the ninth consultative meeting, in London in 1977, with the adoption of Recommendation IX-1, which looked forward to a regime for antarctic minerals and, in doing so, endorsed a number of principles about such a regime, which had been adopted at a special preparatory meeting held in Paris the previous year. The same recommendation made more specific an idea emerging from the previous consultative meeting, namely, that the nationals of consultative parties and other states should be urged to refrain from all exploration and exploitation of antarctic mineral resources "while making progress towards the timely adoption of an agreed regime concerning antarctic mineral resource activities. They will thus endeavor to ensure that, pending the timely adoption of agreed solutions pertaining to exploration and exploitation of mineral resources, no activity shall be conducted to explore or exploit such resources." This contingent measure of restraint has been confirmed at subsequent meetings and is still extant.

Recommendation X-1, adopted by the consultative parties' meeting in Washington in 1979, represented a further accretion of consensus on the contents of a minerals regime; and an informal meeting held in Washington in the following year made further progress in the same direction.

The formal decision to begin detailed negotiations on a minerals regime was taken in Buenos Aires in 1981 at the eleventh consultative meeting. That decision was recorded in Recommendation XI-1, which, once again, picked up and enlarged areas of agreement reached at previous meetings.

Recommendation XI-1 determined that a minerals regime should be concluded "as a matter of urgency." That determination needs to be viewed against the origins of the discussion on minerals. The most important reason for deciding to do the job quickly was that, for so long as a minerals question remained unresolved, it presented a potential threat to the Antarctic Treaty and the Antarctic Treaty System. It was, I believe, recognized in

Buenos Aires in 1981, that to postpone the negotiations until such time as a significant mineral deposit was found in Antarctica and there was a serious economic interest in exploiting it would very greatly complicate, if not make entirely unmanageable, the task of producing agreed answers to the complex issues raised.

Since Recommendation XI-1 constitutes the basis for the negotiations now under way, it is worth noting the main areas of agreement that it recorded:

- The minerals regime should be negotiated by the consultative parties, and they should continue to play an active and responsible role in dealing with the question;
- The Antarctic Treaty must be maintained in its entirety;
- The protection of the unique antarctic environment and of its dependent ecosystems should be a basic consideration;
- The consultative parties, in dealing with the question of mineral resources in Antarctica, should not prejudice the interests of all humankind in Antarctica;
- The delicate balance contained in Article IV of the Antarctic Treaty should not be affected by the regime, and the principles embodied in that article should be safeguarded;
- The regime should include means for assessing the possible impact of mineral resource activities on the antarctic environment, determining whether such activities will be acceptable, and, if they are, regulating them;
- The area of the regime should encompass the continent of Antarctica and its adjacent offshore areas but without encroachment on the deep seabed;
- The regime should cover all mineral resource activities at every stage;
- The regime should be open in the sense that it should include provisions for adherence by states other than the consultative parties on the understanding that adhering states would be bound by the basic provisions of the Antarctic Treaty;
- The regime should include provision for cooperative arrangements with other relevant international organizations;
- The regime should protect the special responsibilities of the consultative parties with

respect to the environment of the whole Antarctic
Treaty area, taking into account responsibilities that
may be exercised in that area by other international
organizations; and
• The regime should promote the conduct of research
necessary to make the environmental and resource
management decisions that will be required.

Additionally--and this is a central issue to which I
shall return--Recommendation XI-1 stated that any agreement that may be reached on a regime should be acceptable
and without prejudice to those states that claim
sovereignty in Antarctica as well as to states that
neither claim sovereignty nor recognize such claims.

The negotiations foreshadowed by Recommendation XI-1
began in Wellington, New Zealand, in mid-1982. Since
then there have been further negotiating sessions held in
Wellington (January 1983), Bonn (July 1983), Washington
(January 1984) and Tokyo (May 1984). The meeting, at
which, for the first time, all the parties to the
Antarctic Treaty and not just the consultative parties
participated, was held in Rio de Janeiro in February/
March 1985 and a further meeting is scheduled in Paris
September/October 1985.

There are several issues, of uneven importance, which
still have to be sorted out. Substantial progress has,
however, been made to the point where some predictions
about the main features of the agreement that is likely
to emerge can be made. I need to stress that these are
predictions only. Given that the negotiations are taking
place under a rule of consensus, no solution on any point
will be reached if it does not attract the support of all
those participating in the decision-making.

The regime will be expressed in the form of a legally
binding international agreement, which, since its principal purpose is to preserve and strengthen the Antarctic
Treaty, will, by one means or another, be closely linked
with that treaty. Following the precedent of the Antarctic Treaty itself and also that of the Convention on the
Conservation of Antarctic Marine Living Resources
(CCAMLR), the agreement will be open for accession by all
interested states.

The agreement will not be a complete and comprehensive
mining code containing a detailed resolution of all the
issues that may arise in future. A code of this kind,
which is common in the national legislation of many
countries, is ruled out for a number of reasons. The

most obvious one is that we are negotiating in a state of considerable ignorance, which has both the advantage that I have referred to (i.e., the absence of pressure resulting from economic interest in an identified resource) and disadvantages. The fact is that it is simply not possible to foresee now the whole range of mineral resource activities that might one day be undertaken in Antarctica. Moreover, our present knowledge of the sensitivities of the antarctic environment is incomplete. The regime must, therefore, constitute a framework within which, as our knowledge increases, the details can be filled in as may be required. That framework will, however, have to be of a kind that will provide all concerned with a clear picture of the way in which decisions will be taken under the regime and the standards against which those decisions will be made. It will also need to ensure that, if mining does take place in Antarctica, those who undertake it have security of title, have their investments protected, and are not subject to the arbitrary disruption of authorized activity.

The regime will create new institutions, in some respects similar to, and other respects differing from, those established by the CCAMLR. There will be a central body or commission with authority relating to the whole area covered by the regime and the regulatory powers necessary to fill in the framework created by it. There will be a subordinate advisory committee, responsible for tendering scientific, technical and environmental advice. There will be a secretariat. And, by way of departure from the CCAMLR precedent, there will be a series of smaller so-called "regulatory committees," with responsibility for certain aspects of the regulation of mining that may take place in an area identified by the commission. Representation on regulatory committees will include, inter alia, the state or states, if any, claiming sovereignty in the identified area and the state wishing to undertake mineral resource activities or to sponsor such activities by its nationals. It will be within these bodies that, consistently with the principles established by the regime and further measures laid down by the commission, many of the detailed rules and regulations governing individual mining projects that might be deemed acceptable would be negotiated and adopted.

The precise composition and functions of the institutions to be created and their method of taking decisions is still under negotiation. But there is, I think, by now, a rather generally shared view that the key to the problem of finding an accommodation between states claiming sovereignty in Antarctica and states neither making nor recognizing such claims will be found largely within the institutional structure of the regime.

As was anticipated in Buenos Aires, it became evident early on that the search for this accommodation between claimants and non-claimants was the single most intractable issue that had to be dealt with. It was also apparent that neither claimants nor non-claimants would be satisfied with a regime heavily loaded in one direction or another but accompanied by a disclaimer based on Article IV of the treaty (although a parallel to that article, as in the CCAMLR, will undoubtedly have its place in the regime). Both claimants and non-claimants demanded, with good reason, that, as foreshadowed in Recommendation XI-1, any minerals regime <u>as a whole</u> would have to be acceptable to them and without prejudice to their conflicting positions on the sovereignty question.

At the first two negotiating sessions, much of the debate about this aspect arose from an examination of the broad question of where the powers required to regulate antarctic minerals should be located. The initial nonclaimant answer to that question was that, for the most part, regulation would have to be undertaken by the parties to the regime, acting collectively and without distinctions of any kind, through its institutions. The initial claimant response was that, for the most part, regulation would have to be undertaken by states acting individually, even if in a manner envisaged by and consistent with the regime or possibly, in certain instances, pursuant to authority delegated by institutions of the regime. With respect to areas of Antarctica subject to claims to sovereignty, the individual states having the major regulatory role would be those claiming sovereignty.

At the meeting in Bonn in July 1983 the consultative parties accepted as a starting point for further negotiations a text that attempted to break this deadlock by switching attention from the question of location of the necessary regulatory powers and focusing, instead, on the institutions of the regime, their composition, and the method by which they would take decisions. The basic assumption underlying this approach was that it was not necessary and almost certainly not possible either to attempt to spell out in detail all the elements of an accommodation on the sovereignty question in the text of the agreement to be adopted or, in particular, to try, in the text itself, to reconcile strongly conflicting views about the substantive powers of claimant and nonclaimant states. The alternative—also in part determined by the lack of knowledge of all the issues that would require regulation—was to leave a significant number of important decisions to be hammered out after the regime had been adopted, through a continuing process of negotiation on a case-by-case basis. Power to take those decisions would be given to the institutions of the regime, but the regime would endeavor to ensure that the composition of those institutions and their method of taking decisions were such as to give claimants and nonclaimants alike a reasonable assurance that their divergent interests would be protected. As in the case of the Antarctic Treaty itself, and consistently with the spirit of forbearance that it embodied and that has informed all the work undertaken since its adoption, the sovereignty issue will not be confronted but stepped around.

The regime will cover the area, offshore as well as onshore, referred to in Recommendation XI-1, and it will cover every stage of mineral resource activities within that area. A distinction will, however, be drawn between, on the one hand, prospecting, defined narrowly to include only activities presenting a minimum of risk to the environment and, on the other, exploration and development. Prospecting, which will not confer title on the prospector, will not require prior authorization. It will, however, be subject to environmental controls applicable to all stages of mineral resource activity. There will be power vested in the commission to impose additional constraints on prospecting, that could have the effect of preventing or limiting any category of prospecting found to present hazards. Provision is also likely to be made to allow for review of prospecting activity that, for example, because of the concentration of prospecting activity in a particular region or because of conflict with other legitimate uses of Antarctica, is thought to present a special problem.

By contrast, in recognition of the very much larger environmental hazards that could be involved, exploration and development will be expressly prohibited until such time as they are authorized by the institutions of the regime. Initially this will require a decision by the commission, acting on the advice of the advisory committee, as to whether or not it should authorize the submission of applications for exploration and development of an identified area. That determination would constitute a critical but not a final decision. A positive determination by the commission would still require that individual applications for exploration and development be scrutinized by the advisory committee and the relevant regulatory committee. Positive results there would still, in turn, require a final approval by the commission before either form of activity could begin.

It is of some importance to note that if the regime is structured in this way with respect to exploration and development, it will stand in sharp contrast to the marine living resources regime established by the CCAMLR. The basic proposition underlying the CCAMLR--that activity is permitted until restrained--will be reversed: the only mineral resource activities that would confer title on the operator and might confer economic benefit will be prohibited unless specifically authorized.

The regime will implement the requirement set out in Recommendation XI-1 that the protection of the antarctic environment and of its dependent ecosystems must be a basic consideration. This will be done in several ways. There will, first, be a set of environmental principles set out near the beginning of the agreement. These principles will cover every phase of activity, including prospecting, and they will be applicable to both operators and institutions created by the regime. They will be relevant at every point to test the acceptability of proposed activity; neither states nor the institutions of the regime will be free to ignore them. Among the principles that it will embody are the following: that no mineral resource activity may take place until adequate information is available to assess its possible impact and it can be judged that such activity will meet rigorous criteria; and that no such activity shall take place until the technology and procedures (including contingency plans in the event of accidents) are available to ensure compliance with these criteria and there exists the capacity to monitor key environmental and ecosystem components. Judgments about the acceptability of proposed activities will be required to take account, inter alia, of the cumulative impact of mineral resource activities and other uses of Antarctica.

A second environmentally important aspect of the regime will involve setting aside, as off limits to all mineral resource activities, existing Specially Protected Areas, Sites of Special Scientific Interest, and any further area that for historic, ecological, environmental, scientific, or other reasons the commission has designated a prohibited area.

The fact that exploration and development will be forbidden until they are determined to be acceptable on environmental grounds will be a significant bar to hazardous activity--the more so if, as seems likely, provision is made for the determination to be made by consensus.

The advisory committee will play a key role in decision making under the regime. All the views expressed by it on a particular proposal, i.e., not merely majority views, will be made known to the bodies to which it gives advice. Provision is likely to be made for the advisory committee to give public notice of its meetings, to receive submissions from concerned international organizations, and to make public its views. Further consideration will be given, in the ongoing negotiations, to alternative techniques for ensuring that environmental considerations and advice are given proper weight and are not overridden by an interest in promoting and deriving benefit from mineral resource activities.

Finally, there will be provision for the suspension, modification, or cancellation of authorized exploration and development if a determination is made that there has been a failure to comply with the rules applicable to it or that new and unforeseen risks to the environment have arisen.

The regime will require the states parties to take measures to ensure compliance with it. It will deal with liability in the event of accident. It will make provision for the settlement of disputes concerning the interpretation or application of the regime. These are among the matters that require further attention in the negotiations.

Finally, the regime will, in several ways, take account of the interests in Antarctica of the international community at large. A general point needs, first, to be made. The regime is intended to, and will, preserve and strengthen the Antarctic Treaty. The benefits to <u>all</u> countries from a guarantee that one sixth of the world will continue to be demilitarized, that scientific research in that region will continue to be conducted under conditions of unique freedom and cooperation, and that a globally important environment will continue to receive protection are considerable. That point has been spelled out at some length in the U.N. debates; I do not need to elaborate on it further here.

The specifics are also important. The consultative parties have not lost sight of the fact that other international agreements extend to Antarctica and other international organizations have, or will have, competence south of 60°S latitude. As I have indicated, the regime will implement that part of Recommendation XI-1 that says the definition of the area will be such as to involve no encroachment on the deep seabed. As also foreshadowed in Recommendation XI-1, the regime will protect the special responsibilities of the consultative parties with respect to the environment throughout the whole of the Antarctic Treaty area. But, in this respect as well, the responsibilities of other international organizations will not be ignored. The regime is likely, in fact, to contain an all-embracing provision requiring the commission to cooperate with all interested international organizations and, more specifically, to develop a cooperative working relationship with any international organization having competence in mineral resources in areas adjacent to those covered by the regime.

Then, of course, there is the basic point that the regime under consideration is not a closed system. Subject only to a willingness to subscribe to the Antarctic Treaty or to its basic provisions (the final choice has yet to be made between these alternatives), every state will be able to adhere to the agreement setting up the regime. Every state party will be entitled to participate in mining found to be acceptable in Antarctica. If it decides to exercise that right, it will participate in all the decision-making relating to its actual or proposed activity.

Are there other elements that might be included in a regime and that would enhance its status vis-a-vis countries not participating in negotiations? Participation by the nonconsultative parties to the treaty in the minerals regime negotiations should provide further insight into this aspect of the regime. So should this workshop. I shall not discuss the question further here except to note, in conclusion, one avenue that I believe to be closed.

Suggestions have been made, in the U.N. debates and in the literature, that Antarctica or its resources or some part of them should be regarded as part of "the common heritage of mankind." The concept of the common heritage of mankind could be applied to outer space and the deep seabed because of a consensus on the status of those areas. Each had been the subject of extremely limited human activity, and neither had been the subject of any claim of sovereignty. It is quite another matter to attempt to apply this notion to an area that, by now, has generated substantial history of human activity, that has been subject to claims of sovereignty dating back more than 75 years, and that is the subject of a preexisting legal regime among the states most directly concerned. It is neither reasonable nor realistic to suppose that the common heritage concept can make headway with respect to Antarctica. There is no consensus on the legal status of the continent that would permit that.

There is another comment that needs to be made about the common heritage concept. As it has developed in the U.N., that concept has a strong exploitation orientation. It looks, essentially, to the development of resources for the benefit of all, especially of the developing countries, rather than to the protection of the environment in which those resources are located. It is, I believe, significant that in the 1982 Law of the Sea Convention the environmental controls on deep seabed mining in the common heritage area are notably weak. Against this background, those who advocate, on the one hand, stringent environmental controls in Antarctica and, on the other hand, the adoption within the U.N. of a common heritage minerals regime are pursuing incompatible goals.

I should like to elaborate on my prepared remarks about the origin of the antarctic minerals regime negotiations by drawing a hypothetical analogy to the discovery of a mountain of gold in the area of Antarctica claimed by New Zealand. Once the gold were discovered there would be considerable pressure on the New Zealand government to develop rules to regulate its exploitation and to treat the deposit as any other deposit of gold located in New Zealand. This would provoke reactions from non-claimant countries, rejecting New Zealand's contention that its rules should govern exploitation, which would in turn create the possibility of unregulated development and destabilization of Antarctica similar to that which existed before the Antarctic Treaty and led to its negotiation.

I should also like to respond to several comments made during the workshop. First, in 1975 the government of New Zealand proposed to prohibit minerals development in Antarctica and declare it a world park. But it did not then make the statement that it would abandon its claim to Antarctica. The reaction from other Antarctic Treaty consultative parties at that time indicated that they were not willing to forgo minerals development in

Antarctica for all time, so the New Zealand proposal was not pursued.

Second, the decision not to reject the possibility of antarctic minerals exploitation does not necessarily indicate that the regime under negotiation must be a "prodevelopment" regime as it is characterized by some nongovernmental organizations. There is a third possibility, which is currently being pursued: a regime constructed objectively to address the question, once it arises, of whether specific types of mineral development activity should be permitted in specific areas of Antarctica, subject to stringent safeguards for safety, conservation, and protection of the environment.

Third, the rationale for the provision in Antarctic Treaty Recommendation XI-1, that "A regime on antarctic mineral resources should be concluded <u>as a matter of urgency</u>" (emphasis added) is not directly related to recent interest in Antarctica. The recommendation was concluded in 1981, two years before U.N. interest was expressed. At that time, the experts were saying that minerals development in Antarctica was not commercially viable. While there are different reasons for the need for urgency, I favor the "gap" theory; that is, that as long as the regime for minerals development remains unresolved, it will pose a threat to the Antarctic Treaty System and cooperation within that system.

With respect to the negotiations themselves, there are four differences among states involved in the negotiations that complicate these negotiations and must be taken into account:

(1) Countries claiming territory in Antarctica and those that do not recognize such claims,
(2) Countries with centrally planned economies and those without,
(3) Countries with access to the requisite technology and those without, and
(4) Countries close to Antarctica whose ecosystems might be affected by antarctic minerals development and those farther away, whose ecosystems are unlikely to be directly affected by such activities.

Finally, the primary matters yet to be settled in the negotiations are the exact methods and procedures by which decisions will be taken. Inspection will be an important element in the minerals regime.

NOTES

1. Zumberge, J.H., 1982. Potential Mineral Resource Availability and Possible Environment Problems in Antarctica. In J.I. Charney, ed. New Nationalism and the Use of Common Spaces, Allenheld, Osmund and Company (Totowa, N.J.); pp. 115-154.
2. Tessensohn, F. 1983. Present Knowledge of Nonliving Resource in the Antarctic, Possibilities for Their Exploitation and Scientific Perspectives. In R. Wolfrum, ed. Antarctic Challenge: Conflicting Interests, Cooperation, Environmental Protection, Economic Development: Proceedings of an Interdisciplinary Symposium, Duncker & Humblot (Berlin); pp. 189-210.

20.

Panel Discussion on Nonliving Resources

The panel consisted of Robert H. Rutford (Moderator), Roger Wilson, Geoffrey F. Larminie, and Adriaan Bos.

REMARKS BY ROGER WILSON

Wilson indicated that he would take a political perspective on Beeby's presentation (Chapter 19). He shared Beeby's view that the "gap" in the Atlantic Treaty System (ATS) on the subject of minerals requires filling but disagreed with the treaty consultative parties' belief that that gap should be filled with a regime that under certain circumstances would permit minerals development. In his opinion, once a regime or a mechanism to govern minerals development existed, minerals development would sooner or later take place.

He questioned the lack of serious consideration given to the nondevelopment option and believed that the reason for this is that antarctic policy is fashioned primarily by civil servants in the foreign affairs ministries of states parties to the Antarctic Treaty who are not directly answerable to the public. This means that no express public mandate, which might challenge conventional wisdom, is sought or derived from discussions within and between political parties. He asked, rhetorically, how many governments had ever made antarctic policy a part of their political platforms, or sought a mandate on antarctic policy, and expressed surprise that few have done so given the serious regard accorded by governments to Antarctica.

He referred specifically to the 1975 New Zealand world park proposal, recalled by Beeby, which received little support from other treaty nations. The 1975 New Zealand government was a Labour Party government, yet today the same party has chosen to continue along the path leading to minerals development. Wilson did not know why this policy change had occurred, but he noted that the New

Zealand Ministry of Foreign Affairs has a significant investment in the existing policy and would have difficulty reversing it once more. The result is a fundamentally different policy direction, taken in the absence of a public mandate, most likely on the advice of civil servants who do not have to answer to the public.

On another point, he noted the acknowledgment by Beeby that perhaps the consultative parties to the Antarctic Treaty had not adequately explained their efforts to negotiate a minerals regime to the rest of the world, which implied that, had they explained, no one would have challenged their efforts. In Wilson's view, the problem is a more fundamental one. To illustrate his point, he quoted the provision of Recommendation XI-1 specifying that one of the principles on which a minerals regime should be based is that "the Consultative Parties, in dealing with the question of mineral resources in Antarctica, should not prejudice the interests of all mankind in Antarctica." He questioned the presumption of the Antarctic Treaty states in appointing themselves to determine what the "interests of mankind" are.

Wilson doubted whether the primary way Beeby had described in which the minerals regime would take account of the interests of the international community at large was indeed sufficient: maintaining good relations with other international organizations having competence south of 60°S latitude. In his view, the fundamental "interests of all mankind" in Antarctica lie in the maintenance of peace in the area and in the results of the science conducted there. The question is then whether these interests are better served by a proexploitation regime or by a protection-oriented regime; Wilson believed humankind's interests could not be guaranteed if minerals exploitation--at best marginal and/or subsidized--were to take place. Nor, clearly, could protection of wildlife or wilderness values be guaranteed.

Wilson raised questions about whether two recent seismic research projects complied with the policy of "voluntary restraint" quoted by Beeby from Recommendation XI-1. He named the Japanese government program carried out with the Hakurei Maru and the 1983-1984 voyage of the U.S. government research vessel, S.P. Lee, noting that once initial investments are made, it would be difficult to halt the momentum toward minerals development. The S.P. Lee program had received early promises of funding from the Circum-Pacific Council for Minerals and Energy Research, a private-industry body whose chairman is an independent U.S. oil company executive, Michael T. Halbouty, but this did not work out in the long run.

On the subject of prospecting, he questioned why prospecting in Antarctica should not be subject to specific authorization under the regime, particularly when many of the consultative parties to the treaty required prior authorization within their own territories.

He ventured that Beeby was optimistic in assuming that countries would not ignore the principles of the regime, including the environmental principles, as he felt had occurred under the CCAMLR, or that political pressures might not ultimately preempt them. He expressed doubts that the advisory committee described in Beeby's presentation would be able to fulfill its task, because the scientists appointed would have numerous commitments on their time, might be appointed on political grounds or to carry out the political aims of their governments, and might not be expert in the disciplines required, or, conversely, to meet the requirement of technical competence they might be those most interested in developing antarctic minerals. He believed that the nonpolitical antarctic environmental protection agency proposed by Greenpeace (see Chapter 14), staffed with full-time professionals, deserves serious consideration.

Wilson also criticized the decision making institutions of the minerals regime as being biased in favor of development, since only those states actually engaged in minerals development activities, during such time as they are engaged in them, can join the consultative parties to the Antarctic Treaty as decision making members in the institutions of the regime; states in favor of environmental protection would not have the same sort of say at the decision making level as those interested in exploiting mineral resources in Antarctica. Clearly the decision making institutions promote minerals activities by giving decision making power to the exploiters.

Wilson cautioned against possible failures to comply with the rules established by the regime and wondered whether members of the decision making commission under the regime would accept the responsibility to discipline one another once a possible disagreement over compliance arose. His lack of confidence stemmed from the difficulty Greenpeace had encountered in seeking consideration during Antarctic Treaty meetings of possible noncompliance by the French government with the Antarctic Treaty Agreed Measures with respect to its construction of an airstrip to serve the French claim in Antarctica.

In conclusion, Wilson advocated a more open system for the minerals regime and that all options be considered, including the possibility of forgoing minerals development completely.

REMARKS BY GEOFFREY F. LARMINIE

Larminie suggested that perhaps a definition of "all mankind" was in order, and noted that by his calculations more than 80 percent of the world's population is already represented within the ATS.

REMARKS BY ADRIAAN BOS

Bos commented that Beeby's presentation dealt mainly with the politics within the ATS. However, the minerals regime negotiations have recently attracted the attention of the wider international community, and for that reason the establishment of the minerals regime has become a global political concern, raising issues new to the ATS framework. He agreed that the consultative parties might have done a better job in explaining their activities to the world at large, but like Wilson he doubted whether this would have prevented questions and tensions arising today about the system.

He cited the language in Recommendation XI-1, which refers to a minerals regime "elaborated by the Consultative Parties," to illustrate how quickly things have changed since 1981; in 1984 the decision was taken to invite the nonconsultative parties to the Antarctic Treaty to take part in these negotiations. He added that it remains to be seen whether the involvement of the outside world can be further enhanced.

He questioned the meaning of the Recommendation XI-1 stipulation that "the Antarctic Treaty must be maintained in its entirety," for, while the achievements of the treaty are admirable and should be maintained, it might be difficult to ask states party to the minerals regime that are not party to the Antarctic Treaty to accept this. He suggested that they might be asked to respect the achievements of the Antarctic Treaty without being asked to maintain the treaty in its entirety.

Bos found the phraseology of Recommendation XI-1, where it refers to "not prejudic[ing] the interests of all mankind," too negative. He stressed the need to find ways to guarantee that the regime be in "the interests of mankind," since this will be the test for acceptance of the regime by the international community. Moreover, he noted that one should not think only in terms of economic benefit; a continuing moratorium on minerals development activities for the time being might also be seen to be in "the interests of all mankind."

SUMMARY

The discussion elaborated on a number of points raised by Beeby (Chapter 19) and by the panelists. Many of these addressed the role of states and organizations currently not involved in the antarctic minerals regime negotiations:

- Their involvement in policy formulation at the negotiation stage;
- Their involvement in the implementation of the minerals regime once completed, and specifically in its decision making processes; and
- Their involvement in potential minerals development activities and benefits therefrom.

(The questions of how the minerals regime will take account of international interests at large and the relationship of the ATS to the United Nations system are discussed further in the next section.)

Other points referred to the earlier technical discussion and the relationship between the timing for possible antarctic minerals development activities and political considerations of urgency in completing the regime. The proposal for a moratorium on antarctic minerals development was also debated.

Several questions produced more detailed descriptions of how the minerals regime, as now contemplated, would operate in practice. Among the points covered were the following: area of application; the framework of the regime; institutions; environmental protection and safety; effects on scientific research; prospecting; environmental protection and confidentiality of data; and enforcement and reporting requirements.

PARTICIPATION

Various approaches to widening international participation with respect to antarctic minerals were considered. In lieu of basing a minerals regime on any one of the four different "theologies" that arise in antarctic minerals debates--world park, claimant, nonclaimant, or common heritage of humankind--it was argued that the ATS should remain 'ecumenical'. The contentious issue, however, was: Who should reconcile these theologies and how? Some participants believed that those appointed to define the "interests of all mankind" should not have a material interest in their trust and that they should be responsible to a higher authority. Equally important was the question of right: Who should appoint those undertaking the definition? Others asked whether a basis for management other than the ATS would likely represent more effectively and equitably "the interests of all mankind"; they stated that "right" derived, among other things, from the fact that the ATS worked; and that the rights and obligations generated by the Antarctic Treaty were exclusive only as between states parties and neither granted rights to nor obligated third parties without their consent.

COMMON HERITAGE OF MANKIND

Beeby's comments on the applicability of the common heritage of mankind concept to Antarctica provoked opposing views from those who believed that Article IV of the Antarctic Treaty suspends the applicability of the claims as long as it is in effect; they did not accept, as debated earlier in the workshop (Section II), that the claims are still very much alive, nor did they necessarily believe that these claims will remain alive in perpetuity (citing historical examples of reversals of territorial claims and the fact that in the Law of the Sea negotiations a number of states had in effect reduced their claims to offshore rights). Others repeated that there is a distinction between the deep seabed and outer space, where the common heritage concept has been applied, and Antarctica.

Several speakers criticized Beeby's characterization of the common heritage as having a strong exploitation orientation; they noted that it encompasses concepts of peaceful use, environmental protection and management, and rational use of resources for present and future generations--all of which are fully consistent with the objectives of the ATS. Nor did they agree with Larminie's "percentage-of-world population" definition of mankind, because the political meaning of the concept is widely understood to refer to all countries, whether rich or poor, developed or developing, technologically advantaged or disadvantaged. Others shared Beeby's view that the environmental provisions developed through a global conference, such as those governing deep seabed mining in the 1982 Law of the Sea Convention, would be far worse than those contemplated in the antarctic mineral regime.

PARTICIPATION IN THE MINERALS REGIME NEGOTIATIONS

Beeby's point that the motivation for the minerals regime talks is essentially political rather than economic in nature led one participant to conclude that this is all the more reason to involve more states than the consultative parties in these talks. Moreover, discussions in the United Nations would not necessarily mean that the common heritage concept would be applied; the U.N. could simply provide the forum that would lead to agreement on how to design a minerals regime acceptable to the wider international community.

Another point of view preferred that the minerals negotiations take place within the Antarctic Treaty framework in order to strengthen the system currently working effectively in Antarctica. Within this context, however, numerous speakers supported opening the negotiations to a broader circle than the small group demonstrating substantial interest in antarctic science. This would avoid the negative ramifications of presenting to the world at large, at some point in the future, a fait accompli.

These participants noted that the invitation to the nonconsultative parties to take part, for the first time, in the fifth meeting in these negotiations, February 26-March 8, 1985, in Rio de Janeiro, meant that any country that acceded to the treaty could take part in the negotiations: the negotiations are no longer closed. Nevertheless, they encouraged making the role of the nonconsultative parties an attractive one, so that additional states will accede to the Antarctic Treaty. For this reason, it would be important to involve the wider group of states fully in these forums and to avoid excessive heads-of-delegation meetings that would be restricted to representatives of consultative parties.

Another suggestion to broaden the dialogue on an antarctic minerals regime returned to the idea of a strategy for environmental management in Antarctica, along the lines of the antarctic conservation strategy suggested by Heap and Holdgate in Chapter 13. A series of sessions on this topic could include scientists, diplomats, and individuals knowledgeable about resource management issues. These sessions could also develop a research agenda that would respond to the needs of those with management and decision-making responsibilities.

Improving the information policies with respect to the minerals negotiations would also help to dispel feelings of exclusion. One participant saw no reason why the basic text of the minerals regime should not be available both to help inform interested outside parties and to allow the negotiations to benefit from relevant comments and criticisms.

PARTICIPATION IN THE ADOPTION OF THE MINERALS REGIME

Returning to the question of who has the right to determine the future of Antarctica, one participant questioned whether, if not in the negotiations, then at the stage when the final product of the minerals negotiations would be formalized, a wider circle might then be involved. He suggested, however, that the final diplomatic conference would also be dominated by those who had negotiated the agreement and that the invitations and rules of procedures would be controlled by them. Others acknowledged that such a diplomatic conference is likely to be held and indicated that at a minimum it will include all those who had participated in the negotiations.

PARTICIPATION IN IMPLEMENTATION OF THE MINERALS REGIME

Participation in the implementation and decision making processes of the completed minerals regime was distinguished from participation in the negotiation of the regime. The ongoing talks clearly contemplated accession by states other than the consultative parties, as noted in Beeby's presentation of the areas of agreement in Recommendation XI-1, so the regime would not be a closed one. It was also noted that majority participation by international community members in the regime would be a factor in avoiding the possibility that minerals operators would seek sponsorship from a state not party to the minerals regime and operate outside its regulatory framework.

The feasibility of broadening participation beyond governments and potential operators in the institutions of the regime produced the comment that further effort is required to determine how to institutionalize the concerned public in debates under the regime in a fruitful manner. There was support for the idea that should problems arise, in one way or another decisions taken on them should be accepted not only by Antarctic Treaty states but also by the outside world. The problem is how to put this into practice. Participation by observers from international organizations in the institutions of the minerals regime was clearly contemplated in the negotiations, and it was pointed out that these organizations have constructive contributions to make.

PARTICIPATION IN ACTIVITIES AND BENEFITS

Several speakers addressed possibilities for wider involvement in potential minerals activities and benefits therefrom. It is critical that the regime reflect principles of justice and effectiveness, which does not necessarily mean application of the common heritage concept. Justice in the minerals regime would mean that it must provide ab initio opportunities for all countries concerned with future resources exploitation to participate, including interested developing states. But justice and effectiveness would also mean taking account of the experience of the ATS. Moreover, while the 1982 Law of the Sea Convention is a good source of principles and norms that could be applied in Antarctica, care should be taken not to fall into the excessive detail found in that convention.

Various options for breathing life into the concept of humankind's interest in Antarctica have been considered in the minerals regime negotiations and these will receive further attention in forthcoming meetings. These options include joint ventures, particularly if they encourage participants from developing counties, and revenue sharing, although lack of knowledge about the economics of antarctic minerals development would make it difficult to develop the kind of precise formula for revenue sharing found in Article 82 of the Law of the Sea Convention.

Caution was expressed about providing encouragement or incentives for minerals activities that might actually stimulate such activities and imply benefits where none in fact exist.

Finally, the establishment of a fund for scientific research was proposed, to help interested countries lacking financial resources and appropriate organizations to take part in scientific activities in Antarctica. Although numerous details remain to be worked out, the funds could be drawn from fees on minerals activities and they would be available only to countries within the ATS, because those countries would have demonstrated some commitment to the treaty. (See Chapter 27 for further comments on the fund.)

URGENCY AND TIMING OF MINERALS ACTIVITIES

Debate over the reasons for urgency in negotiating the minerals regime revealed additional views to that of the preventive approach ("plugging a gap," "filling a vacuum," "putting out a fire") as a means to avoid renewed conflicts over territorial sovereignty should prospects for minerals development improve.

One such justification for the effort to conclude a minerals regime at this time was that the regime should be balanced between environmental protection and development considerations and that it would be easier to accomplish this before the identification of concrete minerals interests in Antarctica.

Some speakers disagree with an earlier remark that the urgency came about as a move to preempt possible pressure for a universalist regime and U.N. involvement.

Others questioned how the evident lack of immediacy with respect to potential minerals activities could be reconciled on the one hand, with the adoption of the policy of "voluntary restraint" quoted by Beeby from Recommendation IX-1, and, on the other hand, with the ongoing seismic research projects named by panelist Wilson. They wondered if characterizations of lack of immediacy were not tailored to allay the concerns of states not party to the Antarctic Treaty. They also challenged the durability of the preventive approach, which once faced with the motivation and technology to conduct minerals activity in Antarctica might not be able to withstand the pressures.

There was some discussion about the deterrent effects of the long lead time, referred to in Croasdale's presentation (Chapter 17), needed to develop the requisite technology for antarctic operation, in the face of a jump in the price of oil or government incentives for investors.

It was pointed out that, while there is no current interest in commercial minerals activities in Antarctica, there is some interest today in prospecting, and this is another factor behind the urgency of completing a minerals regime; the regime would have to be in place in order to regulate prospecting and to provide mechanisms and procedures to deal with the eventuality of a discovery by a prospector.

With respect to the long lead time, a minerals expert described present technological systems as fairly coarse. There is no direct sensing mechanism to find oil and gas in Antarctica, and the search for hydrocarbons is long, slow, expensive and extremely uncertain. As was noted earlier, the logistics of antarctic operations are formidable: First, a base of operations with a suitable communications system would have to be established. Then it would be necessary to ferry rig crews in and out, since the average driller is used to two weeks on and two weeks off duty and would not be attracted by a three month stay offshore Antarctica. In addition, there would have to be a logistic/supply base with an airstrip on land, and some idea of the size and cost of such a facility could be gained by a quick comparison with, for example, the existing U.S. base at McMurdo Sound for the support of scientific study in Antarctica.

On the other hand, there is no doubt that Arctic offshore minerals exploitation technology will continue to be advanced and that the economics of minerals development will depend on factors outside humanity's control, such as the size of deposits eventually discovered, the cost of money, and minerals markets. For these reasons, one should not assume that antarctic minerals exploitation will be uneconomic for many years to come. In fact, the minerals regime negotiations should presume that antarctic minerals development will be economic and to establish a regime to effectively govern commercial operations.

There was some consideration of Wilson's nondevelopment option and whether it was practical, realistic or politically feasible--given the range of nations involved in Antarctica--to set aside minerals development until one is faced with the possibility some time in the future of, say, discovery of gold in Antarctica.

The alternative of a binding moratorium on minerals development for a specified period of years was contrasted with the current policy of voluntary restraint adopted in Recommendation IX-1, which is a tenuous one, contingent on progress in the minerals regime negotiations. The moratorium option would permit time to learn more about the antarctic environment and the nature and effects of possible minerals operations, so that sounder judgments could be made. This possibility, however, is in effect what those negotiating the regime are contemplating, with the difference that there would be no specific term in years. It was suggested that if a formal moratorium on minerals activities were adopted, it should be reviewed halfway through.

Government officials participating disagreed with Wilson's challenge to the bona fides of governments' decisions to proceed with the minerals regime negotiations and noted that their governments had in fact sought and received support for these policies.

THE REGIME

Area of Application

That the area of application of the International Seabed Authority contemplated in the 1982 Law of the Sea Convention might overlap with that of the antarctic minerals regime, as noted in Section II, brought the response that there is a reason for the vague language in the minerals regime with respect to the area of application of the regime; nevertheless, in the end, the antarctic minerals regime will not extend beyond the continental shelf as defined by the 1982 Law of the Sea Convention. Moreover, the minerals regime will have to coordinate with the International Seabed Authority in the future with respect to environmental protection south of 60°S latitude.

The Framework

Provision of additional details about the framework nature of the regime described by Beeby's presentation--which because of the present lack of knowledge about the area and possible operations there would avoid including

detailed terms and conditions until they are required--
was prompted by one participant's statement that as soon
as the first contract is concluded, it would in fact
specify detailed terms and conditions that will serve as
a precedent for future contracts.

In response, it was noted that a contract concluded
for the development of one type of mineral resource in
Antarctica would not likely be relevant to the development
of another mineral resource in another part of Antarctica.
It is more probable that certain areas in Antarctica--as
elsewhere in the world--because of their physical, environmental, and geological nature and characteristics,
should be treated as units. It would simply not be possible to write detailed regulations applicable to all or
even some of these potential resource management
provinces. In addition, at this point it would be
impossible to predict whether interest would emerge in
mineral resource exploration and development in Antarctica, and, if so, where, when, and for what specific
resource(s).

For these reasons, it is not possible at this stage to
seek to elaborate detailed mining codes for Antarctica.
What is needed is to construct a framework that would
identify the decisions that will be necessary to determine the acceptability of possible mineral resource
activities and the basic criteria against which such
decisions will be made and to provide for the establishment of the institutions necessary to make these
decisions and oversee any activities that might be
permitted.

More specifically, such a framework regime would

- Prohibit mineral resource exploration and
development unless specifically authorized through the
institutions of the regime;
- Provide as general criteria that no mineral
resource activity take place unless

   (1) there is sufficient information to judge its
   possible impacts,
   (2) assessment of its possible impacts indicates
   that it would not pose unreasonable risk to the
   environment, and
   (3) technology and procedures exist to permit
   safe operations;

- Provide for the establishment of decision making machinery, along with an advisory body to provide expert scientific, technical, and environmental advice, to apply the general criteria to all decisions about possible mineral resource activities;
- Provide, if and when there were sufficient information to define a resource province in which exploration and development activities could be considered, that the machinery would establish the general terms and conditions to which such activities must conform, including exclusion of such activities in any areas that had been or should be protected; and
- Provide for the consideration of specific proposals for exploration and development activities at specific sites and for elaboration by the machinery of the detailed conditions that would govern any proposals that were approved.

Such a framework regime would involve no presumption about whether mineral resource activities should or should not take place in Antarctica and would allow the necessary specific regulations to be developed if and when necessary.

Institutions

Because the more precise details of the regime would be developed only following the adoption of the regime, the machinery and procedures for taking these decisions would be very important. This would stimulate states' interest in becoming parties to the future minerals regime so that they could take part in the decision making process.

Various speakers stressed that to achieve balance in the decision making institutions of the regime, both kinds of states should be represented, those active in research or minerals activities and those merely interested in the regime without being primarily concerned with benefits from exploitation. This recalled the discussion of the International Whaling Commission membership on the basis of Gulland's contribution to the workshop (Chapters 15 and 16).

The smaller regulatory committees described by Beeby's paper would not be composed solely of the state or states claiming sovereignty in an identified resource province of Antarctica and the state wishing to undertake or sponsor minerals activities in that area; there would be additional representatives on these committees. Thus, there is no possibility that if, say, an Australian were to apply to conduct operations in the Australian claimed sector, Australia could sit alone on the regulatory committee.

Environmental Protection and Safety

Many speakers noted that concern for environmental and safety aspects of the minerals regime is being given paramount importance by those negotiating the minerals regime as well as by those outside. The work of the Scientific Committee on Antarctic Research (SCAR) in assessing the potential environmental implications of minerals activities was referred to (SCAR groups of specialists produced the 1977 Preliminary Assessment of the Environmental Impact of Mineral Exploration/ Exploitation in Antarctica and the 1981 and 1983 reports on Antarctic Environmental Implications of Possible Minerals Exploration and Exploitation), as was the 1984 report of the SCAR Working Group on Logistics, which states with respect to the proposed minerals regime:

> The group discussed the possible impact on scientific support services of those activities likely to be involved in any commercial activities related to exploration for or exploitation of mineral resources. It was agreed that there is a need to ensure that within the documentation of a minerals regime there should be included a statement that all planned commercial activities in Antarctica should give special attention to all aspects of safety and have appropriate search and rescue resources adequate to meet any emergency. No commercial activity should rely to any extent on those services maintained by operations agencies in support of national antarctic research programs.

There were suggestions that the SCAR/International Union for the Conservation of Nature and Natural Resources seminar (Bonn, April 1985), discussed in Chapters 10-14, could contribute to the definition of areas that should be protected from minerals activities in Antarctica as well as to the definition of the resource provinces that would be identified under the regime were interest in actual minerals activities to develop.

Representatives of environmental organizations argued that they do not deserve to be characterized as single-issue constituencies because they work on a wide range of long-term issues and questions that affect the international community and have followed antarctic matters for nearly 15 years. In the context of the antarctic minerals regime, they have supported freedom of scientific investigation and the free exchange of results from Antarctica, the maintenance of the demilitarized status of Antarctica, and the importance of protecting wilderness values and wildlife there. Members of these organizations feared that the threat of minerals activities in Antarctica could jeopardize these values.

Members of the scientific community also expressed concern about the impacts of minerals activities on scientific research programs and the scientific value of Antarctica, and they raised again the question of the long-term value of scientific activities in relation to the short-term value of potential minerals activities noted in Section III. They were worried in addition that funds would be diverted to research related to possible minerals activities and away from scientific research, but they differed on the extent to which they believed this has already occurred in different countries.

Prospecting

Two subjects arose with respect to prospecting: environmental considerations and confidentiality of data. Several participants echoed Roger Wilson's comment that prospecting should be subject to regulation, but it was pointed out that there <u>would</u> be controls over prospecting in the minerals regime described by Beeby. Any prospecting would have to comply with the environmental principles set forth in the regime and would thus be subject to a judgment that prospecting was taking place in accordance with them. In addition, a notification of prospecting would have to include an environmental impact assessment. Moreover, it appears likely that in the final regime the decision making commission will be given power to formulate and apply controls with respect to prospecting activities, and it is also possible that a form of review of prospecting activities could be initiated. On the other hand, it is unlikely that the final regime will require <u>prior</u> authorization of prospecting, not least because prospecting would be indistinguishable from scientific research under the Antarctic Treaty.

This discussion led to one on the importance of ensuring that preminerals development activities do not jeopardize that openness of results guaranteed under the Antarctic Treaty; that is, free exchange of scientific results should not be obstructed by efforts to retain prospecting data as proprietary data, for this information would be of value to the whole international community. One participant suggested that in order to find out what resources exist in Antarctica, governments could adopt a noncommercial approach and pool efforts and results, enhancing the Antarctic Treaty's emphasis on scientific cooperation. Others warned that the prospector should not be required to compromise his investment by turning over data of a proprietary nature and insisted that the rules for data collection in the minerals regime should be clear. The distinction between prospecting and the subsequent stage of exploration would be based on the concept of right; prospecting does not convey exclusive rights to an area, whereas exploration does.

One technical expert ventured that the concept of confidentiality might be being introduced too early in Antarctica. Many more surface geological data would be required, as will results from several stratigraphic information holes. The early introduction of confidentiality could mean that the only entities that would undertake this work would be commercial organizations. He also stressed that current Antarctic Treaty practices should be improved with respect to seismic data exchange; that is, that the seismic tapes, not just summaries of work done, should be made freely available. He noted that the important element is not the data tapes themselves but the interpretation of the raw data. He cited the example of the North Sea, where jurisdictional lines were drawn before 1959 and there are now voluminous bibliographies of freely available data. In fact, industry has made its data available more quickly than is common among the community of academic scientists.

ENFORCEMENT AND REPORTING REQUIREMENTS

As discussed in Chapters 5, 7 and 11, the absence of agreed national jurisdiction in Antarctica requires a strong alternative enforcement mechanism.

On the other hand, referring to the doubts expressed by Gulland (Chapter 15) about the effectiveness of the CCAMLR enforcement system, one participant maintained that the site-specific nature of an oil operation would make that activity far easier to monitor over time than would be the case with fishing activities. One would also need to carry out an environmental impact assessment and collect the baseline data required to complete it, because this would not have been done in Antarctica, and it would be necessary to make provision for monitoring during and after the operation as well as for remedial measures required or contemplated as a result of the operation.

Reporting requirements were also seen as an important part of the future regime, and several speakers expressed interest in further discussion of inspection under the minerals regime. (See Chapter 14 for additional comments on inspection.)

INSTITUTIONS

21. The Antarctic Treaty System from the Perspective of a State Not Party to the System
    Zain Azraai

22. The Antarctic Treaty System from the Perspective of a Non-Consultative Party to the Antarctic Treaty
    Peter Bruckner

23. The Antarctic Treaty System from the Perspective of a New Consultative Party
    L. F. Macedo de Soares Guimaraes

24. The Antarctic Treaty System from the Perspective of a New Member
    S. Z. Qasim and H. P. Rajan

25. The Interaction Between the Antarctic Treaty System and the United Nations System
    Richard A. Woolcott

26. The Evolution of the Antarctic Treaty System— The Institutional Perspective
    R. Tucker Scully

27. Panel Discussion on Institutions of the Antarctic Treaty System

21.

The Antarctic Treaty System from the Perspective of a State Not Party to the System

Zain Azraai

The debate on Antarctica in the 1984 United Nations General Assembly was quite revealing and three elements in particular seem relevant to the present discussions. These are the following:

First, the Antarctic Treaty consultative parties (ATCPs) were clearly determined not to widen or deepen any involvement of the United Nations (U.N.) in dealing with Antarctica;

Second, the Nonconsultative Parties (NCPs) were not consulted in any meaningful way by the ATCPs; and

Third, there was little "debate" as such, but delegations made statements at each other, which often consisted of restatements of well-known positions with little reference to opposing points that had been made.

These three elements bring out in a graphic way certain fundamental problems in relation to the subject of Antarctica.

The first of these elements relates to the assertion by the ATCPs, which emerged clearly during the debate, that they--and they alone--have the right to make decisions pertaining to Antarctica ("exclusive"), that these decisions will cover all activities in Antarctica ("total"), and that these decisions are not subject to review or even to discussion by any other body ("unaccountable"). Here in fact is the fundamental point at issue between the ATCPs and the nontreaty parties (NTPs).

The second element, which is linked to these, is the role of the NCPs; it was clear from the recent debate that the NCPs made no input whatever to the position taken by the ATCPs in dealing with a question that cannot be said to be scientific or technical in nature, requiring special expertise, but that was concerned with how there can be greater international involvement in decision-making and management regarding Antarctica. Whatever may be the rights on paper of the NCPs, this was a disquieting illustration of the two-tier membership of the Antarctic Treaty System in practice.

The third element is the attitude of the ATCPs, which clearly regarded the debate as an irrelevant nuisance to their pursuit of the proper management of Antarctica; indeed, a number of them made the sentiment explicitly clear during the course of the debate that the sooner U.N. discussion of the subject was terminated the better.

From all the above, it is clear that the recent U.N. debate did not make a constructive contribution to a dialogue on Antarctica. It did not examine questions such as the following: Who has the right to decision-making and management of Antarctica? And on what basis? What precisely is the role of the NCPs? How should the international community proceed to deal with the differences of view between the ATCPs and NTPs? Instead, the debate was, for the most part, little more than a replay of assertions and statements that have been made frequently in the past. But perhaps the present workshop will demonstrate that diplomacy is too serious a subject to be left only to diplomats!

On the first question, while it is obvious that I should not pretend to be able to make the case for the ATCPs, it is necessary for me to do so for the purposes of this discussion because the present system of decision-making and management in relation to Antarctica is so anomalous in terms of normal contemporary practice that it, rather than any questioning of it, needs to be justified. At the risk of appearing presumptuous and simplistic, therefore, I will assert that the basic case that the ATCPs advance for insisting on their rights to decision-making in Antarctica, which are exclusive to themselves, which are total--to cover all aspects of Antarctica--and which are not subject to review by any authority, is based on the argument of their expertise and experience in Antarctica. The management of Antarctica, it is said, is sophisticated stuff, and only those states that have real knowledge, based on actual experience in Antarctica, should have the right to make decisions relating to it. By the same token, it is obvious that these states cannot be answerable to that vast majority in the international community who have no (or insufficient) knowledge or expertise of Antarctica.

This has to be the core of the ATCP case, because no other argument can seek to justify their claim to extra-

ordinary and unique rights over Antarctica. The other
argument that is sometimes advanced, that the current
system "works," cannot be the foundation of an exclusive
"right," because it opens itself to the response that it
would require sanction from a higher authority, which
could subject it to review to ensure that it works better.
Nor can the justification be based on the claims asserted
by the ATCPs for the obvious reason that these claims are
not recognized by an overwhelming majority of the international community. Sometimes reference is also made to
the concept of "trusteeship," which the ATCPs are said to
exercise on behalf of the international community in
Antarctica. But, of course, trustees cannot be self-appointed, they should have no material interest in the
trust property, and they must be "accountable." It is
clear, therefore, that the ATCPs' case is a right based
on "expertise." Put in that way, such an assertion must
surely be seen as extraordinary, particularly in the
context of contemporary international relations.

THE RESPONSE OF THE NONTREATY PARTIES (NTPs)

The response of the NTPs to the assertion of this "right"
by the ATCPs can take either of two forms. One consists
of simply denying that there exists or can exist any such
right unless it is conferred by the international community; by implication, this leads to the concept of an
international, universalist regime to replace the current
Antarctic Treaty System.
   The other consists of accepting the reality of the
present situation, legal and factual, but asking nevertheless: Who gave the ATCPs the right that they assert?
And by what authority? What is the justification for the
notion of right based on expertise? Is this privileged
status for the ATCPs acceptable to the international
community today? It then goes on to suggest that, as a
practical matter, the whole subject should be examined in
a forum in which the interested parties would be on an
equal footing, with the implication that the present
system may be changed or amended, which would make it
more acceptable to the international community as a
whole. It is unfortunate that every questioning of the
current Antarctic Treaty System is read, deliberately or
otherwise, by the ATCPs as meaning only the first alternative. In fact, one of the disquieting features of the
recent debate was the reaction of extreme sensitivity and

resentment on the part of the ATCPs regarding any questioning of the current antarctic system. Implicit in that attitude is that any such questioning can come only from irresponsible or ignorant elements.

The case for this first alternative is in fact quite straightforward. Its proponents simply assert that the days when rights may be asserted on grounds such as discovery, occupation, contiguity, inherited rights, geological affinity, possession--or of "expertise"--are past; they go on to point out that Antarctica has no permanent human habitation, that the sovereignty claims are not recognized by the overwhelming majority of the international community (not to mention the problem of the overlapping claims, the existence of an unclaimed sector, and the assertion by two states of a basis of claim); they then conclude that Antarctica must therefore be the common heritage of humankind to be governed by an international regime duly constituted by the international community. This case is as logically self-contained and complete as that of the ATCPs in asserting rights based on expertise and experience.

The second alternative referred to above takes the approach not so much of "rights" as of "interest." This approach begins from the basic question: Does humankind as a whole have a legitimate interest in Antarctica? If so, how might humankind's interest be best served? More specifically, what should be the objectives of a regime which would best serve humankind's interest? What should be the nature of a regime that would best achieve these objectives? And, to bring the discussion to a more concrete level, does the present Antarctic Treaty System meet those objectives? If there are deficiencies, what are the possible remedies? Linked to all these questions is a procedural issue: How best might these questions be discussed? Its proponents respond to these questions as follows:

First, they assert as indisputable that humankind as a whole has a legitimate interest in Antarctica, in how it is governed and managed. This follows from the fundamental fact that the overwhelming majority of the international community does not recognize the sovereignty claims. Furthermore, Antarctica is not a minute atoll of no significance; it occupies one tenth of the surface of the globe. Its location, vastness, fragile ecosystem, and rich marine and possibly mineral resources have great significance for international peace and security, economy, environment, scientific research, meteorology, telecommunications, and so on. These are clearly matters of global interest and fall within the ambit of concern of the international community.

Second (in response to the question: How might humankind's interest be best served or, more specifically, what would be the objectives and the nature of a regime that would best serve this interest, bearing in mind the special characteristics of Antarctica?), they assert that, among other things, the regime should preserve international peace and security, it should promote and facilitate scientific research and exchange, it should protect the environment, and it should ensure that the fruits of any exploitation of Antarctica's resources be equitably shared by humankind. They assert further that a regime serving these objectives should be one in which member states of the U.N. as well as the relevant specialized agencies and other international organizations are able to play an appropriate role and one that is accountable to the U.N. as the most universal and representative international organization.

Third (in testing the current regime of the Antarctic Treaty System against these objectives), they assert that the rights of the ATCPs--exclusive, total, unaccountable-- are the major flaws of the current system, while at the same time they do recognize its many practical virtues. There is simply no denying this deficiency in the current system. Indeed, as has been pointed out earlier, the ATCPs make no attempt to do so; rather, they seek to justify the system on grounds of expertise and experience and, at the same time, to cushion it by asserting that it works (but without explaining why it cannot work in a more open or universal system) and by pointing to the participatory role of the NCPs and of certain international organizations. (The latter justification is asserted, however, with some limitations and only after 23 years of the operation of the Antarctic Treaty and-- dare one suggest it?--after the validity and fairness of the present systems have been called into question by a number of interested NTPs.)

The issues thus stated may be examined further by considering the notion of expertise itself, based on experience and technological know-how, which serves as the essential justification for the rights that the ATCPs enjoy. If this notion was applied in relation to the recent negotiations leading to the 1982 U.N. Convention on the Law of the Sea (UNCLOS), what was the expertise of the overwhelming majority of the member states in relation, let us say, to seabed mining to justify their participation in those negotiations? Whatever views may be held on UNCLOS (and the 159 signatories cannot all be marching out of step), it has not been generally denied that the member states of the U.N. had a "right" to participate in those negotiations based not on their "expertise" but on natural justice and their material "interest" in the issue. Likewise, it can be asked without, I hope, pressing this argument too far: What was the expertise of the colonial peoples to be involved in the decolonization process? They certainly had no expertise in the complexities of governing or managing a country; yet I believe that no one today would seriously deny their right to do so. Indeed, developments in the norms of international life in the past four decades have advanced us from the concept of right based on expertise in dealing with questions of peace and security, disarmament, international trade and finance, and decolonization, to a right based on natural justice and interest. Why cannot this conceptual progress in the management of international affairs be extended to Antarctica?

In asserting this, one is not necessarily asserting a universalist dogma (one country, one vote), which, after all, does not apply in a number of important international institutions such as the U.N. Security Council, where the veto exists; the World Bank and the International Monetary Fund, which have weighted voting; the International Seabed Authority and various commodity arrangements, where the interests of specific groups--producers, consumers, and other parties most directly affected--are taken into account. This takes us into the area of "special" rights as distinct from "exclusive" rights.

Accepting, without necessarily admitting, that some special expertise is required to make decisions affecting peace and security, the environment, and scientific research in Antarctica, should it be exercised exclusively by the ATCPs? And what is the justification for extending this right to cover every aspect of activity in Antarctica, including the possible exploration and exploitation of its mineral resources, which involves, but is not limited to, the question of the equitable sharing of its benefits, on which the expertise--let alone the exclusive expertise--of the ATCPs is not self-evident? And, finally, what is the justification for the exercise of such rights in a manner unaccountable to the rest of the international community?

In the light of all the points made above, the second element mentioned at the beginning of this discussion, namely, the precise role of the NCPs, falls into place. Here recent developments within the Antarctic Treaty itself, in the Convention on the Conservation of Antarctic Marine Living Resources (CCAMLR), and in the minerals negotiations allowing the participation of the NCPs must be welcomed. Nevertheless, it must also be noted that these developments appeared to have taken place as reluctant concessions, which had to be extracted painfully after more than two decades of the functioning of the treaty itself and, with regard to the CCAMLR, to require a prior acceptance of the complete validity of the Antarctic Treaty itself. Also, there are specific gaps such as the exclusion of the U.N. Environment Program and some other international organizations, and, even more pointed, the exclusion of the NCPs from the so-called heads-of-delegation meetings, the justification for which is not easy to see. But perhaps these specific omissions may be remedied over time. Nevertheless, from the point of view of an outsider, the role, or rather the non-role, of the NCPs in providing inputs to the position taken by the ATCPs during the recent U.N. General Assembly debate was not particularly reassuring, and the basic question remains, therefore, of their precise role and effectiveness more generally in the Antarctic Treaty System.

Finally, and most immediately, there is still the question: Where do we go from here? On this point, the experience of the recent General Assembly debate was again disappointing. In this connection, it may be recalled that a proposal was put forward for the creation of a U.N. Committee on Antarctica. This was firmly resisted by the ATCPs, despite lengthy and elaborate explanations that such a committee would not require any of the parties to give up its position on the appropriate system for the government of Antarctica and that the committee was not intended to be a parallel system or to be the thin edge of the wedge to supplant or replace the current system. Rather, it was intended as a forum in which all participants would be on an equal footing and that would examine in depth issues that are not discussed in the current system. Principally these issues would include the following questions: How can the achievements of the Antarctica Treaty System be preserved and, at the same time, meet the legitimate interests of the international community as a whole in Antarctica? What is the status and legal significance of the UNCLOS and the International Seabed Authority on the situation in Antarctica, more specifically in relation to the exploration and possible exploitation of its minerals resources?

This proposal for the creation of a U.N. committee was not pressed, and, instead, the suggestion was made that member states be invited to comment on the U.N. Secretary-General's study on Antarctica. Comments would be circulated in advance of the fortieth session of the U.N. General Assembly of 1985 in order to facilitate discussions. But even this suggestion was adamantly opposed by the ATCPs. It is difficult to resist the conclusion from these developments that the basic attitude of the ATCPs is that the U.N. has no business to be dealing with Antarctica at all. Such an attitude, if true, is surely unfortunate.

The fact is that serious issues are at stake with regard to Antarctica, which involve recognizing the realities of the situation, both legal and factual, and the legitimate interests of all parties. These cannot be resolved either by a simplistic, universalist approach or by an adamant attitude that the ATCPs are better informed. All sides need to exercise forbearance, refrain from casting aspersions on one another's motives, and examine issues with an open mind, in the full knowledge that there are no easy answers to the complex question of how Antarctica might best be governed and managed in the interest of all humankind.

I should like to elaborate on my prepared remarks by stressing the strong desire of Malaysia and of other countries raising the question of Antarctica in the U.N. to be consulted seriously on the subject. I object to the characterization of the Malaysian initiative by some representatives of the ATCPs as having as its sole aim the dismantling or replacement of the Antarctic Treaty System. The fact is that in Antarctica, the ATCPs enjoy extraordinary rights, which are exclusive, total, and unaccountable. While I am aware of the ATCPs' arguments for the assertion of these rights, I am not convinced that the arguments justify the unique character of the rights in the circumstances of Antarctica; that is, there is no agreement regarding sovereignty and developments in Antarctica, which are matters of global concern.

My final point relates to the future of the antarctic debate. I was disappointed at the rejection of the Malaysian initiative to establish a special U.N. committee on Antarctica during the fall 1984 U.N. General Assembly. Malaysia's alternative effort met with equally strong resistance from the ATCPs: a call for comments on the U.N. antarctica report, which could have served as a basis for discussion at the 1985 U.N. General Assembly. My question is: How can governments find a forum in which they can, on equal footing, examine in depth the issues raised in the U.N.? I do not believe that it is appropriate simply to ask Malaysia and others to put forward specific proposals to be discussed at some unspecified time in the future, because that could freeze Malaysia into the stated position and she would be unable to negotiate effectively.

This contribution is presented by Ambassador Zain in a personal capacity.

22.

The Antarctic Treaty System from the Perspective of a Non-Consultative Party to the Antarctic Treaty

Peter Bruckner

INTRODUCTION

According to the provisions of Article XIII(1), the Antarctic Treaty shall be open for accession by any state that is a member of the U.N. or by any other state that may be invited to accede to the treaty with the consent of all the contracting parties whose representatives are entitled to participate in the meetings provided for under Article IX of the treaty, the so-called consultative parties (CPs).

The Antarctic Treaty, signed on December 1, 1959, by the 12 participants in the International Geophysical Year (IGY): Argentina, Australia, Belgium, Chile, France, Japan, New Zealand, Norway, South Africa, the USSR, the United Kingdom and the U.S. of America, entered into force on June 23, 1961, when it was ratified by all 12 signatories.

The following states have subsequently acceded to the treaty at the times indicated:

Poland	June 1961
Czechoslovakia	June 1962
Denmark	May 1965
The Netherlands	March 1967
Romania	September 1971
German Democratic Republic	November 1974
Brazil	May 1975
Bulgaria	September 1978
Federal Republic of Germany	February 1979
Uruguay	January 1980
Italy	March 1981
Papua New Guinea	March 1981
Peru	April 1981

Spain   March 1982
People's Republic of China   June 1983
India   August 1983
Hungary   January 1984
Sweden   April 1984
Finland   May 1984
Cuba   August 1984

According to the treaty, these states are also contracting parties; they are subject to the same general obligations and enjoy the same general rights under the treaty as the original participants.

However, the original twelve states enjoy a special status under the treaty. They are born members of the consultative meetings provided for under Article IX of the treaty. Any acceding state may--pursuant to Article IX(2)--be entitled to appoint representatives to the meetings "during such time as that contracting party demonstrates its interest in Antarctica by conducting substantial scientific research activity there, such as the establishment of a scientific station or the dispatch of a scientific expedition." The following acceding states have been recognized as CPs: Poland (1977), the Federal Republic of Germany (1981), and Brazil and India (1983).

The purpose of this chapter is to present the outlook of an acceding state--a nonconsultative party (NCP)--on the Antarctic Treaty System. This system is now composed of the 1959 Antarctic Treaty, the 1972 Convention for the Conservation of Antarctic Seals, the 1980 Convention on the Conservation of Antarctic Marine Living Resources and the nongovernmental Scientific Committee on Antarctic Research (SCAR). The system also covers all the recommendations approved by the CPs.

This chapter will focus mainly on the Antarctic Treaty and will essentially be based on the history of Denmark's participation in the treaty since 1965. The opinions expressed are the personal views of the author and do not necessarily represent those of his government.

Furthermore, this discussion does not purport to reflect the views of other acceding states, except where official statements have provided a sufficient basis for expressing more generalized NCP opinions. In this respect, the report that the U.N. Secretary-General was requested to submit to the 39th U.N. General Assembly would have constituted a valuable source of updated information on other NCP views. However, this material was not available at the time of drafting the manuscript.

MOTIVES FOR ACCESSION

Several factors influenced the Danish decision on accession to the Antarctic Treaty. Danish scientists participated in research activities, together with colleagues from other nonsignatory countries, during the IGY. Scientific experience from Greenland--the polar regions of the Kingdom of Denmark--undoubtedly influenced the scientific interest in Antarctica. Danish scientific circles strongly supported the principles of the treaty, in particular the principle of free scientific research.

In 1952 the vessel Kista Dan of the Danish company J. Lauritzen made a voyage to Antarctica, thereby starting an annual series of calls that the company's ships have since maintained.[1]

The treaty provides that the contracting parties inform one another of certain activities, such as shipping related to expeditions to and within Antarctica. Difficulties concerning the implementation of these obligations were eventually overcome. The fact that NCPs had no influence on recommendations adopted at consultative meetings also sparked certain hesitations. In this respect, however, it was argued that the recommendations would not become binding on any country unless that country had given its express consent.

An essential factor in the Danish decision-making process was general interest in supporting a treaty in which East and West had been able to join in cooperation based on laudable principles.[2]

In retrospect, the Danish expectations have not been disappointed. In its contribution to the study of the U.N. Secretary-General made pursuant to Resolution 38/77 of December 15, 1983, the Danish government stated, inter alia:

> For more than 20 years the 1959 Antarctic Treaty has provided a legal regime in Antarctica which has removed the potential for disputes relating to the exercise of sovereignty and guaranteed peace and stability in the region. In the view of the Government of Denmark it is of particular importance that the Treaty prohibits any military use of the region and guarantees its

status as a nuclear weapons free area. Furthermore, the Treaty has provided an exemplary framework for free scientific research and has created the basis for an extensive international cooperation to protect the extremely fragile ecosystem of Antarctica.

There is a comprehensive system of on-site inspection, with observers being guaranteed complete freedom of access at any time to any and all areas of Antarctica. The Scientific Committee on Antarctic Research (SCAR) also forms part of the system and has served to initiate, promote and coordinate scientific activities in Antarctica.

The Antarctic Treaty has so far proved its value for the benefit and interest of mankind as a whole. It has set an example of international cooperation which has succeeded according to its purpose.

FUNCTIONING OF THE TREATY SYSTEM

As a general rule the treaty system has functioned smoothly in practice, seen from a Danish point of view. One of the important aspects has been _information_ on developments within the system. Article III provides that in order to promote international cooperation in scientific investigation in Antarctica the contracting parties shall--to the greatest extent feasible and practicable--exchange information regarding plans for scientific programs and scientific observations and results from Antarctica. The information received ex officio from contracting states, whether CPs or NCPs, through the official channels has not been abundant. The official documentation received deals essentially with plans and programs in scientific research activities rather than with the results thereof. However, it is our impression that interested parties, scientists, etc. have been able to obtain all the scientific material that they need through other channels.

In this respect, the international cooperation within SCAR plays a significant role. SCAR is open to all countries actively engaged in antarctic research, to scientists appointed by the International Council of Scientific Unions (ICSU), of which SCAR is a component, and to each of the ICSU-federated international scientific unions. Canada, which is not a contracting party, took part in working groups and groups of specialists in 1982 and 1983. Thus, SCAR seems to serve admirably the principle of freedom of scientific research, in particular of making the results thereof freely available to interested parties.

The increasing interest in and concern about all activities in Antarctica have prompted certain questions that have been difficult to answer. In 1983, the Danish Parliamentary Committee on Foreign Affairs asked certain questions relating to seismic investigations by Japan, Norway, and the United States in Antarctica, to the French airstrip proposal for Pointe Géologie, and to the environmental impact of bases in Antarctica. At the time, the Danish authorities were not able to offer entirely satisfactory replies.[3]

The question is whether the provisions of Articles III and VII(5) of the treaty are sufficiently wide to guarantee the availability of all relevant information on activities in the Antarctic. The current interest and concern seem in particular to focus on environmental issues. The treaty system does not seem to contain instruments likely to ensure satisfactory responses to such concerns.

NCPs receive no official reports on the deliberations during the meetings of CPs. Until fairly recently the texts of the recommendations have been difficult to obtain. After the eighth consultative meeting, in Oslo in 1975, a collection of recommendations was issued. In conjunction with the tenth consultative meeting, in Washington in 1979, the U.S. Department of State brought out a Handbook of Measures in Furtherance of the Principles and Objectives of the Antarctic Treaty.

During the twelfth consultative meeting, the information system was further improved. The Handbook of the Antarctic Treaty (renamed) is now to include final reports of consultative meetings. The NCPs also benefit from these information measures. Furthermore, an NCP should, as an observer, be entitled to make use of the new rules concerning public availability of its own documentation and that of other NCPs and CPs.

Inspection is an area that involves a certain discrimination of NCPs and gives rise--at least in theory--to some particular legal problems. Pursuant to Article VII(1), only CPs are entitled to designate observers to carry out inspection. Observers must be nationals of the CPs that designate them. All areas of Antarctica, including all ships at points of discharging or embarking cargoes or personnel in Antarctica, are to be open at all time to inspection by any observer. In practice, few inspections of any sort are conducted.[4] In principle, ships of an NCP state may be inspected only by CP observers. As the rule is drafted, NCPs are not entitled to conduct inspection, not even of their own ships.

A related problem, which has so far remained fairly theoretical, is that of <u>jurisdiction</u>. This issue, which is intimately connected with the delicate problem of sovereignty, was not properly solved at the Washington conference in 1959.[5] Article VIII of the treaty provides that scientists and officially designated observers, as well as their staffs, remain under the jurisdiction of the countries of which they are nationals, regardless of where they may be in Antarctica. This rule applies to all contracting parties. However, it does not apply, for instance, to the members of the crew of the ship carrying a scientific expedition to Antarctica.

Furthermore, what would be the legal situation in the case in which a Danish sailor were arrested in the Antarctic Peninsula, where jurisdiction is contested. In theory, at least, he might be subject to Argentine, British, and Chilean law--but perhaps, according to the treaty, not to Danish law. This situation, however, is not peculiar to NCP citizens; it may arise also with regard to CP nationals other than scientists and observers.[6]

RIGHTS AND OBLIGATIONS OF NCPs UNDER THE TREATY

As stated in the Introduction, the term "contracting states" also covers NCPs. In principle, CPs and NCPs as contracting parties enjoy the same rights and are subject to the same obligations under the Treaty. However, in many respects the treaty has established a two-tiered system of participation. Certain aspects of the differential system of rights and obligations under Article VII in regard to inspection have already been dealt with above. CPs also enjoy a privileged status under Article XII concerning amendments and pursuant to Article XIII(1) concerning accession.

By far the most important example of "privileged status" is the fact that the day-to-day management of the treaty system has been entrusted to the CPs. Until 1983, NCPs were not even consulted.

Article IX provides that the CPs meet periodically to exchange information, to consult together on matters of common interest pertaining to Antarctica, and to formulate, consider, and recommend to their governments measures in furtherance of the principles and objectives of the treaty. Among the categories of measures specifically mentioned are those relating to questions of facilitating scientific research and international scientific cooperation, to the exercise of the rights of inspection and jurisdiction, and to the preservation and conservation of living resources in Antarctica. Any such measures recommended by the representatives of CP meetings to their governments become effective when approved by all of the consultative parties.

So far 138 recommendations have been adopted and more than 130 approved. NCPs have had no opportunity to influence the drafting of these recommendations. Conversely, the recommendations are not legally binding on NCPs unless expressly approved by them.

In view of this situation, in which different treaty parties might be bound by different sets of rules, the CPs in 1975, in Recommendation VIII-8, urged "the States that have or will become Parties to the Antarctic Treaty to approve the recommendations adopted at consultative meetings." In 1977 the final report of the special CP meeting emphasized that CPs might urge a state that considered itself entitled to CP status to make a declaration of intent to approve the recommendations in force and might also invite such a potential CP to consider approval of the other recommendations.[7]

Certain issues have been made subject to separate agreements negotiated under the authority of the treaty: the 1972 Convention for the Conservation of Antarctic Seals (Seals Convention) and the 1980 Convention on the Conservation of Antarctic Marine Living Resources (CCAMLR).

The Seals Convention is open for accession by states invited with the consent of all parties. The CCAMLR is open for accession by states interested in research or harvesting activities related to antarctic marine living resources and to certain regional economic integration organizations (such as the European Economic Community). Last but not least, the ongoing negotiations on the future minerals regime of Antarctica have so far been reserved to the CPs.

These features, in particular, must have been in the minds of the critics of the Antarctic Treaty System during the 38th U.N. General Assembly debate, where terms such as "exclusivity of the treaty system," "enhancement of interests of the privileged few," and the "secrecy" of the meetings were frequently used.

One way of attempting to remedy this "exclusivity" and "secrecy" might be to modify the criterion for qualifying as a CP. As noted above, Article IX(2) provides that each contracting party that has acceded to the treaty be entitled to participate in the consultative meetings during such time as that contracting party "demonstrates its interest in Antarctica by conducting substantial scientific research activity there, such as the establishment of a scientific station or the dispatch of a scientific expedition." A recommendation adopted during the special consultative meeting in July 1977 contains the procedures giving effect to Article IX(2) (see Appendix 22-A.):

> Firstly, the criterion of Article IX(2) only applies to acceding States, not to signatories.
> Secondly, it may be asked whether the criterion is reasonable measured by a yardstick of the 1980s.

As stated by Bilder, "This differential status [of CPs and NCPs] is ostensibly related to a rational and 'neutral' criterion--demonstrated interest in Antarctica, and any Party may in theory obtain consultative status by engaging in 'substantial' activities in Antarctica. In practice, however, many nations may not have sufficient wealth or technical skill to mount and support such activities."[8]

Assuming that all interested nations possess the necessary wealth and skill, it may even be questioned whether the present criterion for qualifying as a CP is likely to enhance--in a longer-term perspective--the objectives and principles of the treaty.[9] Indeed, in a statement made by Lee Kimball on September 24, 1984, before the subcommittee on Science, Technology and Space of the U.S. Senate Committee on Commerce, Science and Transportation, it is said, inter alia:

> As long as the ticket to antarctic decision-making remains something on the order of the establishment of a scientific station or the dispatch of a scientific expedition to Antarctica, we can expect additional strains on the antarctic

environment and its pristine value as more countries seek to meet this qualification. Growing tourism and fishing efforts will increase congestion and the potential for accidents in the area.

Kimball adds that the onslaught of antarctic minerals development and supporting operations would add a whole new dimension to the possibility of accidents and cumulative environmental impacts in Antarctica.

Translated into the present context, an admission ticket based on performance criteria might in certain respects lead to situations that would raise serious objections on environmental grounds.

In a reply to a question put by the Parliamentary Committee on Foreign Relations of Denmark in February 1984, the Danish foreign minister stated, inter alia, that it would be appropriate first to assess how the newly established observer arrangement for NCPs is functioning in practice before considering any attempt to seek a modification of the criteria for qualifying as a CP. Furthermore, it was stated that in view of the unanimous opposition by CPs to amendment of the treaty, any attempts in this regard would be counterproductive.

On the other hand, it must be recognized that the minerals issue gives rise to specific considerations. Luard observes that the treaty powers must recognize that the situation today is by no means the same as it was 25 years ago: "A Treaty that functioned well when its main purpose was to provide for a system of peaceful and cooperative scientific research will not necessarily work well for the quite different purpose of establishing a viable minerals regime."[10]

Among the three broad measures discussed by Luard as the means for fulfilling certain basic conditions, the first is to maintain the existing treaty system and establish a new minerals regime within it.[11] This appears to be a reasonable approach.

Indeed, it seems necessary to make a proper distinction among separate issues: participation in the Antarctic Treaty and in the regular CP meetings; participation in the negotiations on the minerals regime; and the question of participation in decision-making under the future minerals regime.

There is no apparent need to change the rules on participation in the regular and special CP meetings—that is, to modify the criteria for qualifying as a CP—in order to meet the increasing concern regarding participation in the minerals regime. This problem should, in any case, be solved within the minerals regime itself. It should, however, be ensured that all legitimately interested states may participate in the negotiations leading to this regime. An effective regime requires the consent by a large majority of the world community. Otherwise, those outside the regime could quite legally undertake mining activities.

The increasing interest in these negotiations in particular, and in the management of the Antarctic in general, should and could be met--at least as a first step--by an appropriate observer-status arrangement in the regular and special consultative meetings (see below). Attempting to modify the criteria for qualifying as a CP at this stage may be tantamount to opening the whole Antarctic Treaty for revision with the inherent risk of prejudicing the basic principles of the treaty and present international cooperation on antarctic issues. There is no reason to destroy the successful features of the Antarctic Treaty in order to find a satisfactory solution to the novel problems concerning a future minerals regime.

The CCAMLR, which was negotiated under the provisions of the Antarctic Treaty, has in fact established its own participation rules which differ from those of the Antarctic Treaty. Membership in the convention is open to states--or competent regional organizations-- interested in research or harvesting activities related to antarctic marine living resources. Participation in decision-making meetings in the commission established by the convention is open to all original signatories that are parties and to acceding parties during such time as they are engaged in antarctic living-resources research or harvesting.[12] The convention has also established its own observer regime.

Mentioning CCAMLR as a precedent does not necessarily mean that it is a perfect one in all respects. Criticisms have been voiced on two important points in particular. First, it has been questioned on what grounds the CPs felt entitled to negotiate in secret a treaty regarding high seas resources and then to present the document to the rest of the world to endorse a fait accompli. Second, the criteria for accession and for participation by acceding states in the decision-making do not apply to the signatories. Even if some of these do not carry out any fishing activities in antarctic waters, they remain permanent members of the decision-making commission.

Furthermore, the relationship of the CCAMLR to the Antarctic Treaty, which specifically includes the high seas, and to the International Whaling Commission and the Seals Convention has also been criticized.[13]

A minerals regime will be far more difficult to negotiate than the regime on the living resources. It is to be hoped that the lessons from the CCAMLR will be kept in mind during the continuing minerals negotiations.

THE OBSERVER ISSUE

The Antarctic Treaty contains no provisions concerning observers. As far as can be ascertained, the subject was not even discussed at the Washington conference. This may in part be explained by the fact that the treaty does not establish any separate international organization nor for that matter any kind of permanent secretariat.

The observer issue was discussed among the CPs in 1981 without success. A consensus was reached in the spring of 1983, perhaps as a first reaction to emerging international criticism of the secluded character of the Antarctic Treaty System. During the 37th U.N. General Assembly, the Prime Minister of Malaysia in his speech on September 29, 1982, argued for U.N. action on Antarctica. The Malaysian statement during the signing ceremony of the Law of the Sea Convention at Montego Bay in December 1982 echoed the same theme. At their seventh summit conference in New Delhi, March 7-12, 1983, the heads of state of the nonaligned countries considered that in view of the increasing international interest in the Antarctic, the U.N. should undertake a comprehensive study of the subject.

At the preparatory meeting for the twelfth CP meeting, in April 1983, the CPs decided to invite the NCPs to attend the twelfth biennial CP meeting, in Canberra in September 1983, as observers. All NCPs--except Czechoslovakia--were represented at the meeting in Canberra. In their opening statements the CPs welcomed the NCPs, which in turn expressed their appreciation for the invitation.

The new observers were confronted with certain difficulties that made their first attendance somewhat cumbersome. They lacked background information from the preparatory meeting. Moreover, the fact that the session was divided into open plenary meetings, closed meetings of heads of CP delegations, and working groups made it difficult to follow the course of the deliberations.

On request to the chairman, the NCP observers obtained admission to the working group on environmental questions. In practice, observers were not able to participate in a major part of the final discussions, which took place in closed meetings of heads of CP delegations. Statements by observers were not allowed during the plenary debate on the final report nor during the discussions on procedures relating to the negotiation of the minerals regime. Observers could speak on the question of rules of procedures concerning their status, but the actual negotiations on this issue were conducted in closed meetings.

The CPs accepted the inclusion in the final report of a formal statement made in common by the NCPs (See Appendix 22-B). In this statement, the NCPs noted with satisfaction that the CPs were receptive to more meaningful and substantive participation by the NCPs, which would undoubtedly contribute to strengthening the system. They also stressed the need for prior background information, which would facilitate their participation in the various antarctic meetings.

As far as is known, the draft rules of procedure on observers considered by the CPs in Canberra provide that observers may speak freely, receive documentation, submit information documents, and attend all plenary and formal committee and working group meetings. They may not take part in decision-making.[14]

The CP meeting in Canberra decided to invite the NCPs as observers to their next meeting, in Brussels in 1985, and to the preceding preparatory meeting. It was also agreed to consider, on a case-by-case basis, inviting as observers at future CP meetings representatives of international organizations having a scientific or technical interest in Antarctica.

During the special consultative meeting on the minerals regime in Tokyo in May 1984, the NCPs were kept informed at regular intervals. It was decided to invite NCPs as observers to future meetings on this theme. Their status will be the same as that enjoyed in the regular CP meetings.

Some CPs include nongovernmental organization representatives in their delegations to CP meetings. In regular meetings, the United States has followed this practice since 1977; Australia did so for the first time in 1983. In special meetings on the minerals regime, the United States has done it since 1982, Australia since 1983, and New Zealand since 1984. The Danish observer

delegation to the minerals regime meeting in Rio de
Janeiro in February-March 1985 included a representative
from a nongovernmental organization.

ANTARCTICA AND THE U.N. GENERAL ASSEMBLY

The inclusion of the question of Antarctica on the agenda
of the U.N. General Assembly was viewed with much sympathy
by the Danish government. In its reply to the U.N.
Secretary-General's <u>note verbale</u> made pursuant to
Resolution 38/77, it was stated that the Danish
government--recognizing the legitimate interest of the
world community--was

> prepared to support the efforts aiming at
> introducing greater openness in the international
> cooperation concerning Antarctica, provided that
> neither the basic principles of the Treaty nor the
> positive results of the present international
> cooperation are jeopardized.

The Danish government furthermore recalled that the
Antarctic Treaty is in conformity with the principles and
purposes of the U.N. Charter and that it is open for
accession by all members of the United Nations. In the
view of the Danish government, the international
cooperation concerning the Antarctic should be pursued
within the framework of the treaty. However,

> accession to the Treaty becomes meaningful only if
> the acceding parties are entitled to participate
> in the international antarctic cooperation in a
> manner which corresponds to the obligations they
> have undertaken according to the Treaty.

The Danish government therefore welcomed the invitations to participate as observers in the future regular
CP meetings and in the special consultative meetings on
the minerals regime. Finally the hope was expressed that
these moves

> may lead to a generally acceptable permanent
> arrangement which will ensure that NCPs may
> participate fully and effectively in the entire
> range of international cooperation and management
> concerning the Antarctic. Progress in this

direction will doubtless serve to rally the ample
support for the Treaty which appears imperative in
order to preserve it as the international
framework for cooperation in Antarctica for the
benefit and interest of mankind as a whole.

CONCLUDING REMARKS

The final words of the Danish contribution to the U.N.
study reflect what others have more elegantly labeled
"external accommodation" or "accountability": "It is up
to those who would prefer to build on the Antarctic
Treaty system to determine how far they are willing to go
in the area of accountability in lieu of being confronted
by major institutional changes."[15]

The major--but not the only--challenge to the treaty
system is the minerals issue.

Whether in the light of these challenges the new
observer arrangement will serve the purpose of increased
accountability remains to be seen. The comment has been
made that since decisions in CP meetings are determined
not by voting but rather by discussion and consultation
leading to consensus, the influence of NCP observers on
the decision-making process need not be far different
from that of the CPs.[16]

This, of course, is to be hoped. However, if owing to
such increased incentives the total membership of the
Antarctic Treaty is substantially increased, the current
management mechanisms may need to be further developed.
Holding meetings among representatives of 20 or 30 countries is one thing; organizing meetings of representatives
from 60 countries or more is a more complex task. The
question of the establishment of a more permanent infrastructure should therefore be considered an urgent
matter.[17]

Together with the other NCPs, Denmark will be prepared
to contribute as constructively as possible to the development of international cooperation under the treaty
system. Denmark and the two other Nordic NCP countries,
Finland and Sweden, all have polar regions and a long-
standing interest mainly in the Arctic region. Denmark
has a long experience in the area of exploration and
exploitation of minerals in polar regions (Greenland).
Although this experience cannot, of course, be applied
automatically to all aspects of similar issues in
Antarctica, it may prove to be of some value, for

instance, for comparative purposes. One aspect that most
certainly will be strongly emphasized is the need for
strict safeguards aimed at protection of the sensitive
antarctic environment. In this regard the words from the
Swedish contribution to the study of the U.N. Secretary-
General seem particularly pertinent:

> In view of the great importance of Antarctica to
> global climate and oceanic conditions in general
> it is clear that disturbances in the antarctic
> environment can have consequences that are both
> unpredictable and hazardous. These important
> problems have to be confronted with great
> seriousness and full openness.

In elaborating on his written presentation, Bruckner
noted that it is difficult to identify any objective
criteria used for the determination of CP status and
wondered whether the qualification cited in Article IX(2)
of the Antarctic Treaty is reasonable today. Moreover,
it might even be discriminatory, since not all states can
afford scientific expeditions.

He noted that Denmark has a history of demonstrated
interest in Antarctica by virtue of its involvement in
ship transport, construction work, and scientific research
there, but the first two types of activity do not seem to
be considered in the determination of CP status.

Bruckner added that if CP status were to be relevant
to the minerals regime, any of three changes in that
status could make it more acceptable to those outside:
modify the criteria, interpret them more flexibly or
reduce the differences between CP and NCP status. He did
not see the need to modify the criteria at this time, but
he could envisage the second or third options or a combination of the two.

In this context, he noted that the Antarctic Treaty in
effect forces countries to allocate scarce funds for
scientific research in a manner that is not necessarily
rational. He suggested that it would make sense for a
number of countries to pool resources for antarctic
science and spread the costs out over several years. The
question that remained, however, was whether at the end
of that period the contribution of each country would be
considered sufficient for each to qualify for consultative
status. On the other hand, if the difference between CP
and NCP status disappeared, this would reduce the pressure
on governments to engage in questionable scientific
activities.

With respect to the observer role of the nonconsultative states in antarctic meetings, Bruckner noted that "participation short of the right to take part in decision-making" is not a fixed concept. Whatever rules of procedure had been adopted in Canberra in September 1983 to govern NCP participation had not been conveyed to Denmark and to other NCPs by January 1985. In his view, the effect of NCP participation should be to allow the NCPs to exercise reasonable influence on decisions, to allow them to be heard and have their views taken into account, and to allow them to submit their views orally and in writing, whether as proposals or suggestions. He also believed that it would be useful if "heads-of-delegation meetings" could include representatives from both consultative and nonconsultative states. In the end, he believed that the role of observers in practice will depend on the quality and constructive nature of their contributions rather than on the application to the letter of the observer rules of participation.

In conclusion, he noted that in addition to the development of the role of nonconsultative states as a means to increase support for the Antarctic Treaty System, it will be important to increase the flow of information on antarctic affairs and make it more easily accessible, to consider the contributions that members of the U.N. family--such as the U.N. Environment Program--could make to the system, and to develop means to take account of the views of the concerned public.

NOTES

1. The J. Lauritzen polar vessels have throughout the years carried Australian, French, British, Belgian, and Dutch scientific expeditions and their supplies to the antarctic continent and returned with parties who had wintered there; see Thorsoe, S. 1984. *J. Lauritzen 1884-1984*. World Ship Society, 1984. Approximately 25 localities in the Antarctic have been named after J. Lauritzen ships or members of their crews. The Danish company A. E. Sorensen has also carried expeditions to Antarctica. The German subsidiary company of Christiani & Nielsen has carried out construction and other work in connection with the establishment of the scientific station and other activities of the Federal Republic of Germany.

2. See Frivagten, November, 1965.
3. A similar situation arose in the Belgian parliament. See question No. 413 from Mr. Daras of April 20, 1984 (Brussels).
4. See The Future of the Antarctic - Background for a Second U.N. Debate. Greenpeace International, October 22, 1984, p. 8.
5. See Quigg, P.W. 1983. A Pole Apart. The Emerging Issue of Antarctica, New Press, McGraw-Hill Book Company (New York); 1983, p. 150.
6. Quigg, op. cit., p. 151.
7. Quigg, op. cit., p. 152. Denmark has not officially approved any CP recommendation.
8. See Bilder, R.B. 1982. The Present Legal and Political Situation in Antarctica. In J.I. Charney, ed. The New Nationalism and the Use of Common Spaces. Allanheld, Osmun (Totowa, N.J.), 1982, p. 173. Belgium and Norway maintain no permanent scientific stations in the Antarctic. Belgium had undertaken little scientific research work for some time. Some new CPs had undertaken a rather limited amount of independent scientific work before being admitted.
9. According to Quigg, op. cit., p. 148, one of the most time-consuming matters at the Washington conference was conditions and procedures for new consultative memberships.
10. See Luard, E. 1984. Who owns the Antarctic? Foreign Affairs, p. 1184.
11. The two subsequent measures related more specifically to the minerals regime, which as such is not a subject for discussion in this chapter.
12. The CCAMLR has now been ratified by the 15 original signatories. All are members of the decision-making commission, together with the European Economic Community, which acceded in 1982. Sweden, Spain, India, the Republic of Korea, and Uruguay have also acceded to the convention and Brazil plans to.
13. See Quigg, op. cit., pp. 189-193.
14. See Report of the Twelfth Consultative Meeting, Polar Record 22(136):109.
15. See paper delivered by L. Kimball in Antarctic Politics and Marine Resources: Critical Choices for the 1980s. 1985. Center for Ocean Management Studies, University of Rhode Island (Kingston, R.I.), 1985, p. 247. See also Beck, P.J., The

United Nations and Antarctica. Polar Record, 22:137-144; Sollie, F. Polar Politics: Old Games in New Territories, or New Patterns in Political Development, p. 26: "Here, mutual accommodation clearly is needed in that the privileged few must adapt the operation of the Treaty and the regime for resources to take all due account of the broader international interest, while those who emphasize the doctrine of the common heritage must adjust their demands to established rights and to actual possibilities for development."

16. See Report on Antarctica, November 1, 1984. International Institute for Environment and Development (Washington, D.C.), 1984, p. 6.
17. See Report of the Twelfth Consultative Meeting, Polar Record 22(136):108.

APPENDIX 22-A:
EXTRACT FROM THE FINAL REPORT OF THE
FIRST SPECIAL ANTARCTIC TREATY CONSULTATIVE MEETING

"The Representatives of the Consultative Parties (Argentina, Australia, Belgium, Chile, France, Japan, New Zealand, Norway, the Republic of South Africa, the Union of Soviet Socialist Republics, the United Kingdom of Great Britain and Northern Ireland, and the United States of America) met in London on 25, 27 and 29 July 1977.

"The Meeting considered in Plenary Session the question of procedures to be adopted to give effect to Article IX, paragraph 2, of the Antarctic Treaty ... and decided as follows:

"The Representatives of the Consultative Parties Recognizing the need for a procedure of consultation to be adopted between them in the event that another State, having acceded to the Antarctic Treaty, should notify the Depositary Government that it considers it is entitled to appoint Representatives to participate in Antarctic Treaty Consultative Meetings;

"Recalling that Recommendations which became effective in accordance with Article IX of the Treaty are, in terms of that Article 'measures in furtherance of the principles and objectives of the Treaty';

"Recalling their obligation under Article X of the Antarctic Treaty to exert appropriate efforts, consistent with the Charter of the U.N., to the end that no one engages in an activity in Antarctica contrary to the principles or purposes of the Treaty;

"Recognizing that the entitlement of an acceding State to appoint Representatives to participate in Antarctic Treaty Consultative Meetings under Article IX, paragraph 2, of the Treaty depends on such a State demonstrating its interest in Antarctica by conducting substantial scientific research activities there, such as the establishment of a scientific station or the dispatch of a scientific expedition;

"Unanimously decide:

1. An acceding State which considers itself entitled to appoint Representatives in accordance with Article IX, paragraph 2, shall notify the Depositary Government for the Antarctic Treaty of this view and shall provide information concerning its activities in the Antarctic, in particular the

content and objectives of its scientific programme. The Depositary Government shall forthwith communicate for evaluation the foregoing notification and information to all other Consultative Parties.

2. Consultative Parties, in exercising the obligation placed on them by Article X of the Treaty, shall examine the information about its activities supplied by such an acceding State, may conduct any appropriate enquiries (including the exercising of their right of inspection in accordance with Article VII of the Treaty) and may, through the Depositary Government, urge such a State to make a declaration of intent to approve the Recommendations adopted at Consultative Meetings in pursuance of the Treaty and subsequently approved by all the Contracting Parties whose Representatives were entitled to participate in those meetings. Consultative Parties may, through the Depositary Government, invite the acceding State to consider approval of the other Recommendations.

3. As soon as possible, but in any case within 12 months of the date of the Communication by the Depositary Government to the other Consultative Parties referred to in paragraph 1 above, the Government which is to host the next Consultative Meeting shall convene a Special Consultative Meeting in order that it may determine, on the basis of all information available to it, whether to acknowledge that the acceding state in question has met the requirements of Article IX, paragraph 2 of the Antarctic Treaty. The adequate preparation of the Special Consultative Meeting shall be undertaken through diplomatic channels.

4. With the agreements of the Representatives of all the Consultative Parties, the Special Consultative Meeting shall record this acknowledgment in its report. The acceding State shall be so notified by the host Government of the Special Consultative Meeting.

5. The procedure hereby established may be modified only by a unanimous decision of Consultative Parties."

APPENDIX 22-B:
STATEMENT OF NONCONSULTATIVE PARTIES

"The delegations of the Nonconsultative Parties to the Antarctic Treaty having been present at the Twelfth Consultative Meeting express appreciation to the Government of Australia and to the other Consultative Parties at having been invited to this Meeting.

"Our presence reflects the interest of our Governments in the development of the antarctic system and our willingness to contribute to the maintenance and further development of the principles and objectives of the Antarctic Treaty.

"We all recognize the achievements of the Treaty, for example with regard to cooperation in the field of scientific research, the protection of the environment, and demilitarization. We noted with satisfaction the recognition by the Consultative Parties of the difference in position between the Nonconsultative Party and observers.

"We have noted, furthermore, with satisfaction that the Consultative Parties are receptive to a more meaningful and substantive participation of Nonconsultative Parties, which would undoubtedly contribute to strengthening the system. Likewise the delegations of the Nonconsultative Parties fully endorse statements of Consultative Parties which have been made during the Twelfth Consultative Meeting regarding the importance of the availability of information to the Nonconsultative Parties so as to facilitate their participation in the various antarctic meetings.

"We believe that the participation of Nonconsultative Parties in the various activities of the antarctic system is important for the strengthening of the system and for the contribution thereto by the Nonconsultative States.

"We request that this statement be attached to the Final Report of the Twelfth Consultative Meeting.

"Canberra, September 27, 1983."

23.

The Antarctic Treaty System from the Perspective of a New Consultative Party

L. F. Macedo de Soares Guimaraes

In principle, a new consultative party should see the Antarctic Treaty System (ATS) from the same perspective as do the other Antarctic Treaty consultative parties (ATCPs), since it adheres to the spirit as well as to the letter of the treaty and carries out activities with the same purposes in mind. There are, however, some differences that relate to the specific situation of the new ATCP regarding its economic and scientific development as well as to the evolution of the ATS since its inception. In this sense there is not much to say about the treaty itself, its positive qualities and possible deficiencies. It would be more interesting to examine how this new ATCP inserts itself into a system that has attained maturity but is still evolving. This leads me to describe the way the new ATCP (in this case, Brazil) has taken to reach that position and its views on how it interacts with the other ATCPs inside the system. Let us, then, start with a historical resumé.

Brazil conducted research projects in the south Atlantic during the International Geophysical Year (IGY) but did not send an expedition to Antarctica. I could not find any document stating a clear decision on this point, but it is not difficult to guess the reason. In 1957-1958, the building of Brasilia, the new capital, was at its most feverish. Not only did this construction represent a huge effort in itself; it also engrossed the entire country, its government, and its population as a major concern.

Brazil was going through what could be called a Buddhist phase--it was looking at its own navel. The new capital was the symbol of national integration, the interiorization of the country, the effective economic linkage of the western and northern regions with the industrialized and more densely populated center-south. Anyone who at that moment suggested an expedition to Antarctica would at least have been fined for driving in the wrong direction.

When it was established that antarctic endeavors during the IGY would be the admission ticket to the negotiations for the Treaty of Washington, Brazil was left out. Accordingly, Brazil protested that decision, since it was evident that the country had antarctic interests.

Undeniably, being out of the negotiations and out of the treaty had some advantages. Nevertheless, the advantages did not compensate for the disadvantages. Confronted with the new reality, Brazil had to consider its position in relation to antarctic questions and take a decision on the course to follow.

As stated, there was no doubt about the country's interest in Antarctica. No scientific expertise is needed to understand that the climatic phenomena that powerfully interfere with the economy of the center-south regions of Brazil have their origin in Antarctica. It is also easy to see the importance of antarctic waters in ocean processes along Brazilian coasts. Suffice it to say that Brazil is much closer than most of the ATCPs to Antarctica. Naturally, it had to develop and decide on an antarctic policy.

A first possibility was a national solution. Not having taken part in the negotiations, and not being a party to the treaty, Brazil could choose to claim an area of the continent regardless of possible reactions from the parties. Indeed, Brazil has a sufficient basis for claiming territorial sovereignty in Antarctica, but I will spare my audience a description of that basis. For a long time, a territorial approach to the antarctic question had been advocated in certain circles. Private personalities, university professors, members of parliament, and high officials made statements and wrote articles or books defending that position. A decision was eventually taken not to claim any part of Antarctica; fortunately, I would say. Because, in my view, territorial claims are an old-fashioned, if not primitive, form of policy--primitive both in the historical sense, for the most ancient conflicts in history derived from territorial disputes, and in a psychological sense, for one of the earliest instincts of a child is to take possession of something and refuse to share it.

As a matter of fact, it was recognized--and here I am not strictly obeying chronological order--that involvement in activities in areas not traditionally subject to national jurisdiction would not require the assertion of territorial sovereignty. In other words, to conduct research and eventually to explore and exploit the deep seabed, outer space, or Antarctica, it is not necessary to exercise sovereignty over parts of these areas. On the contrary, in the case of Antarctica, a territorial claim would lead the country to concentrate on the areas claimed. It seemed to be much more advisable to maintain our freedom of action and to perform scientific research in any region of interest to us. A territorial claim, though possible, was not advisable, and it would restrict rather than enlarge the opportunities offered by Antarctica.

In the following years, Brazil had the advantages of observing the performance of the treaty before taking a decision. By the end of the 1960s, Brazil's policy of national integration was bearing fruit and there was rapid development in the natural sciences and in earth studies and atmospheric and marine fields. Scientific capacity, though limited, was strengthened. This fact would be important for Brazilian participation in antarctic activities.

The decision taken in 1975 to adhere to the treaty was based on the following factors, inter alia:

(1) Brazil has a natural and historical interest in Antarctica;
(2) The Treaty was the sole instrument internationally valid and accepted for the whole antarctic region;
(3) Brazil takes part in all international efforts to regulate activities in areas not subject to national jurisdiction; and
(4) Brazil should, consequently, participate in the discussion concerning Antarctica.

A logical consequence of becoming a party to the treaty, because of the relevance of Antarctica for Brazil, was to develop scientific research there and, therefore, to apply for consultative status.

The launching of the Brazilian Antarctic Program suffered from some delay, mostly because of internal controversy concerning its management. In 1982, the government decided to establish a National Commission for Antarctic Affairs responsible for the policy and entrusted a successful Interministerial Commission for the Resources of the Sea with the task of planning and carrying out the program. I shall spare you the description of the program, but it is important to point to its favorable reception by the Brazilian scientific community, proving that there was a natural wish to extend Brazil's activities to Antarctica. Just after the necessary mechanisms were set up, some 60 proposals were submitted. It is also important to stress the enthusiasm shown by public opinion and by all interested circles in support of the program.

At the same time, it cannot be denied that the program is extremely modest. Its budget for 1984 was around U.S. $4 million. Yet the results have been quite encouraging. A national symposium in October 1984 heard some 70 papers, half of them worth publishing in international scientific publications.

It is necessary to refer to the invaluable help of many countries active in Antarctica, beginning with Argentina and Chile.

It is obvious that in joining the consultative group Brazil was not seeking to belong to an exclusive group composed largely of industrialized countries. That group had been recently enlarged by the admission of two new members, although these were also European countries--Poland and the Federal Republic of Germany. In 1983, two developing countries gained consultative status--Brazil and India.

Brazil did not join the consultative group because of its composition. It did so because this is the forum for cooperation among the countries active in the region. I should also mention Brazil's admission to the Scientific Committee on Antarctic Research, as a necessary step toward full integration into the system. Ratification of the Convention on the Conservation of Antarctic Marine Living Resources is under consideration by the National Congress of Brazil, and it is likely that the process for adherence to it will be completed in 1985.

Let me now turn to two questions that are the subject of my intervention. The first one is the role played by the ATCPs vis a vis the international community. The second one is Brazil's position within the consultative group. Both are presented in testimony to our experience rather than as a thesis.

The brilliant paper presented by Ambassador Zain (Chapter 21) is a model for the logical arrangement of ideas. I do not intend to contradict it, since my demarche, as I stated, is not an abstract one. I would say, nevertheless, that it leads to logical conclusions from assumptions that are not accurate. It says that the ATCPs assert <u>rights to decision-making based on their expertise and experience in Antarctica</u>. I would submit that the ATCPs do not speak of rights. Rights, in this sense, suppose a permission to perform certain activities under limits set beforehand.

Until 1961, the ATCPs worked in Antarctica without any legal framework. Nothing could be unlawful because there was no law. After that, they applied to themselves a lawful instrument, thereby restricting their activities. Their rights, if the word can be used, relate to the treaty and to the United Nations Charter, to which the treaty is subordinated. They do not assert rights vis a vis other states or other instruments, except, of course, under the general principles of international law.

The activities of the ATCPs are not based on expertise or experience. Brazil could hardly claim that. The right words are not expertise or experience but involvement and responsibility.

Some countries are now contesting this absence of rights, this self-imposed legal framework. They want to establish a legislation that will then create rights. Why would this be necessary? Their argument is that the present system has not received the sanction of humankind. This leads to the question of whether the ATCPs represent the rights of humankind. Here again the question is not well formulated.

Since the treaty is not the basis of any rights for its parties, it equally does not imply the denial of rights to nonparties. The idea seems to be to establish another system to which the ATCPs would be accountable. Would it not be simpler to join the existing system and be accountable collectively as is the present case?

The term "antarctic club" has been frequently used. Let us take a club as an example. Suppose that some people are not members of a club, which is not closed to them. They say: "Let us eliminate the club's regulations and submit its members to an outside body to which we would belong." In fact, it would be more to the point if the outsiders joined the club and exercised their rights within it. In other words, they propose a meta-club that would replace the original club, so that they could do what they would do if they simply joined the club. Coming back to the ATS, they propose a metainternationalization instead of adhering to the international system.

Some voices have been raised in defense of the internationalization of Antarctica. That is exactly what the treaty ensures, as opposed to the nationalization advocated by the claimants. This new position could be called ultrainternationalization. However, the question here is not one of mere terminology.

It is well known that Brazil has been a champion of the concept of the common heritage of humankind applied to the seabed beyond national jurisdiction. The application of that concept to Antarctica could seem logical if not obligatory. The question was therefore considered in depth, since consistency should be a necessary foundation for any foreign policy.

First, you may have noticed that I used the term concept to identify the nature of the common heritage of humankind. It is not a principle, a slogan, an ideal pointing to an objective to be attained. Common heritage of humankind is a concept newly incorporated into public international law with a clear meaning, a distinct substance. Of course, this is not the place to make a dissertation on this concept. Suffice it to say that it is not a concept to be applied automatically to any area not traditionally subject to national sovereignty. For instance, it should not be applied to Antarctica. The reason for that has nothing to do with the existence there of territorial claims.

The interest of the international community in the deep seabed was drawn by specific resources, the exploitation of which required financial and technological means and legislation suitable to every state regardless of its financial and technological capability. It should be noted that the concept of common heritage of humankind refers not only to the international seabed area but also to its resources.

Antarctica is a completely different case. The antarctic system does not stem from specific interests in resources. The problem there has never been how to devise a system that would ensure an equitable sharing of riches. The treaty and the subsequent recommendations of the ATCP meetings intend to regulate the activities of those countries that perform scientific research in the area. The treaty and the recommendations impose a burden on those countries. The word privilege is misplaced as regards the ATS. The responsibilities bestowed on the ATCPs derive not from their power or prestige but from their real involvement in antarctic activities, their commitment to study and preserve this magnificent area.

In recent years, the problem of economic resources has indeed been raised. However, that a new question arises does not invalidate the treaty. On the contrary, its solution must be subject to the treaty and ensuing recommendations. The ATCPs do not approach the question of resources from the point of view of sharing them among themselves. Their difficult task is to conform any regulation to the principles set by the treaty. In this sense, emphasis must be placed on the interests of all nations, especially the developing ones.

This leads me to the second question, that is to say, how Brazil as a new ATCP views the ATS. I harbor a healthy suspicion concerning the intentions of industrialized countries. On the other hand, I know the interests of developing countries. I will not be challenged on this second point. I do not know the interests of Costa Rica, per se, but I do know the interests of Costa Rica as a developing country. Certainly, Brazil does not represent the interests of developing countries in the ATS. But, necessarily acting as a developing country, Brazil will contribute to the defense of the interests of developing countries.

It is necessary to explain how Brazil can act as a developing country in the consultative group. Years ago I was talking with a member of one of those communities that proliferated in the late 1960s and early 1970s. He told me that when a new member was admitted, the initial idea was that this newcomer would have to adapt himself to the community's system. In fact, the opposite took place. The community had to change its ways in order to adapt to the new member. This apparent paradox is not difficult to understand. The new member brought into the community personal abilities and talents. It was in the interest of the community to take advantage of those abilities and talents instead of forcing the newcomer to perform duties for which the newcomer was not prepared. If the system was to rotate service in the kitchen, and the newcomer was a bad cook and a good gardener, it would be in the interest of the community to change the system.

Brazil can bring to the consultative group some abilities and many shortcomings. This will have to be taken into account in the interest of the group. New forms of cooperation must be established. This is not easy, for the first impression is that the system is a zero-sum game and that Brazil's shortcomings must be compensated for by the other parties in order to come out even. This is just not so. The contributions that

Brazilian research projects can bring to the system will enrich it. Possibly divergent views will not impede progress but, on the contrary, will bring about wider-ranging decisions.

Only many years after the treaty entered into force was the apostolic number of the ATCPs increased; first, by two European countries, and, one year ago, by two non-European countries. The challenge that this represents will bring new life to the system that has to undergo some modifications.

It became clear, for instance, at the last meeting on minerals, that the accommodation is not just between claimants and nonclaimants. It is also between industrialized and developing countries. Certainly, this presents new problems for the negotiations, but the results will be better.

This is only one example. There will be need for more dissemination of information within and outside the system and among its components. Perhaps this will require some institutional arrangements of an administrative nature.

Coming into the consultative group does not mean that Brazil will change or lose its identity. The group will have to cope with that. Some three years ago there was a humorous chronicle published in a Brazilian newspaper, projecting the first Brazilian expedition to Antarctica. This was before the real one took place:

> The Brazilians establish their camp just beside a Norwegian camp. On the very first day, they discover they have forgotten to bring a number of things. So, every now and then, a Brazilian goes to knock on the Norwegians' door to ask for a cup of sugar, one or two pairs of socks, some toothpaste, and so on. During the night, the Norwegians can't sleep, for the Brazilians are playing drums and tambourines, singing and dancing the samba. When Sunday rolls around, the Brazilians decide to do something to thank the Norwegians. They had the idea to invite the Norwegians to a football match. So they go to the Norwegian camp and knock at the door. No answer. They open the door. No one. The Norwegians had fled, disappeared. The Brazilians commented among themselves: 'Strange fellows, these Norwegians.'

Times are changing in the antarctic system.

24.

The Antarctic Treaty System from the Perspective of a New Member
S. Z. Qasim and H. P. Rajan

INTRODUCTION

India acceded to the Antarctic Treaty on August 19, 1983, and was accorded the status of a consultative member on September 12, 1983, that is, within one month after the accession to the treaty. We consider this a very significant event, because in the past, states that acceded to the treaty had to wait for several years before being admitted to consultative status. The grant of consultative status to India, therefore, clearly shows an international recognition of India's scientific accomplishments in Antarctica.

India's research program in Antarctica began with the launching of a first expedition to the continent in December 1981, although India's interest in Antarctica dates back many years. India is separated from Antarctica only by a few islands and a continuous stretch of the Indian Ocean. Unlike the Pacific and the Atlantic oceans, which communicate to both the North Pole (Arctic) and the South Pole (Antarctic), the Indian Ocean is bounded to the north by a landmass; it communicates only to the Antarctic Ocean and the South Pole, from which it derives energy and fertility.

Recent investigations have revealed that the weather over the Indian Ocean is greatly influenced by the Antarctic environment. It is also interesting to note that the ice sheet of Antarctica is in several ways akin to the Himalayan ice column. The Himalayan ice samples have revealed that they are inextricably mixed with the effects of spasmodic uplift. The Antarctic, on the other hand, represents a stable situation affected only by global climate. For scientific purposes, therefore, it is possible to use Antarctica as the southernmost reference point, with the Himalayas as the corresponding northernmost point. Studies on the mass balance of annual glaciers provide data on the current short-term climatic situation in the Himalayas; linking these data with annual changes on the largest freeze/melt operation on the Earth, that is, the sea regions of Antarctica, would provide scientific information on global weather phenomena.

Scientific research in Antarctica is important because

(1) Antarctica is an important location for observing the interaction of the Earth's magnetic field and charged particles from the Sun. It is perhaps the only place in the Southern Hemisphere from where observation of simultaneous activity in the ionosphere and the Earth's magnetic field can be made. It provides relative freedom from man-made sources of electrical interference (noise). Hence, it forms an ideal environment for conducting studies on radiowave propagation and radio-noise levels, both in the ionosphere and in the lower atmosphere.

(2) The North and South poles maintain the heat budget of the world in balance. The heat transferred through the atmosphere and the oceans to the poles is dissipated in space in the form of long-wave radiation. The cold air from Antarctica, on meeting the warm air in the atmosphere of the lower latitudes, changes into moisture-bearing clouds. Antarctica thus regulates the global climate and in particular that of the Southern Hemisphere.

(3) The waters of the Indian, Atlantic, and Pacific oceans merge around Antarctica, forming a distinct body of water that girdles the continent and remains uninterrupted by any landmass. The mixing process between the cold and warm waters demarcates the area of the antarctic convergence that has unique physical, chemical, and biological characteristics.

(4) Antarctica provides a distinct, unpolluted and stable environment for carrying out scientific observations.

(5) Many important oceanographic features of the Indian Ocean are governed by the Antarctic Ocean. Hence, to understand the processes occurring in the Indian Ocean, knowledge of that part of the Antarctic Ocean that joins the Indian Ocean becomes very necessary.

(6) In the Mesozoic era, the supercontinent of Gondwanaland had a common landmass of five continents, namely, Africa, Antarctica, Australia, India, and South America. Later, the continents drifted apart and the oceans came between.

The study of Antarctica, therefore, is of great significance to India.

BACKGROUND OF THE ANTARCTIC TREATY

Polar expeditions date back to time immemorial. The Greeks knew of the existence of Antarctica but had no proof. They had named the brightest star circling the sky "Arktos" (meaning the bear), and they called the pole around which it appeared to revolve the Arctic Pole. To balance the natural order, they felt that there must be a similar opposite pole and named it "Anti-Arctic," which in later years has come to be known as Antarctica.

International scientific cooperation, however, began only during the first International Polar Year in 1882-1883. During this period, the latitude and the longitude of the antarctic region were determined, and studies began to be carried out in the fields of magnetics, meteorology, and glaciology. A second international Polar Year was convened 50 years later, in 1932-1933. With the advances made in the field of geophysics and in various techniques, particularly those relating to investigation in the ionosphere, it was felt that the third International Polar Year should be scheduled after a lapse of only 25 years. Accordingly, this was called for the years 1957-1958.

To undertake the preparatory work for the third International Polar Year, the proposal was formally placed before the Mixed Commission on Ionosphere--a body formed by the International Council of Scientific Unions (ICSU). A resolution was moved in the bureau of ICSU, stipulating that the third International Polar Year be scheduled for the year 1957-1958, and that an International Polar Year Commission be established in 1951 to do the necessary planning for the program. The resolution was approved and the special committee was duly convened in 1952. The World Meteorological Organization (WMO) was also invited to participate. The WMO suggested that an International Geophysical Year (IGY) would be more useful, stressing the need to extend synoptic observations of geophysical phenomena over the whole surface of the Earth. Accordingly, a Special Committee for the International Geophysical Year (CSAGI) was constituted to undertake the preparatory work for the IGY. The CSAGI held four meetings and four conferences prior to the start of the IGY. The work of the CSAGI also led to the establishment of an ad hoc committee in 1957 and thereafter a standing Special Committee on Antarctic Research. The special committee was renamed the Scientific Committee on Antarctic Research (SCAR) in 1961.

Considering the political situation prevailing at that time, the IGY needs to be commended for its contributions. Antarctica for many decades prior to the IGY had been an object of continued interest among many nations. Of these, seven nations had made territorial claims to the continent, some of which overlapped. Many nations, including the United States, the USSR, Australia, New Zealand, and the United Kingdom, had attempted to solve the controversies relating to the antarctic territorial claims, but these attempts did not succeed. The claims controversy by and large did not significantly affect the relationships among the scientists carrying out work in Antarctica. Scientists from various countries, including those having overlapping claims in Antarctica, demonstrated quite convincingly that they could carry out their work together peacefully, without being affected by political differences. This was due partly to common scientific objectives in the most inhospitable continent and partly to the fact that the scientists were not personally responsible for respective national interests in Antarctica, since such interests were not at stake during the IGY.

However, it became clear to all nations involved in antarctic investigations that much could be gained if scientific work that began during the IGY could be continued thereafter. On May 2, 1958, the United States proposed to other participants in the IGY that they should all join "in a Treaty designed to preserve the continent as an international laboratory for scientific research and ensure that it be used only for peaceful purposes."[1]

Thus, in the invitation to the Washington conference, the President of the United States had pointed out that the "present situation in Antarctica is characterized by diverse legal, political, and administrative concepts which render friendly cooperation difficult in the absence of an understanding among the countries involved," and he

proposed the negotiation of a treaty that in addition to
securing freedom of scientific investigation for "citizens, organizations, and governments of all countries"
should also secure the use of Antarctica for peaceful
purposes only and for any other peaceful purposes not
inconsistent with the charter of the United Nations.[2]

The invitation was issued after careful and confidential consultations.[3] All nations accepted the idea in
principle. However, during the preliminary talks in
Washington, the USSR opposed the claims on antarctic
territory by some nations and Chilean and Argentine
delegations were reluctant to agree to international
control.[4] Nevertheless, the advisability of maintaining Antarctica free for scientific investigations,
with particular reference to the studies that started
during the IGY, prevailed, and a formal treaty conference
was opened on October 15, 1959. Finally, the treaty was
signed on December 1, 1959. Japan was the first nation
to ratify the treaty on August 4, 1960. The treaty
entered into force on June 23, 1961.

THE ANTARCTIC TREATY SYSTEM

The Antarctic Treaty is a remarkable instrument, drafted
through a unique negotiating process at a time when the
cold war was at its peak. It has been particularly successful in two aspects, namely, keeping Antarctica free
from military activities, including nuclear weapons, and
persuading the seven states with territorial claims to
put them aside for at least 30 years. It has worked
admirably well administratively and more so scientifically, because scientific investigations have been
continuously in progress irrespective of the political
differences and territorial controversies.

Nevertheless, in the eyes of outsiders, the treaty was
regarded as an exclusive club. The most important provision concerning the operational part of the treaty is
contained in Article IX, which has created a system of
regular, periodic meetings of representatives "for the
purpose of exchanging information, consulting together on
matters of common interest pertaining to Antarctica, and
formulating and considering and recommending to their
governments measures in furtherance of the principles and
objectives of the treaty...." The main objectives of the
treaty as it stands today include measures regarding

(1) The use of Antarctica for peaceful purposes only,
(2) Facilitation of scientific research in Antarctica,
(3) Facilitation of international scientific cooperation in Antarctica,
(4) Facilitation of the exercise of the rights of inspection provided for in Article VII of the treaty,
(5) Questions relating to the exercise of jurisdiction in Antarctica, and
(6) Preservation and conservation of the antarctic environment and of living resources in Antarctica.

The procedure established by these provisions is commonly referred to as the consultative procedure, and the parties participating in the meetings are known as the consultative parties. All the original members of the treaty were automatically regarded as consultative parties, whereas the countries that acceded to the treaty later could become consultative parties after they demonstrated a tangible interest in Antarctica by conducting substantial scientific research activity there, such as the establishment of a permanent scientific station or the dispatch of a scientific expedition. On such a demonstration, the country concerned ipso facto is entitled to consultative status. However, in view of the circumstances under which the Antarctic Treaty itself was negotiated, it has become a procedural formality to bring such a demonstration by a new member to the attention of all other consultative parties to the Antarctic Treaty at an Antarctic Treaty consultative party (ATCP) meeting. It is, however, somewhat interesting to note that while the original members of the treaty retain their consultative status for all time, a new member remains a consultative party only during such time as it continues its scientific interest in Antarctica.

However, despite these limitations, it is erroneous to maintain that the Antarctic Treaty is a closed club and that the decisions are taken only by those who are represented in the ATCP meetings. On the contrary, the Antarctic Treaty is an open treaty. It is open to accession by any state that is a member of the United Nations. It is also open for accession by any state that may be invited to accede to the treaty; but in the case of accession by invitation, the consent of all contracting parties with consultative status is required. Table 24-1 gives a list of countries that have acceded to the Antarctic Treaty together with those that have been given consultative status.

TABLE 24-1 Antarctic Treaty Members

Original Members (All Consultative Parties)

    Argentina
    Australia
    Belgium
    Chile
    France
    Japan
    New Zealand
    Norway
    Union of South Africa
    Union of Soviet Socialist Republics
    United Kingdom of Great Britain
      and Northern Ireland
    United States of America

Acceding Members and Years of Accession

Member	Year
Poland	1961[a]
Czechoslovakia	1962
Denmark	1965
The Netherlands	1967
Romania	1971
German Democratic Republic	1974
Brazil	1975[b]
Bulgaria	1978
Federal Republic of Germany	1979[c]
Uruguay	1980
Papua New Guinea	1981
Italy	1981
Peru	1981
Spain	1982
Peoples Republic of China	1983
India	1983[d]
Hungary	1984
Finland	1984
Sweden	1984
Cuba	1984

(a) Became consultative party in 1977
(b) Became consultative party in 1983
(c) Became consultative party in 1981
(d) Became consultative party in 1983

The Antarctic Treaty provides for meetings of the ATCP members at regular intervals, and the practice has been that the meetings are held once every two years. During the regular ATCP meetings, a large number of recommendations are adopted on matters concerning the various aspects of Antarctica. The number of such recommendations varies greatly from meeting to meeting (from 28 at the fourth to 3 at the eleventh). So far, 12 meetings have taken place and in all 138 recommendations have been adopted. Several of these deal with purely procedural and administrative matters, while many others serve to implement measures on specific matters. By such a procedure, a number of recommendations on important subjects, termed the "Agreed Measures," have been adopted. Some of these Agreed Measures, according to reviewers of the treaty, are tantamount to substantial legislation. The consultative parties have demonstrated a strong desire and a will to avoid conflict and promote cooperation by acting jointly to fulfill the objectives of the treaty.[5]

It may also be noted that the Antarctic Treaty System is not merely confined to the working principles established by the Antarctic Treaty parties and their consultative mechanism. A closer examination of the treaty reveals that it has intrinsic links with other organizations of the scientific community and in particular with the Scientific Committee on Antarctic Research (SCAR).

Article II of the treaty stipulates that the freedom of scientific investigations in Antarctica and the cooperation envisaged toward that end, as enunciated during the IGY, shall continue. A question may be asked: What exactly was the cooperation extended during the IGY? From the start, the IGY unanimously adopted a notion that the overall aim of the antarctic program should be entirely scientific. However, in Antarctica the situation is such that it is difficult to separate science from politics. In 1957, the International Council of Scientific Unions (ICSU) took steps to examine the merits of scientific investigations in Antarctica, and in 1958 SCAR was established. In the same year, it was also agreed to extend the IGY program through the year 1959. Thus, the work carried out by SCAR becomes an integral part of the cooperation, and both the IGY and SCAR have become indispensible elements of the Antarctic Treaty System. It must, however, be made clear that SCAR, by itself, does not conduct scientific programs. All research in Antarctica is carried out and financed by national organizations. The activities of SCAR are governed by its constitution, framed in 1958, which with time has undergone minor changes. According to the SCAR constitution, it is a scientific committee of ICSU charged with the initiation, promotion, and coordination of scientific activity in the Antarctic, with a view to framing and reviewing scientific programs of circumpolar scope and significance.

In establishing the programs, SCAR respects the autonomy of other existing international bodies. Since no infrastructure existed in Antarctica before the IGY, the scientific organizations coordinated by SCAR had to organize logistic bases for different types of operations. To carry out scientific research in Antarctica, many SCAR nations have established special polar research institutes responsible for field operations, the running of permanent bases, and planning and coordinating of scientific programs. India and Brazil were admitted as full members of SCAR at the meeting held in October 1984 in Bremerhaven, Federal Republic of Germany.

It must be mentioned that the geographical area of interest to SCAR extends to the antarctic convergence, where the cold antarctic waters meet the warmer currents from the north; it thus includes some of the islands lying north of the convergence. Thus, the area of interest to SCAR is wider than that covered by the Antarctic Treaty, which covers the area south of 60°S latitude only.

SCAR has been active in planning and coordinating antarctic research since the end of the IGY. At the twelfth ATCP meeting, held in Canberra in 1983, a resolution was adopted that recognized that SCAR has a unique assemblage of knowledge and expertise in antarctic science and expressed appreciation for the advice provided by SCAR to the ATCPs in response to various requests.[6] Thus, the ATCP meetings and their recommendations, along with the work of SCAR and the contributions of other scientific bodies, such as the Scientific Committee on Oceanic Research and the International Union for the Conservation of Nature and Natural Resources, which are concerned with antarctic matters, form a close-knit system governing the Antarctic.

What exactly does this system seek to accomplish? The first 10 to 15 years of the existence of SCAR and the treaty have been sometimes referred to as the "honeymoon" time in Antarctica.[7] Science and conservation were the only two activities that mattered. The treaty made no substantive reference to resources. It merely provided a forum for discussion on the future elaboration of measures to preserve and conserve the living resources of Antarctica. The question of nonliving resources was not mentioned at all. No activity was in progress or even contemplated at that time for the extraction of mineral resources. Thus, although it was deemed necessary to make a special reference to living resources as a subject for consultation and conservation, it was thought that there was no practical or urgent need for a similar provision on nonliving resources. It is also significant to note that, with respect to living resources, measures have been adopted that go far beyond the question of conservation alone. In other words, the conservation clause in the Antarctic Treaty has been effectively used even for the regulation of activities that may very often be regarded as commercial in nature. This is particularly true for the Convention for the Conservation of Antarctic Seals adopted in 1972. This process of regulating activities under the conservation clause of the Antarctic Treaty has been referred to as a policy of "indirection" by some writers.[8]

Convention on Seals

The consultative parties have accorded a high priority to the preservation and conservation of antarctic living resources. The very first ATCP meeting adopted a recommendation that

- Recognizes the urgent need for measures to protect living resources from uncontrolled destruction by man;
- Calls for the establishment of such measures in a suitable form;
- Proposes, as an interim measure, the adoption of general rules of conduct based on guidelines developed during the first meeting of SCAR; and
- Encourages cooperation in promoting scientific studies of Antarctica.

Similar concerns were expressed at the second consultative meeting, and the draft of the Agreed Measures for the Conservation of Antarctic Fauna and Flora was improved and later adopted at the third consultative meeting. After the fourth consultative meeting, the consultative parties turned their attention to the conservation of antarctic seals. Recommendations of the fourth meeting designated the fur seals and the Ross seals as specially protected species. Thus, these animals, which were in great danger of extinction in the nineteenth and early twentieth centuries, started to multiply again because of suspension of economic exploitation. In an attempt to fill in the gaps left by the Agreed Measures that were adopted, the governments of the consultative parties were called to regulate seal hunting at the national level. During the fifth consultative meeting, the parties that reviewed the SCAR report prepared for that purpose concluded a draft convention regulating pelagic sealing. There were two reasons for the choice of a convention rather than a set of "agreed measures" such as those adopted in 1964. First, the Agreed Measures on the antarctic fauna and flora are in the form of a recommendation, which, since it had not yet been approved at that time by all parties represented at the consultative meetings had not become binding. Second, the scope of a recommendation was considered limited to the consultative parties, and it was desirable to try to increase the effectiveness of Antarctic Treaty recommendations by inviting other states to observe them. It was therefore decided to convene a special conference to negotiate a convention on seals in the Antarctic. The 12 consultative parties participated in the conference. On June 1, 1972, the Convention on the Conservation of Antarctic Seals (Seals Convention) was finally adopted in London.

The Seals Convention is open for accession by any state that may be invited to accede to it with the consent of the contracting parties. The area covered by the convention lies south of 60°S latitude. The convention maintains a close link with SCAR. To avoid possible conflicts relating to sovereignty claims, Article IV of the Antarctic Treaty of 1959 is specifically mentioned in Article I of the Seals Convention. The convention provides that the contracting parties may from time to time adopt other measures with respect to conservation, scientific study, and rational and humane use of seal resources. It also includes provisions specifying measures to be adopted for protection of the antarctic seals, such as permissible catch, open and closed seasons, seals reserves, exchange of information, and consultation among the contracting parties. Article I specifies the species to which the convention will be applicable and indicates the specially protected species that cannot be killed or captured. The convention entered into force in 1978.

Convention on the Conservation of Antarctic Marine Living Resources

By 1975, the ATCPs became concerned with the inadequacy of information concerning the stocks of marine living resources in Antarctica. At the eighth consultative meeting, held at Oslo in June 1975, a recommendation was adopted to encourage studies that could lead to the development of effective measures for the conservation of antarctic marine living resources in the treaty area. It was realized during this meeting that the backbone of the antarctic ecosystem is the krill, on which most fish, birds, seals, and whales are dependent. It was therefore felt that extensive exploitation of krill might have a detrimental effect on the entire ecosystem. Increasingly, nations, particularly the USSR and Japan, had begun to exploit krill. As a result of this, much concern was expressed at the ninth consultative meeting, held in London in 1977, about such exploitation, and the parties recommended to their governments that a definite conservation regime should be established before the end of 1978. The recommendation outlined the basic elements of a convention and contained general interim guidelines. It also provided a mandate to prepare a draft for the regime of living resources to be discussed at a special consultative meeting.[9] The special consultative meetings were held at different stages and prepared the final draft in 1978. Informal negotiations were held in 1979 that related, inter alia, to French sovereignty over Kerguelen and Crozet islands and to the participation of the European Economic Community. The draft formed the basis for the negotiations, and a convention was adopted on May 20, 1980, in Canberra. It was ratified by eight of the original signatories and entered into force on April 7, 1982. Thus, the entire process from conception to adoption of the Convention on the Conservation of Antarctic Marine Living Resources (CCAMLR) was completed in the short span of about five years.

The convention not only applies to antarctic marine living resources in the area south of 60°S latitude, which is the Antarctic Treaty area, but also extends north to the area between that latitude and a defined boundary line around the continent that comes close to the natural antarctic convergence. The purpose of the enlargement of the area of application is to extend protection to the antarctic marine ecosystem as a whole. Thus, like the Seals Convention, this convention maintains a close link with the Antarctic Treaty. However, this convention is more open that the Antarctic Treaty or the earlier Seals Convention because it stipulates that any state interested in research on or harvesting of the marine living resources to which the convention applies may accede to it. Using the principle of conservation, including rational use, the convention governs the exploitation of the marine living resources of the area, mostly krill.

The convention establishes a permanent commission, a scientific committee and an office of the executive secretary. The commission is an intergovernmental organization with its own legal framework. Its functions include, inter alia, the facilitation of research; formulation, adoption and revision of conservation measures; and implementation of a system of observation and inspection. Decisions in the commission are taken by consensus. The commission can adopt conservation measures, which will become binding on all members 180 days following notification, although the option is given to members to indicate that they cannot accept any conservation measure within 90 days of its adoption. Thus, by providing that the measures are adopted by consensus, and by allowing members to opt out of any measures within 90 days, a kind of double-veto system operates.

The scientific committee is composed mainly of the representatives of the CCAMLR parties. The committee may invite other scientists or experts, for example, from the United Nations Food and Agriculture Organization (FAO), to participate as observers. Its functions include making recommendations on the establishment of conservation measures, the assessment of the state of living resources and the effects of proposed measures, submission of studies to the commission, and formulation of proposals for research. Some writers believe that the convention is in most aspects similar to an international fisheries commission adapted especially for the antarctic waters,[10] while others are of the opinion that the convention can be regarded as a model instrument for the protection of the ecosystem, which differs greatly in both substance and form from fisheries agreements.[11] It will soon be put to the test, by which its effectiveness will be able to be assessed fully. India acceded to this convention in June, 1985.

Mineral Resources

CCAMLR is viewed at times as a model or a standard for a future regime on mineral resources in Antarctica. The question of mineral resources, however, is far more complex, as these resources are finite and are found and extracted in fixed locations. The development of mineral resources also depends on the availability of technical personnel and installations that require security in situ, as well as security in investment. These issues obviously raise questions relating to title and property. Thus, by its nature and character, minerals exploitation can reopen and bring to the fore the conflict between the claimant and nonclaimant groups.

The subject of mineral resources was first raised informally at the sixth consultative meeting, held in 1970, and it was put on the agenda for the first time at the seventh meeting, held in 1972. At that meeting it was recommended that the subject of antarctic mineral resources and minerals exploitation be carefully studied and included in the agenda for the eighth consultative meeting. Thereafter, the subject has been regularly discussed at every session, at experts' meetings, in informal working groups, and at extraordinary meetings. Sessions of the special consultative meeting on minerals have been held in Wellington, Bonn, Washington, Tokyo, and Rio de Janeiro. The next one meets in Paris in September/October 1985. India will be actively participating in all future meetings relating to mineral resources. At the twelfth consultative party meeting, in which India participated, it was noted with appreciation that the negotiations on mineral resources were taking place in different forums, and the ATCPs welcomed the progress achieved in negotiations that had taken place so far and expressed the hope for an early completion of negotiations.[12]

A number of other organizations are also involved in the question of mineral resources of Antarctica. SCAR's international working groups on geology and geophysics as well as a group of specialists have held a number of meetings and have produced numerous documents that not only evaluate the available information on potential resources but also assess the impacts that mining and exploitation activities might have on the antarctic environment. One group has recommended measures for environmental protection. It may, however, be mentioned that the present state of technology is not adequate to permit minerals exploitation activities to be undertaken in Antarctica for economic gain, and this is likely to remain so for quite some time in the future. There are, however, some indications that in the next decade or so it may become possible to exploit hydrocarbon resources commercially. Perhaps it might become desirable to work out guidelines for a regime of hydrocarbons and thereafter modify it for other minerals.

Protection of the Antarctic Environment: Specially Protected Areas and Sites of Special Scientific Interest

Protection of the antarctic environment has been the prime concern of the ATCPs, and large numbers of areas in Antarctica, which have colonies of penguins, other birds, seals, etc., have been declared Specially Protected Areas (SPAs). No activities of any type which are detrimental to the environment, such as putting up either temporary or permanent stations, are permitted in SPAs. Similarly, a large number of areas in Antarctica, including marine regions that have rich fauna and flora, have been proposed as Sites of Special Scientific Interest (SSSIs). A proposal for such a designation is first sent to SCAR, and on approval by SCAR it is referred to the ATCP meeting, which formally designates the area as a SPA or a SSSI.

SPAs and SSSIs are measures designed to protect designated areas of Antarctica from harmful interference. SPAs are areas of biological importance and entry into these areas is allowed only by permits issued for compelling reasons. SSSIs protect areas of ongoing scientific interest and may also include nonbiological regions. Their main purpose is to protect a particular program of research as well as the area itself as identified in the management plan. Tables 24-2 and 24-3 contain lists of SPAs and SSSIs, respectively.

Scientific Research Program under the Treaty System

The scientific research program in Antarctica falls under five main areas, which can be summarized briefly as follows:

(1) <u>Ocean Studies</u>: Ocean studies include antarctic oceanography and studies in antarctic marine geology and marine biology. Biologists have already gathered

Table 24-2 Specially Protected Areas

Site Number and Name of Area	Coordinates	ATCP Recommendation	Reason for Designation as SPA
1. Taylor Rookery Mac. Robertson Land	Lat. 67°26'S Long. 62° 50'E	IV-1	For protection of emperor penguins
2. Rookery Islands, Holems Bay (Mawson Coast)	Lat. 67°37'S Long. 62°33'E	IV-2	For protection of colonies of six species of birds, two of which, namely, giant petrel and cape pigeon, are not found anywhere else
3. Ardery Island and Odbert Island, Budd Coast	Lat. 66°22'S Long. 110°22'E and Lat. 66°22'S Long. 110°33'E	IV-3	For protection of several species, particularly Antarctic petrel and Antarctic fulmer
4. Sabrina Islet, Balleny Islands	Lat. 66°54'S Long.163°20'E	IV-4	For protection of flora and fauna
5. Beaufort Island, Ross Sea	Lat. 76°58'S Long.167°63'E	IV-5	For protection of substantial and varied avifauna
6. Cape Crozier, Ross Island	Lat. 77°32'S Long.169°19'E	IV-6 but terminated by VIII-12	For protection of microfauna and microflora. Now SSSI by Recommendation VIII-4.

7. Cape Hallett, Victoria Land	Lat. 76°18'S Long.170°19'E	IV-7	For protection of diverse vegetation and terrestrial fauna
8. Dion Islands, Marguerite Bay, Antarctic Peninsula	Lat. 67°52'S Long. 68°43'W	IV-8	For protection of emperor penguins
9. Green Island, Berthelot Islands, Antarctic Peninsula	Lat. 65°19'S Long. 64°10'W	IV-9	For protection of rich vegetation
10. Byers Peninsula, Livingston Island, Southern Shetland Islands	Lat. 62°38'S Long. 61°05'W	IV-10 but terminated by VIII-12	Diverse plant and animal life. Now SSSI by Recommendation VIII-4.
11. Cape Shirreff, Livingston Island, South Shetland Islands	Lat. 62°28'S Long. 60°48'W	IV-11	For protection of plant and animal life including invertebrates and elephant seals
12. Fildes Peninsula	Lat. 62°12'S Long. 58°58'W	Created by IV-12 modified by V-5 and terminated by VIII-12	Biologically diverse region with numerous lakes; ice-free in summer. Now SSSI by Recommendation VIII-4.

Table 24-2 (Continued)

Site Number and Name of Area	Coordinates	ATCP Recommendation	Reason for Designation as SPA
13. Moe Islands, South Orkney Islands	Lat. 60°45'S Long. 45°41'W	IV-13	For protection of the maritime Antarctic ecosystem. Moe Island provides representative sample of the maritime Antarctic ecosystem.
14. Lynch Island, South Orkney Islands	Lat. 60°40'S Long. 45°38'W	IV-14	For protection of the dense areas of grass
15. Southern Powell Island and adjacent islands, South Orkney Islands	Lat. 60°45'S Long. 45°01'W	IV-15	For protection of substantial vegetation, birds, and mammal fauna and fur seals
16. Coppermine Peninsula, Robert Island	Lat. 62°23'S Long. 59°42'W	VI-10	For protection of rich vegetation, terrestrial fauna, and avifauna
17. Litchfield Island, Arthur Harbor, Palmer Archipelago	Lat. 66°16'S Long. 64°06'W	VIII-I	For protection of rich collection of marine and terrestrial life; native place for six species of birds

Table 24-3 Sites of Special Scientific Interest (SSSI)

Site Number and Name of site	Reasons for Designation as SSSI	ATCP Recommendation	Date of Expiry	Whether Extended
1. Cape Royds, Ross Island	This area supports the most southerly Adelie Penguin	VIII-4	June 30, 1981	Extended up to June 30, 1985 by Recommendation X-6. Further extended up to December 31, 1985 by Recommendation XII-5.
2. Arrival Heights, Hut Points Peninsula, Ross Island	This area is an electromagnetically and natural "quiet site"	VIII-4	June 30, 1981	Extended up to June 30, 1985 by Recommendation X-6. Further extended up to December 31, 1985 by Recommendation XII-5.
3. Barwick Valley, Victoria Land	Environmentally unique area and possesses extreme polar desert ecosystem	VIII-4	June 30, 1981	Extended up to June 30, 1985 by Recommendation X-6. Further extended up to December 31, 1985 by Recommendation XII-5.
4. Cape Crozier, Ross Island	Penguin colonies are found here	VIII-4	June 30, 1981	Extended up to June 30, 1985 by Recommendation X-6. Further extended up to December 31, 1985 by Recommendation XII-5.

5.	Fildes Peninsula, King George Island, South Shetland Islands	Biologically diverse region with numerous lakes; ice free in summer. Contains representative sequences of tertiary strata	VIII-4	June 30, 1981	Extended up to June 30, 1985 by Recommendation X-6. Further extended up to December 31, 1985 by Recommendation XII-5.
6.	Byers Peninsula, Livington Island, South Shetland Islands	Diverse plant and animal life. Fossils of this area provide evidence of link with other southern continents	VIII-4	June 30, 1981	Extended up to June 30, 1985 by Recommendation X-6. Further extended up to December 31, 1985 by Recommendation XII-5
7.	Haswell Island, Queen Mary Land	Exceptionally prolific and representative breeding locality for many species of birds: five species of petrel, one species of skua, and one species of penguin	VIII-4	June 30, 1981	Extended up to June 30, 1983 by Recommendation X-6. Further extended up to December 31, 1985 by Recommendation XII-5 (voluntary)
8.	Western Shore of Admiralty Bay, King George Island, Southern Shetland Islands	This area supports an exceptional assemblage of antarctic birds and mammals	X-5	March 31, 1985	Extended up to December 31, 1985 by Recommendation XII-5

much information on antarctic marine life. Research is now being conducted to study interactions among species, their life histories, abundance, distribution, and their special adaptation to the environment.

(2) <u>Glaciology</u>: Antarctic glaciers constitute about 90 percent of the Earth's glaciation. Glaciologists study the structure and dynamics of the antarctic ice sheet. Ice coring and isotopic analysis have provided information on climatic conditions that existed thousands of years ago.

(3) <u>Atmospheric Sciences:</u> Scientists in Antarctica are investigating physical processes occurring in different regions of the atmosphere, their interaction with the ocean and their influence on global climate. The remoteness of the continent from anthropogenic sources of air pollution makes it an ideal place to gather data on atmospheric-aerosol transport and precipitation and the variability of trace-gas constitutents and their effect on climate. Upper-atmospheric research in Antarctica is concerned with geomagnetism, cosmic rays, and auroral and ionospheric physics.

(4) <u>Earth Sciences</u>: The earth science group consists of scientists working in the fields of cartography, geodesy, geology, and solid earth geophysics. A number of other related disciplines such as seismics, magnetics, gravimetry, geochemistry, volcanic geology, glacial geology, and paleontology are also encompassed within the scope of their work.

(5) <u>Biological and Medical Sciences:</u> Terrestrial life in Antarctica is meager. Certain bacteria, algae, and lichens survive just below the surface of some rocks. The inland animal world in Antarctica is represented by several insect species. The largest antarctic terrestrial animal is a 5-mm-long wingless insect found in the Antarctic Peninsula. Biologists are studying these organisms and their responses to human intervention. From the biomedical point of view, the human being is also an object of research. Human physiology and behavioral reactions in extreme antarctic conditions and in isolation are being studied.

INDIA'S SCIENTIFIC EXPEDITIONS

In view of the importance of Antarctica to India, as noted earlier, it was decided in 1981 to undertake expeditions to this remote continent. After careful consideration and planning, a team of 21 members was selected for the first expedition, with clear-cut scientific objectives to be undertaken.

The expedition left Goa on December 6, 1981, on board M.V. Polar Circle, a chartered ship from M/S G.C. Reiber & Co., Bergen, Norway. The members of the team were drawn from seven different institutions in the country and included oceanographers, meteorologists, biologists, geologists, geophysicists, radio-communications experts, and naval personnel. The participants were acclimatized to the cold in a vigorous training program in the Himalayas and to the sea on board a ship. Before departure, the entire team was briefed by several top national scientists. After a successful cruise, they landed in Antarctica at 30 minutes past midnight January 9, 1982.

The main objectives of the first expedition were as follows:

(1) To initiate studies and build facilities and expertise in different oceanographic disciplines with reference to antarctic waters,
(2) To continue and strengthen a program of routine data collection and studies, and
(3) To identify programs of significance in scientific and economic terms and pursue these as thrust areas in order to establish the position of Indian science in Antarctica.

The expedition set up a base camp on the ice shelf and another, which was named Dakshin Gangotri, in the hilly terrain.

The scientific report on the first expedition has already been published. It includes 30 original papers on different scientific disciplines.

A second expedition was organized in December 1982. This expedition consisted of 28 members drawn from nine different organizations in the country. The same vessel, Polar Circle, was chartered again for this expedition. The second expedition left Goa on December 1, 1982, and landed on December 28, 1982.

The objectives of the second expedition were as follows:

(1) To survey the area and select a site for setting up a permanently manned station in Antarctica,

(2) To work out the logistics for setting up and servicing the permanently manned research station,
(3) To survey and identify a suitable airstrip and prepare it for landing an aircraft,
(4) To establish direct communications links between the base camp in Antarctica and India and between the base camp and the mobile parties on the landmass and on the ship, and
(5) To obtain more knowledge in all the scientific disciplines initiated on the first expedition.

The second expedition successfully accomplished these tasks and in particular selected a site for a permanently manned station. It also recovered the cassette from the automatic weather-recording station left behind during the first expedition and fixed a new cassette after fully overhauling the entire system.

The third Indian expedition consisted of a team of 81 persons, including two women scientists, chosen from 13 different organizations. The significant objectives of this expedition were to build the permanent station and to leave a team of 12 members to winter in Antarctica and continue experiments during the antarctic winter.

In view of the size of the third team, a larger Finnish vessel was chartered for the expedition. It left Goa on December 3, 1983, and landed in Antarctica on December 27, 1983. The building material for the station consisted of prefabricated material and containers specially designed to expand the existing Indian barracks erected during the earlier expeditions. Four helicopters to carry personnel and material from the ship to the base camp were also taken.

A fourth Indian expedition was launched on December 4, 1984. This expedition was composed of 82 persons.

India has defined its short-term and long-term objectives in Antarctica for the next decade.

POLITICAL ISSUES

The possibility in the future of exploiting mineral resources for commercial gain has resulted in a sudden increase of interest in Antarctica among many other countries. Thus, between 1961 and 1974, during a period of 15 years, only six additional states acceded to the Antarctic Treaty. Between 1975 and 1980, four states deposited their instruments of accession, but between

1981 and 1984, 10 states acceded to the treaty. In addition to the possibility of minerals exploitation, this may be attributed to other important events that have taken place in the last two decades. First, technological breakthroughs and the emergence of newly independent nations in the 1960s have brought demands for change in existing international legal rules and regulations. These demands arose as a result of participation by the newly independent states in the law-making process. This in turn has resulted in calls for a new international economic order and for the utilization of all natural resources beyond national jurisdiction for the common benefit of humankind.

The longest and most widely attended conference in the history of the United Nations, the third Law of the Sea Conference (UNCLOS), witnessed the great enthusiasm with which these demands were projected. The day the U.N. General Assembly resolved to convene the third UNCLOS, it passed a resolution with 108 votes in favor, none against, and 14 abstentions, which declared, inter alia, that the resources of the seabed and the subsoil thereof beyond the limits of national jurisdiction were the "common heritage of mankind" (CHM). The CHM principle, as it is commonly known, is regarded as the cardinal principle of the new law of the sea. The CHM resolution has been regarded as law making, and attempts have been made to apply the same principle to other natural resources; for example, the 1970 Agreement Governing the Activities of States on the Moon and other Celestial Bodies declared the resources of the moon to be a "common heritage of mankind."

The 15-point CHM declaration on the seabed envisaged the establishment of international machinery for the administration of the principle. However, the UNCLOS witnessed the erosion, at least to a large extent, of the CHM principle. The machinery that it established did not exactly conform to the lines envisaged in the CHM Declaration; instead, it was based on a pragmatic approach, incorporating a "parallel" system of exploitation.

In fact, the entire Law of the Sea Convention has to be viewed in two parts. One part includes the codification of existing rules and customary law covering matters falling within national jurisdiction. The provisions of the other part relate to the resources beyond the limits of national jurisdiction (the international seabed area). They are in a more or less contractual form, which takes into account the interests of the countries that have the technology and the capital necessary for seabed mining.

Law of the Sea Conference Resolution II, adopted along with the convention, recognizes a special status known as "pioneer" status, for nations that have already made substantial investments in seabed mining. In contrast, the 1970 CHM Resolution on the seabed had incorporated the principles of an earlier resolution passed in 1969 known as the moratorium resolution, which had prohibited activities relating to seabed resources pending the establishment of international machinery.

Thus, without going into much detail, it becomes quite clear that the CHM principle as contained in the Law of the Sea Convention under the section entitled "Principles Governing the Area" is contradictory in terms. It is therefore very difficult to accept the theory that the CHM principle has attained the status of universal international law, or <u>jus cogens</u>, insofar as the natural resources beyond the limits of national jurisdiction are concerned. It would thus be erroneous to superimpose the CHM principle on antarctic resources without understanding how the system as a whole has worked so far, however noble the principle may appear to be at first sight.

It may be recalled that at the initiative of Malaysia and Antigua and Barbuda, the United Nations produced in October 1984 a comprehensive study on Antarctica and in debating it in the General Assembly many countries referred to the CHM principle. Yet in light of the principles discussed above as to how the antarctic system has worked so far, it is doubtful whether the CHM concept, however relevant in the context of the law of the sea, is appropriate or workable if applied to antarctic resources.[13]

The consultative parties, in various informal meetings, have already expressed their grave concern at attempts to internationalize the Antarctic without appreciating the openness of the Antarctic Treaty as well as how the system has worked so effectively thus far. Even politically, the Antarctic Treaty has undergone many crucial tests. When the treaty was signed in 1959, there were seven states that claimed sovereignty over Antarctica. The United Kingdom's claim dates back to 1908, based on discovery and formal acts of possession. The Chilean claim to territorial sovereignty was made in 1940, but Chile did not establish a station until 1947. The claims of Chile, Argentina and the United Kingdom overlap. The Australian claim (1933) covers the largest area, about

6.5 million square kilometers. France did not make a territorial claim until 1924. New Zealand's claim dates back to 1923. The Norwegian claim was made in 1939, presumably to prevent German activity in the area. The Argentine claim was first decreed in 1946. Only the five claims of Australia, France, New Zealand, Norway and United Kingdom are mutually recognized. About 15 percent of the continent--Marie Byrd Land--is free from any claims.

One of the most important contributions of the treaty system has been to freeze the territorial claims and thus put to rest controversies surrounding the claims.[14] It may also be mentioned that even the claimants could be divided into the northern and southern groups. The southern group would consist of Argentina, Australia, Chile, and New Zealand and the northern group would comprise France, Norway, and the United Kingdom. The southern claimants can practically control all ports and other land-based facilities that may serve as communications links with Antartica. Major operations to Antarctica could turn out to be impossible or prohibitively expensive if logistic support from bases in at least one of these countries were not available.[15]

The nonclaimant parties to the Antarctic Treaty are Belgium, Brazil, the Federal Republic of Germany, India, Japan, Poland, South Africa, United States and the USSR. It may, however, be mentioned that Brazil, Peru, and Uruguay have informally speculated about making claims in Antarctica.

The role of the two major powers, the United States and the USSR, also deserves special mention. Although the United States has not made any official claim, many attempts have been made unofficially regarding the option of the United States' staking a claim. In fact, the document entitled "The Political Legacy of the International Geophysical Year," prepared for the Subcommittee on National Security Policy and Scientific Developments of the Committee on Foreign Affairs of the U.S. Congress, includes the United States among the claimant nations, although the word "unofficial" is used in brackets.[16] The USSR, on the other hand, has reserved all rights, based on the Russian discoveries, which include the right to present territorial claims if need be. The role of these two countries, which have, over the years, made massive investments in Antarctica and have contributed enormously to antarctic science, becomes significant in light of the attempts to internationalize the issue in the U.N. forum. It may be mentioned that in the negotiations for a mineral regime, the Beeby draft text has sought to give a special status to these two countries among the treaty parties as "the two states which, prior to the entry into force of the Antarctic treaty, had asserted a basis of claim in Antarctica."[17]

Apart from the recent attempt in the United Nations, there have been other instances when there were efforts to internationalize the antarctic issue. In 1971, the Committee on Natural Resources of the U.N. Economic and Social Council (ECOSOC) was told by Secretary-General U Thant that "the era of systematic exploration for antarctic resources had arrived" and that the work of the committee would be "incomplete and unrealistic if any significant portion of the globe went unreported and excluded."[18] In 1975, the issue was raised in meetings of the U.N. Environment Program (UNEP) and again in the ECOSOC. In 1976, the FAO attempted to develop a $45 million Southern Ocean program with U.N. Development Program (UNDP) assistance for exploration, exploitation, and utilization of Southern Ocean resources for the benefit of the world as a whole and developing countries in particular. Because of the treaty parties' objections, this program was radically reduced to a $200,000 information project.[19]

CONCLUSIONS

In light of the foregoing account, a very important question that needs to be considered is: What is the kind of regime that should be established in Antarctica? In this regard, the role of the consultative parties becomes crucial, as does in particular the role of new members such as India. India's position becomes even more important because of the status that it holds as the chair of the NonAligned Movement.

It is undoubtedly clear that, whatever mechanism is advocated for a future regime, the antarctic system as a whole cannot be replaced. The present discussions in the United Nations and the possibility of a review of the Antarctic Treaty in the year 1991 make it important to consider right now what amendments/modifications are necessary to maintain scientific activity and cooperation in Antarctica. Already it can be noticed that many conditions in the ATCP meetings are now relaxed. The openness of the system is further demonstrated by recent decisions allowing parties to the treaty that are not consultative parties to attend the consultative meetings as observers. Similarly, in all meetings and discussions related to the minerals regime in Antarctica, the nonconsultative members will henceforth be invited to attend as observers. The number of consultative party members, which remained at 12 from 1959 to 1977, has now increased to 16. It is hoped that many more states will want to join the group.

To sum up, it seems that an overhaul of the Antarctic Treaty System is neither possible nor advisable. Such an overhaul would affect not only the Antarctic Treaty provisions but also an entire scientific system developed through scientific cooperation over the past 100 years or so. This scientific cooperation can neither be ensured through another structure nor governed by provisions of any new treaty. What is important is the will of the nations, and in particular the dedication of individual scientists, who have braved their way to the remote continent and conducted investigations and experiments there for the benefit of mankind as a whole. The findings and results of these investigations are what humankind looks forward to. That, in short, is the true essence of the CHM principle. The scramble for economic gain coupled with notions of sovereignty will seriously hamper scientific progress in Antarctica.

Internationalization of the Antarctic as an issue in political forums cannot be expected to produce any substantial results. Regrettably, it may perhaps lead to another historic conference like the third UNCLOS. While admittedly the Law of the Sea Conference was historic in many senses, it also offers many lessons. To maintain the true essence and spirit of the CHM principle, it is necessary to maintain scientific cooperation and activity unhampered by political or economic interests, and steps should be taken for the dissemination of scientific information obtained thus far for the benefit of mankind as a whole.

NOTES

1. Bullis, H. 1973. The political legacy of the International Geophysical Year, in hearings before the Foreign Affairs subcommittee on National Security Policy and Scientific Development, U.S. House of Representatives (November 1973), p. 57.

2. Sollie, F., 1984. The future of Antarctica. Mazingira, p. 5.
3. Daniels, P.C., 1973. The Antarctic Treaty. In R.S. Lewis and P.M. Smith, eds. Frozen Future, (New York), p. 36.
4. Bulls, H., 1973. The political legacy of the International Geophysical Year, in hearings before the Foreign Affairs subcommittee on National Security Policy and Scientific Development, U.S. House of Representatives (November 1973), p. 57.
5. Sollie, F., 1984. The development of the Antarctic Treaty System--trends and issues. In R. Wolfrum, ed. Antarctic Challenge: Conflict, Interests, Cooperation, Environmental Protection, Economic Development, Dunker & Humblot (Berlin), p. 20.
6. Recommendation XII-8, adopted at the 1983 meeting of the ATCPs. Text of the recommendation is contained in SCAR Bulletin No. 76, January 1984, p. 121.
7. Gjelsvik, T., 1984. Scientific research and cooperation in Antarctica, in R. Wolfrum, ed. Antarctic Challenge, Dunker & Humblot (Berlin), p. 46.
8. Sollie, F., 1984. The development of the Antarctic Treaty System--trends and issues. In R. Wolfrum, ed. Antarctic Challenge: Conflict, Interests, Cooperation, Environmental Protection, Economic Development, Dunker & Humblot (Berlin), p. 32.
9. For the text of the recommendation, see SCAR Bulletin No. 58, January 1978, p. 90.
10. Sollie, F., 1984. The development of the Antarctic Treaty System--trends and issues. In R. Wolfrum, ed. Antarctic Challenge: Conflict, Interests, Cooperation, Environmental Protection, Economic Development, Dunker & Humblot (Berlin), p. 33.
11. Couratier, J., 1983. The regime for the conservation of Antarctica's living resources. In F. Orrego Vicuna, ed. Antarctic Resources Policy, Scientific, Legal and Political Issues Cambridge University Press (London), pp. 147-148.
12. Report of the Twelfth Consultative Meeting, Canberra, 13-17 September 1983, in SCAR Bulletin No. 76, January 1984.
13. A report of the U.N. Secretary-General on the question of Antarctica was submitted to the U.N. General Assembly on October 31, 1984. This study is in two parts. U.N. Doc. A/39/583.

14. Article IV of the treaty reads as follows:

"1. Nothing contained in the present Treaty shall be interpreted as:
"(a) a renunciation by any Contracting Party of previously asserted rights of or claims to territorial sovereignty in Antarctica;
"(b) a renunciation or diminution by any Contracting Party of any basis of claim to territorial sovereignty in Antarctica which it may have whether as a result of its activities or those of its nationals in Antarctica, or otherwise;
"(c) prejudicing the position of any Contracting Party as regards its recognition or non-recognition of any other State's right of or claim or basis of claim to territorial sovereignty in Antarctica;
2. No acts or activities taking place while the present treaty is in force shall constitute a basis for asserting, supporting or denying a claim to territorial sovereignty in Antarctica or create any rights of sovereignty in Antarctica. No new claim, or enlargement of an existing claim, to territorial sovereignty in Antarctica shall be asserted while the present treaty is in force."

15. Antonsen, P., 1984. On the balance of power in Antarctic Treaty System. Nansen Newsletter No. 2:8.
16. Bullis, H. 1973. The political legacy of the International Geophysical Year, in hearings before the Foreign Affairs subcommittee on National Security Policy and Scientific Development, U.S. House of Representatives (November 1973), p. 56.
17. See Appendix 8, The Future of the Antarctic: Background for a U.N. Debate. October 1, 1983. Greenpeace International (Lewes, United Kingdom), 1983, p. 4.
18. Mitchell, B. 1983. Frozen stakes: the future of Antarctic minerals. International Institute for Environment and Development (London), p. 41.
19. Mitchell, B. 1983. Frozen stakes: the future of Antarctic minerals. International Institute for Environment and Development (London), 1983, p. 41.

25.

The Interaction Between the Antarctic Treaty System and the United Nations System
Richard A. Woolcott

When in 1773 the great navigator Captain James Cook became the first known person to force a small vessel beyond the Antarctic Circle, he recorded in his journal that "I can be bold enough to say that no man will ever venture further than I have done and that the land which may lie to the south will never be explored."

Well, here just over 200 years later, we have actually conducted a seminar on the Beardmore Glacier of the continent that Cook predicted would never be explored. Hobson's theory of imperialism suggested that the trader followed the missionary and the flag followed the trader. In Antarctica it has been different; the scientists followed the explorers and seminar attendees have followed the scientists.

For me, personally, this visit fulfills a lifelong ambition. For Australia and Australians, Antarctica is a neighboring continent of great importance. Indeed, the earliest maps of the Southern Hemisphere show a gigantic mass, usually called Terra Australis (the southern land) and, in those maps, Australia and Antarctica were connected. While physically they are separated by sea, they remain connected by a tradition of proximity, interest, exploration, and research.

So I am very glad to have had the opportunity to participate in the Workshop on the Antarctic Treaty in three capacities: in my personal capacity, as the chairman of the New York group of the Antarctica Treaty consultative parties, and as the representative of Australia at the United Nations.

As I am sure is well known, the United Nations General Assembly (UNGA) considered "the question of Antarctica" in 1983 and again in 1984. At the conclusion of the debate in 1984, it was agreed to inscribe the item again on the provisional agenda for the fortieth session of the UNGA in 1985.

This is not, of course, the first time that consideration has been given to the United Nations' playing some role in antarctic affairs. In the years before the signature of the Antarctic Treaty in 1959, a number of individuals and organizations, and even some states, made suggestions for some relationship between the United Nations and Antarctica. As long ago as 1956, India, whose scientists are now conducting research at a new antarctic station on the other side of this continent, proposed that the question of Antarctica be included on the agenda of the UNGA. In the event, that did not happen, and it was not until 1983 that Antarctica became a subject of discussion in the General Assembly.

To some this might suggest that there has been little interaction between the United Nations system and the Antarctic Treaty System. Some might even see this lack of discussion of Antarctica in the United Nations as evidence of what has been alleged to be "secrecy" or "exclusivity" on the part of the Antarctic Treaty consultative parties. It would be a serious mistake to reach such conclusions.

The fact that states have not, until recently, discussed the subject of Antarctica in the General Assembly indicates to me a lack of interest among the majority of countries not involved in the continent as well as a measure of acquiescence on the part of the world community in the system of management established by the Antarctic Treaty. This has occurred both because the practical framework of cooperation operating under the treaty system has promoted the principles and purposes of the U.N. system and because, in appropriate areas, there has been practical cooperation between the two systems.

PROMOTION OF PRINCIPLES AND PURPOSES OF UNITED NATIONS CHARTER

The United Nations Charter defines a number of important purposes of that organization, including the development of friendly relations between nations, the achievement of cooperation between nations in solving international problems, and, most importantly, the maintenance of peace and security. The United Nations Charter also emphasizes a number of principles governing the actions of states, including respect of the sovereign equality of states, encouragement of the peaceful settlement of international disputes, the avoidance of the threat or use of force, and the promotion of economic and social advancement of all peoples.

I should like to concentrate, first, on the ways in which the Antarctic Treaty has given practical effect to these important principles and purposes of the U.N. system and, in particular, how it has encouraged international cooperation in Antarctica and how it has insulated the continent from international conflict and rivalries over the last quarter of a century.

In order to measure the achievements of the Antarctic Treaty System and to relate them to U.N. principles, we should reflect briefly on the situation in Antarctica before the signature of the Antarctic Treaty in 1959. Territorial claims, and nonrecognition of these claims by other nations active in Antarctica, had led to political tensions. Rivalry among certain of the claimants had led to incidents and protests. One country took steps to bring its claim before the International Court of Justice, and the other countries involved declined to accept the jurisdiction of the court. In short, a difficult international situation had developed that posed a potentially serious threat to peace in the region.

The unprecedented cooperation that underlay research activities in Antarctica in the fields of geophysics and geography during the International Geophysical Year (IGY) of 1957-1958 marked a turning point in relations between countries with interests in the continent. Because of the lack of basic data on Antarctica and the extremely difficult operating conditions there, it was imperative that there be maximum cooperation among scientists engaged in common research programs, including the free exchange of views and personnel and free access to stations and installations. There was also a strong sense, on the part of the scientists involved, that their studies in Antarctica were important to the entire world community and that the results of their efforts should be freely available to all who were interested.

This cooperative spirit led the 12 nations that had conducted research in Antarctica during the IGY to reflect on the earlier unsatisfactory situation and to consider a more permanent framework for continuing their joint efforts, in a way that would allow all nations wishing to be active on the continent to do so without threatening the interests of others or destabilizing the area.

The result was the Antarctic Treaty of 1959. This was the starting point; and today it is still the core of what is known as the Antarctic Treaty System (ATS). The latter comprises a cooperative framework and a wide range of carefully considered recommendations and legally binding instruments that address living and working in Antarctica, exchanging information, protecting the environment, and conserving and making rational use of resources. In this regard, I should refer in particular to the 1964 Agreed Measures for the Conservation of Antarctic Fauna and Flora, the 1972 Convention for the Conservation of Antarctic Seals, the 1980 Convention on the Conservation of Antarctic Marine Living Resources, and the current negotiations toward an Antarctic minerals regime.

Let us now look more closely at how the Antarctic Treaty, in its language and in its practical application, relates to the fundamental purpose of the United Nations Charter: the maintenance of international peace and security.

The preamble of the treaty states that "it is in the interest of all mankind that Antarctica shall continue forever to be used exclusively for peaceful purposes and shall not become the scene or object of international discord." It continues that "a treaty ensuring the use of Antarctica for peaceful purposes and the continuance of international harmony in Antarctica will further the purposes and principles embodied in the Charter of the United Nations."

What could be a more forthright commitment to the aims of the United Nations with respect to peace and security? The treaty also goes beyond this general commitment by including a number of specific provisions designed to ensure that this commitment is upheld.

Article I of the treaty prominently and firmly establishes the position of the treaty partners that "Antarctica shall be used for peaceful purposes only." It prohibits "any measures of a military nature, such as the establishment of military bases and fortifications, the carrying out of military manoeuvres, as well as the testing of any type of weapons."

Other provisions relating to the demilitarization of Antarctica include Article V, which prohibits nuclear explosions; Article VII, which provides full access for purposes of inspection by consultative party observers to promote the objectives and ensure the observance of the provisions of the treaty; and Article IX (1a), which permits consultative parties to adopt measures regarding the "use of Antarctica for peaceful purposes only."

Special mention should also be made of the ingenious device contained in Article IV of the treaty, which puts to one side the question of territorial sovereignty, the source of earlier tensions, as an obstacle to peaceful cooperation.

Article IV represents the practical application of specific exhortations of the United Nations, including respect for the sovereign equality of states and the encouragement of peaceful settlement of disputes. The "freezing"--no pun is intended--of positions with respect to claims removed the basis for the earlier contention between countries active in Antarctica.

Collectively these provisions, and the fact that they have worked in practice over a long period, establish the Antarctic Treaty as the only effective post-World War II international agreement ensuring the demilitarization of a large geographic region. The treaty has also provided a precedent for numerous subsequent initiatives and proposals for regional disarmament measures, including nuclear-weapons-free zones. In this way, as in others, the treaty has made an important contribution in furthering the principles and purposes of the United Nations Charter.

With the advantages of hindsight and knowledge of the extreme difficulty of negotiating measures to limit conventional or nuclear weapons today, the achievement of the Antarctic Treaty in this area must be seen both as impressive and of great value to all states--particularly, may I say, to states such as those of Australasia, of the South Pacific, of South America, and of southern Africa, which are within close geographic proximity to Antarctica.

In spite of other differences, the importance of maintaining Antarctica demilitarized, free of nuclear weapons and strategic competition, and a zone of peaceful cooperation was clearly recognized by most delegations that spoke on the antarctic item at both the 38th and the 39th UNGA sessions.

LINKS WITH THE UNITED NATIONS SPECIALIZED AGENCIES

I shall turn now to links between the treaty system and the specialized agencies of the United Nations.

Article III(2) of the Antarctic treaty makes specific provision for the establishment of cooperative working relations with the U.N. specialized agencies and other international organizations with a scientific or technical interest in Antarctica. The anticipated increased international interest in the continent and the need for cooperation and the sharing of the results of antarctic research with the whole world community, including those states that lack the desire or the means to become actively involved in antarctic research.

Although existing links are not highly formalized, practical working relations have already been developed with U.N. family organizations, such as the World Meteorological Organization (WMO), the International Telecommunications Union (ITU), the Food and Agriculture Organization (FAO), the United Nations Environment Program (UNEP), and the Intergovernmental Oceanographic Commission (IOC) of the United Nations Educational, Scientific and Cultural Organization (UNESCO), as well as with organizations outside the U.N. system, such as the Scientific Committee on Antarctic Research (SCAR) and the Scientific Committee on Oceanic Research (SCOR) of the International Council of Scientific Unions (ICSU), the International Union for the Conservation of Nature and Natural Resources, and the International Whaling Commission.

At their very first meeting the consultative parties adopted a recommendation making it an obligation of governments to encourage the work of relevant international organizations, including the U.N. specialized agencies, and to promote on a bilateral basis the development of cooperative relations with those organizations (Recommendation I-V). In considering the degree of cooperation already established with U.N. bodies, particular mention should be made of links established with WMO, ITU, IOC and FAO.

Antarctica has a crucial effect on the atmosphere of the whole Southern Hemisphere, and meteorological data from Antarctica play an important role in predicting global weather and climate. Hence the great value of cooperation between nations active scientifically in Antarctica and the WMO. An important element in this cooperation is the provision of daily antarctic meteorological data via the Global Telecommunications System in the WMO's World Weather Watch program, which has been continuing since the IGY in 1957-1958. The system also involves the cooperation of the ITU.

The WMO has an executive committee working group on Antarctic meteorology, through which antarctic treaty

consultative nations regularly contribute expertise to
the coordination and development of antarctic meteorology. The consultative parties are also making substantial
contributions to the World Climate Research Program
(WCRP), which is coordinated jointly by the WMO and the
ICSU. They were closely involved also in the very successful Global Atmospheric Research Program, which ran
from 1967 to 1982 and which was the forerunner of the
WCRP.

Cooperation with the IOC has included contributions by
the consultative parties to the Program Group on the
Southern Ocean, a coordinating body for marine biology
research. The IOC also has permanent observer status at
meetings of the commission of the Convention for the
Conservation of Antarctic Marine Living Resources
(CCAMLR).

The FAO also has observer status at CCAMLR meetings.
It is working with the CCAMLR secretariat to develop
fishing data for the Southern Ocean area. The FAO,
together with SCAR and other committees of the ICSU, is
also sponsoring the 10-year Biological Investigation of
Marine Antarctic Systems and Stocks program, the aim of
which is to gain a deeper understanding of the structure
and functioning of the antarctic marine ecosystem as a
basis for the future management of antarctic living
resources. This program also relates to the work of the
IOC.

I should also refer to the close links between the
consultative parties and SCAR. This is in keeping with
the facts that the ICSU played a prominent part in
developing the cooperation that occurred during the IGY,
and that cooperation among the consultative parties has
traditionally been most extensive in the field of scientific research. There are also close links with SCOR and
with the Scientific Committee on Problems of the Environment of ICSU. ICSU, whose charter provides for the
encouragement of international scientific activity for
the benefit of humankind as a way of contributing to the
cause of peace and international security, has consultative status with UNESCO, the Economic and Social Council
of the U.N. (ECOSOC), the FAO, and the International
Atomic Energy Agency (IAEA) and working relations with
the ITU, the World Health Organization (WMO), the United
Nations Disaster Relief Organization, and the United
Nations Research Institute for Social Development.
Membership of ICSU consists of science academies,
research councils, and associations or institutions from
some 65 countries.

One can see from this that a wide range of practical interaction already exists among the members of the Antarctic Treaty, the United Nations system, and other relevant scientific bodies.

THE FUTURE

In the light of growing international interest in Antarctica in recent years, reflected in the consideration of it within the United Nations, it is worthwhile to speculate on how future interaction between the ATS and the U.N. system might develop and on what other steps might be taken to accommodate this wider international interest in the affairs of the Southern continent.

In this connection, I am speaking not as the chairman of the Antarctic treaty consultative party coordination group in New York, nor as a representative of the Australian government, but rather on the same basis as all the speakers at the Workshop on the Antarctic Treaty-- that is, in my personal capacity.

I would begin by recalling the maxim, "Diplomacy is the art of accommodation and compromise," and, in this context, I shall attempt to balance the concerns of those who have argued that there is a need substantially to revise, or even to replace, the existing system of management in Antarctica against the views of those who want to see the ATS maintained because of its value and importance but who are prepared to acknowledge that there may be scope for further evolution within the present system. To do this, it is necessary to address some of the main criticisms of the treaty system.

A major concern has been with the decision-making process, which has been criticized as being exclusive, undemocratic, and discriminatory against small states. It is suggested instead that a universal framework based on equality, for example within the United Nations, would be more appropriate. While this latter obviously has appeal for countries that have not previously been involved but now have an interest in Antarctica, I would suggest that, in fact, the existing two-tier system has several important advantages.

The present system is based on the notion that those countries that are directly engaged in antarctic activities are, through their experience in Antarctica, best placed to take decisions affecting such matters as scientific programs and environmental protection in the

continent. Moreover, it is they who will be bound by the
obligations that arise from those decisions, which are
taken continuously by means of consensus. Hence, giving
consultative parties a special role is both logical and
workable from a management viewpoint.

This is not dissimilar to the practice in many other
international bodies, including a number in the U.N.
system. Furthermore, consultative party status is not,
as has sometimes been suggested, effectively restricted
to wealthy countries. The present list of consultative
parties demonstrates that there are no ideological or
economic divisions on north/south or east/west lines
within the treaty framework. The present arrangement
permits practical cooperation in an atmosphere free of
the political rhetoric or posturing that often infects
much of the United Nations system.

Beyond this, I should emphasize that the Antarctic
Treaty--which has now been acceded to by 32 states,
including 6 new signatories in the past two years, is
open to all member states of the United Nations. In 1983
it was decided that all parties to the Antarctic Treaty
may attend consultative meetings and, subsequently, also
future meetings of the minerals negotiations as observers.
Thus any country that is interested in antarctic affairs
may now, by simply acceding to the Antarctic Treaty,
which involves no cost, have the opportunity to participate in and influence decisions taken in these meetings.
Acceding states may speak and circulate papers. Furthermore, as decisions are taken by consensus, their influence
on the decision-making process would in practice be
significant. The distinction between Antarctic Treaty
consultative parties and nonconsultative parties is
therefore likely to become less marked in the future.

This represents an important change from the past. It
shows the capacity and the will of the consultative
parties to adapt the treaty system to changing circumstances and interests. Small states, which do not maintain antarctic programs and which perhaps never intend to
do so, can nevertheless, through acceding to the treaty,
become fully informed about antarctic affairs and bring
considerable influence to bear when decisions are made,
thus ensuring that their interests are taken into account.
I would advocate strongly that such countries consider
carefully the benefits this opportunity provides.

It has also been suggested that access to information
under the present system is poor. It is true that, in
the past, exchanges of information on antarctic matters

occurred largely among those countries directly engaged in Antarctica. This was not the result of any desire to withhold information from the rest of the world. Far from it; the results of scientific research in Antarctica have always been made freely available to the international community. Rather, it was the consequence of the relatively limited interest shown by other countries in antarctic affairs until recent times.

Recognizing the growth of international interest, the consultative parties, at their twelfth meeting in 1983, decided to take steps to disseminate more widely information on antarctic activities, to study further steps, and to consider them in 1985 (Recommendation XII-6). To this end, in the period since, they passed to the United Nations Secretary-General for circulation to all member states the report of their twelfth meeting and, individually, they made substantial contributions toward the United Nations Secretary-General's study of Antarctica. An improved and expanded Handbook of the Antarctic Treaty is being prepared.

This is not to suggest that more cannot be done. I am aware that many countries still believe that information flows could be improved further. An important way of improving their access to information would be for them to accede to the treaty and attend meetings. Another would be for consultative parties to circulate copies of information regularly exchanged among themselves and through SCAR to the U.N. secretariat for information. Means of ensuring effective dissemination of information on Antarctica will be considered at the Brussels meeting in October, 1985, and any ideas others may have will be welcome.

It has been argued that links between the ATS and U.N. specialized agencies and other relevant nongovernmental bodies have not been adequately developed. I have already canvassed the range of links that currently exist. These links are not perhaps so extensive as they might be, and there may well be benefit in seeking to extend them in future in the cases of organizations that have relevant expertise. I believe that the consultative parties might carefully consider the scope for intensifying their existing links with U.N. agencies and nongovernmental bodies and the opening up of new links with other relevant bodies. Where appropriate, such organizations might attend consultative meetings in an observer capacity, as they do meetings of CCAMLR. Also, U.N. specialized agencies and other relevant nongovernmental bodies might be encouraged to consider giving Antarctica higher priority in terms of allocation of resources within their total programs than they have done to date. Close links with such bodies would also assist the process of improving the dissemination of information on Antarctica to the international community.

Judging from statements made at the last two UNGAs, it seems to me that the interests and concerns of nontreaty countries relate mainly to the question of possible mineral resources. This is especially true of the negotiations taking place through special Antarctic Treaty consultative meetings to develop an antarctic minerals regime. It is suggested that the rich, developed countries are seeking to exploit the resource potential of Antarctica for their own benefit, at the expense of others, and that wider international involvement in the negotiation of any minerals regime is necessary. It is also suggested that Antarctica is the "common heritage of mankind" and that any revenues derived from the exploitation of its resources should be shared equitably. In the same context, it is argued that territorial claims in Antarctica are outmoded and that the prospect of future resource exploitation contains the seeds for potential conflict there.

There are many complex issues involved here. First, let me deal with the concept of "common heritage." While many states, including Australia, accept the validity of the concept for the international seabed and outer space, there is no international consensus that it is applicable to Antarctica. Moreover, in the U.N. context, analogies with the Outer Space Convention and the Law of the Sea Convention do not really work. The oceans are global. Outer space surrounds us all. Antarctica, on the other hand, has fixed dimensions and neighbors such as my own country, Australia, with a sustained and serious interest in the continent.

A most important reason why "common heritage" is not applicable to Antarctica is the fact of the existence of long-standing territorial claims in Antarctica. While such claims might not be recognized by the international community, they remain a fact of international life. Differences over claims have been usefully put to one side under the Antarctic Treaty in favor of international cooperation. In my view, to reopen this question will only provoke an unnecessary renewal of tensions, which have been successfully avoided for the past 25 years.

Common heritage also has an exploitive connotation, and it is important to emphasize that the principal purpose of the minerals negotiations is not, as is sometimes suggested, to carve up the resources of Antarctica. It is to ensure that unregulated minerals activity does not take place in future that could prove environmentally harmful, adversely affect other uses of the continent, and lead to renewed contention. The consultative parties are seeking to negotiate an agreement that will lay down the broad rules for future activity and establish arrangements with the objectives of ensuring that the fragile antarctic environment is fully safeguarded and that potential conflict is avoided. It is important to negotiate such a regime now, in the absence of pressures to exploit.

On the question of wider involvement in the antarctic minerals negotiations and the need to ensure that the interests of all countries are adequately taken into account, I mentioned earlier the decision by the consultative parties to open future meetings of the minerals negotiations to all parties to the Antarctic Treaty. This goes a long way toward meeting such concerns. Any country interested in antarctic minerals may now attend minerals negotiating meetings if it first accedes to the Antarctic Treaty. By this means it can exert an influence on the proceedings.

Furthermore, in developing a minerals regime, negotiators have been conscious of the need to ensure that it serves the interests of all humankind in Antarctica, that it is open to accession by all states, and that there is fair opportunity for all states to participate in future minerals activities. The negotiations have been based from the outset on the principle of nondiscrimination among states. It is also widely recognized by those engaged in the minerals negotiations that in order to arrive at an internationally accepted minerals agreement there will be a need, among other things, to accommodate the interests of the wider international community, including the many countries that may not become engaged in any possible future mining activity in Antarctica.

This leads me to a third and related point, the question of revenue sharing from antarctic resource activity on an equitable basis. This is a difficult area to address, largely because so little is known about the resource potential of the Antarctic continent, much less how it might be exploited. Many exaggerated notions of the resource wealth of Antarctica have been expressed.

The reality is that most of the information available is speculative and based largely on geological hypothesis. The few known, low-grade, on-shore deposits that are accessible in the two percent of the continent not covered by deep moving ice will not be exploited for a very long time indeed. Much prospecting and exploration would need to be done before it could be determined whether exploitation is feasible or economic. Although prospecting might commence in the next decade or so, minerals exploitation, if it ever takes place in Antarctica, even on its continental shelf, will not occur until well into the next century. The higher costs that would be involved in operating in Antarctica, as well as the long distances and inhospitable climate, make it likely that developers will concentrate on other areas of the world that remain to be exploited before turning their attention toward Antarctica. Thus, in my view, notions of taxation and revenue sharing in the Antarctic context seem very premature and uncertain indeed.

Nevertheless, it seems to me inconceivable that an antarctic minerals regime that ignored notions of equity could be successfully negotiated and find international acceptance in the current political climate. That would be politically naive. Accordingly, in the longer term, if exploitation were to take place in Antarctica, it seems only reasonable to expect that all states would have rights of access to antarctic mineral resources and that they should not be excluded from the economic benefits derived from them, after the costs of recovery and of the minerals regime's institutions have been covered. How these concepts find expression in the minerals regime will be a matter for discussion and negotiation. Realistically, the details are likely to be left to the institutions created by the minerals agreement to sort out in the future.

In the same context, I believe that there is scope for those already engaged in antarctic activities to offer greater cooperation to developing countries interested in antarctic scientific research within their national programs. In particular, they might share their antarctic experience with scientists of other countries, undertake joint research where appropriate, and advise countries interested in conducting their own Antarctic programs on scientific and logistic support matters. They might also extend opportunities, within the limits of their own national programs, for scientists from interested countries, especially developing countries, to conduct research in Antarctica.

RELATIONSHIP TO THE UNITED NATIONS SYSTEM IN THE FUTURE

I would note that during discussion on the antarctic item at the 1984 U.N. General Assembly, most speakers--even the critics of the present treaty system--paid tribute to the virtues of the treaty, particularly its provisions for demilitarization, denuclearization, scientific cooperation, and environmental protection. Their criticisms were aimed at a number of other areas, which I have addressed. The various suggestions that I have made in response to these concerns amount to saying that I believe there is scope <u>within</u> the treaty system to address many of these concerns. Some can be resolved by encouraging closer interaction between the ATS and the U.N. system; others will need to be resolved by different means. Such matters will need to be addressed at the 1985 thirteenth Antarctic Treaty consultative meeting as well as at the fortieth UNGA session, where Antarctica will again be on the agenda.

I return to my earlier reference to diplomacy as the art of accommodation and compromise. If we are to avoid dispute and confrontation, there is a responsibility on the part of all concerned to show realism and flexibility. Attempts to replace the ATS, or to set up new institutional arrangements for the management of antarctic affairs within the United Nations, will be neither practicable nor productive; on the other hand, Antarctic Treaty consultative parties should continue to find ways of overcoming the reasonable concerns of other countries by improving the operation of the treaty system.

An increase in public interest, or the launching of a particular initiative in the UNGA by one or two countries, should not lead us to the conclusion that a new set of arrangements should be developed to replace the ATS. As Evan Luard wrote in an article on Antarctica in the Summer 1984 issue of <u>Foreign Affairs</u>, "as long as the Treaty is progressively opened to new members, as their scientific capacities increase, there would seem to be little advantage in replacing it with new mechanisms...." In my own view, there would be no advantage in seeking "new mechanisms."

I am not opposed to U.N. interest in Antarctica or to the discussion of Antarctica in the United Nations. There is already a measure of constructive interaction between the treaty partners and the U.N. system, which I have noted. What I am opposed to is pressure--to a large extent artificially stimulated--for some institutionalized U.N. political involvement in an area in which there are no threats to the peace nor any substantial political, economic, or social problems. What I would oppose, for example, would be attempts to establish yet another U.N. committee, which is not needed, which would not be effective, and which could lead to attempts to undermine or replace an existing, sound, flexible, and open instrument in the Antarctic Treaty.

When I became Australia's ambassador at the United Nations and, then, chairman of the New York group of the Antarctic Treaty consultative parties, I knew little about the inner workings of the U.N. system and less about the processes of the ATS--although I had, of course, been interested in a general way in both matters.

In earlier days--in the 1960s and 1970s--I was attracted to international approaches to questions such as that of Antarctica and to the concept of the "common heritage of mankind." My personal experience in New York, over the last two and a half years, and in Antarctica itself, however, has demonstrated to me that one system--the U.N. system--works less effectively than I had hoped, while the other system--the ATS--works more effectively than I had expected. This experience has turned me away both from attempts to apply the common heritage principle to Antarctica and from efforts to institutionalize U.N. involvement in Antarctica, however well intentioned these attempts may be.

In the Antarctic Treaty we have something of value. It is a sound and proven basis on which to build in the future. I am confident of that. As is well known, I am a strong supporter of the United Nations and of the multilateral process, but in this particular case I cannot perceive a useful role for the United Nations beyond the production of the Secretary-General's study on Antarctica and the dissemination of information about Antarctica to interested countries.

The fundamental issue is this. Could the United Nations provide a practical alternative to the treaty or a more effective framework to regulate future activities in Antarctica? On the basis of my own experience I would have to answer, no.

The United Nations does have the potential to pass resolutions promoted by Third World countries, because of the majority that these countries can normally command on certain political and economic issues. However, the passage of some such resolutions, without consideration of whether a resolution can or will be implemented, or of its practicability, can have the result of undermining the credibility of the United Nations itself. While, as I have said, I personally am a strong supporter of the multilateral diplomatic process, I do not wish to see the United Nations drawn into areas in which it is bound to prove unable to act effectively, thereby risking the further erosion of its international credibility.

In conclusion, I believe that the ATS is an evolving and continuing experiment that has served the international community well. As Philip Quigg wrote in his recent book, A Pole Apart,[1] "Just as Antarctica's unique environment must be protected from exploiters, so must its political and economic future be protected from ideologues." The treaty, he continued, "deserves the opportunity to prove again its adaptability and the capacity of its members for adjustment and compromise, not only among themselves but with the rest of the world."

I should like to finish my remarks by quoting an old American adage which goes, "If it ain't broke, don't fix it." Well, in my opinion the Antarctic Treaty is not a broken instrument. It is a successful, open, and effective instrument. So I would urge its critics not to try to dismantle it but to accept it and to build on the flexible framework it provides. This would be--to use the language of the treaty--"in the interests of all mankind."

NOTE

1. New Press. A Twentieth Century Fund Report, McGraw-Hill Book Company (New York), 1983.

26.

The Evolution of the Antarctic Treaty System—The Institutional Perspective

R. Tucker Scully

INTRODUCTION

The Antarctic Treaty and the agreed recommendations and separate conventions developed pursuant to the treaty compose a system for managing activities in Antarctica--those activities that were taking place in 1959 when the treaty was concluded as well as activities that have developed since then. Analysis of the treaty system logically proceeds from two general perspectives: first, consideration of the substantive provisions of the treaty and the content of the steps undertaken under the aegis of the treaty--an analysis of what the treaty system has accomplished; and second, consideration of the process by which these steps have been accomplished--an analysis of how the treaty system operates. The emphasis of this chapter is on the latter perspective. However, the substantive elements of the Antarctic Treaty System are closely linked to the manner in which the system operates, and a summary of these elements serves as a necessary backdrop.

THE ANTARCTIC TREATY

The basic obligations in the Antarctic Treaty relate to the dedication of Antarctica exclusively to peaceful purposes and to maintenance of the freedom of scientific research there. In its preamble, the treaty recognizes that "it is in the interest of all mankind that Antarctica be used exclusively for peaceful purposes." The preamble further expresses the conviction that "a treaty ensuring the use of Antarctica for peaceful purposes only and the continuance of international harmony in Antarctica will further the principles and purposes embodied in the Charter of the United Nations."

The preamble articulates objectives with respect to scientific research that reflect the stimulus to conclusion of the treaty provided by the cooperative scientific programs undertaken in Antarctica during the International Geophysical Year. It acknowledges "the substantial contributions to scientific knowledge resulting from international cooperation in scientific investigation in Antarctica" and expresses the conviction that "establishment of a firm foundation for such cooperation on the basis of freedom of scientific investigation as applied during the International Geophysical Year accords with the interests of science and the progress of all mankind."

To give effect to the goals encompassed in the preamble, the treaty provides that Antarctica shall be used exclusively for peaceful purposes. It bans any measures of a military nature, including establishment of bases or fortifications, military maneuvers, and the testing of weapons. These prohibitions do not preclude use of military personnel or equipment for scientific research or any other peaceful purpose. The treaty also bans nuclear explosions in Antarctica as well as the disposal there of radioactive waste material.

The treaty provides for the continuation of the freedom of scientific research in Antarctica and international cooperation therein as applied during the International Geophysical Year. Further, the parties to the treaty agree, to the greatest extent practicable and feasible, to exchange information regarding plans for scientific programs in Antarctica, to exchange scientific personnel among expeditions and stations in Antarctica and to exchange and make freely available scientific observations and results from Antarctica. Finally, in the promotion of international cooperation in scientific research in Antarctica, every encouragement is to be given to the establishment of cooperative working relations with those U.N. specialized agencies and other international organizations having a scientific or technical interest in Antarctica.

Relatedly, the treaty obliges each party to provide advance notice of (1) all expeditions to Antarctica on the part of its ships or nationals and all expeditions to Antarctica organized in or proceeding from its territory, (2) all stations in Antarctica occupied by its nationals, and (3) any military personnel or equipment to be introduced into Antarctica in support of scientific research or other peaceful uses.

To ensure compliance with its provisions and promote its objectives, the treaty establishes a right of on-site inspection of all stations and installations in Antarctica. Each consultative party to the Antarctic Treaty has the right to designate observers, whose names shall be communicated to all other consultative parties. Each observer so designated enjoys complete freedom of access at any time to any or all areas of Antarctica. The treaty further stipulates that all areas of Antarctica, including all stations, installations, and equipment within those areas and all ships and aircraft at points of discharging or embarking cargoes or personnel in Antarctica, shall be open at all times to inspection by designated observers.

To establish these basic obligations relating to the peaceful use of Antarctica and to scientific research, the Antarctic Treaty also had to deal with basic legal and political differences over the status of Antarctica. Seven of the original signatories--Argentina, Australia, Chile, France, New Zealand, Norway, and the United Kingdom--assert claims to territorial sovereignty in Antarctica. The other five original signatories--Belgium, Japan, South Africa, the USSR, and the United States--neither assert nor recognize such claims. Article IV of the treaty incorporates an imaginative juridicial formulation through which the parties agree to disagree over sovereignty in Antarctica. Article IV provides that nothing contained in the treaty be interpreted as

- A renunciation by any party of previously asserted rights of or claims to territorial sovereignty in Antarctica,
- A renunciation or diminution by a party of any basis of claim to territorial sovereignty in Antarctica, or
- Prejudicing the position of any party as to recognition or nonrecognition of any other state's right of or claim or basis of claim to territorial sovereignty in Antarctica.

Further, Article IV establishes that no acts or activities taking place while the treaty is in force "shall constitute a basis for asserting, supporting or denying a claim to territorial sovereignty in Antarctica or create any rights of sovereignty in Antarctica. No new claim, or enlargement of an existing claim, to territorial sovereignty shall be asserted while the present Treaty is in force."

Article VI of the treaty delineates its area of application as "the area south of 60° south latitude, including all ice shelves." It adds, however, that "nothing in the present Treaty shall prejudice or in any way affect the rights, or the exercise of rights, of any state under international law with regard to the high seas within that area."

The juridical provisions of the Antarctic Treaty contained in Articles IV and VI are of central importance. They incorporate the accommodation between claimant and nonclaimant that permitted the conclusion of the treaty. These provisions create the means for parties to the treaty--starting from different assumptions--to apply common obligations to activities in Antarctica. They establish the basis for international cooperation in Antarctica and thus for the evolution of the Antarctic Treaty System.

As has often been pointed out, the Antarctic Treaty is a limited-purpose agreement. It sets out far-reaching obligations, which maintain Antarctica as a zone of peace and free and cooperative scientific research. It incorporates an imaginative juridical formulation to permit effective implementation of these obligations. However, the treaty did not deal with all possible activities in Antarctica and did not extend the juridical accommodation to those activities with which it did not deal.

At the same time, the treaty provides a mechanism for addressing new activities and new circumstances in Antarctica. Article IX of the treaty provides that the 12 original contracting parties meet within two months of entry into force of the treaty, and at suitable intervals thereafter, for the purpose of exchanging information, consulting together on matters of common interest pertaining to Antarctica, and recommending to their governments measures in furtherance of the principles and objectives of the treaty.

In addition to representatives of the 12 original signatories, participation in the meetings referred to in Article IX is open to representatives of any party that later accedes to the treaty during such time as that party demonstrates its interest in Antarctica by the conduct of substantial scientific research activity there, such as the establishment of a scientific station or the dispatch of a scientific expedition. (It should be noted that the Antarctic Treaty is open to accession by any member of the United Nations or by any other state so invited by the consultative parties to the treaty.)

The meetings provided for in Article IX are known as consultative meetings. Since the Antarctic Treaty entered into force in 1961, there have been 12 regular consultative meetings held at approximately two-year intervals. The next is planned to be held in Brussels in October 1985. There are now 16 consultative parties entitled to participate in these meetings. Four acceding parties have met the activities criterion of Article IX and joined the twelve original signatories in participating in consultative meetings--Poland in 1977, the Federal Republic of Germany in 1981, and Brazil and India in 1983.

The consultative mechanism outlined in Article IX of the treaty lies at the heart of what is known as the Antarctic Treaty System. It has been the vehicle for the development of agreed recommendations, to give effect to the substantive provisions of the treaty and, equally important, to delineate and respond to new issues and situations that have arisen since 1961. At the same time, the consultative mechanism has itself demonstrated a considerable development as it has been applied to respond to the substantive requirements of evolving activities in Antarctica.

THE ANTARCTIC TREATY SYSTEM--SUBSTANTIVE CONTENT

The Antarctic Treaty System, thus, refers to the Antarctic Treaty and the recommendations, measures, and agreements developed pursuant to the treaty, taken together with the consultative mechanism itself. There have been 138 agreed recommendations to governments developed at the 12 regular consultative meetings held since 1961.

The agreed recommendations have dealt with a wide variety of subject areas, spanning the range of possible human activities in Antarctica. These subject areas include the following:

- Facilities of scientific research, including the designation of Sites of Special Scientific Interest, where human activity is strictly limited to facilitate particular kinds of scientific observation;
- Cooperation in meteorology and in the exchange of meteorological data, including procedures for integrating antarctic data into worldwide analytical systems;

- Cooperation in telecommunications, including procedures for contact between stations in Antarctica;
- Cooperation in air transport and logistics, including search and rescue and emergency assistance;
- Tourism, including development of guidance for visitors to Antarctica to ensure observation of the measures adopted to protect the antarctic environment from harmful impacts;
- The impact of humans on the antarctic environment, including a recommended code of conduct for stations in Antarctica and recommendations to develop procedures for assessing impacts of operations in Antarctica;
- The preservation of historical sites and monuments;
- Exchange of information, including procedures for elaborating the advance notification and data sharing obligations of the treaty;
- The operation of the Antarctic Treaty System; and
- The preservation and conservation of wildlife and living resources in Antarctica, including the Agreed Measures for the Conservation of Antarctic Fauna and Flora, which established a system of specially protected species and specially protected areas to ensure that the impacts of human activity on native species of wildlife in Antarctica are properly controlled and regulated.

Equally important, there have been agreed recommendations concerning antarctic resources. With respect to living resources, consultative meeting recommendations have led to the negotiation and conclusion of the 1972 Convention for the Conservation of Antarctic Seals and the 1980 Convention on the Conservation of Antarctic Marine Living Resources.

The Convention for the Conservation of Antarctic Seals, which entered into force in 1978, applies to the area of the Antarctic Treaty and prohibits the killing or capturing of Ross seals, southern elephant seals, and southern fur seals. It establishes quotas for the takes of crabeater seals, leopard seals, and Weddell seals as well as closed seasons for sealing, seal reserves, and sealing zones. The convention is designed to create a means of controlling commercial sealing in Antarctica should such activity emerge (or re-emerge). It calls upon the Scientific Committee on Antarctic Research (SCAR) to perform at least for an initial period scientific and advisory functions, but also provides for the establishment of permanent machinery should commercial sealing be initiated. To date, commercial sealing has not developed.

The Convention on the Conservation of Antarctic Marine Living Resources (CCAMLR) entered into force in April 1982. As is indicated in its title, CCAMLR has as its objective the conservation of antarctic marine living resources, with conservation understood to include rational use of resources. CCAMLR rests on an "ecosystem approach" to management of living resources and requires that any harvesting activities be conducted in accordance with conservation principles designed to ensure the health of target and dependent populations, to maintain ecological relationships, and to prevent irreversible changes in the antarctic marine ecosystem.

Consistent with its conservation objectives, the convention applies to a geographic area designed to approximate the full extent of the antarctic marine ecosystem. This area, defined by specific coordinates, extends to those waters found south of the Antarctic convergence, or polar front, which is the transition zone between antarctic waters to the south and warmer subantarctic waters to the north. It should be noted that the convention area is considerably larger than that covered by the Antarctic Treaty (which applies to the area south of 60°S latitude).

The marine area covered by the convention may offer significant potential for harvesting. At the same time, scientific investigations have indicated that the antarctic marine ecosystem--which is characterized by short, simple food chains and dependency upon a single species, antarctic krill--may be particularly vulnerable to uncontrolled harvesting. For this reason, the consultative parties committed themselves to conclusion of a conservation system prior to the initiation of large-scale harvesting activities. Implementation of the convention therefore offers an unusual opportunity to elaborate and apply an effective management framework to these resources before they become the object of significant exploitation pressure.

The convention deals with basic differences of view over the existence and nature of marine jurisdiction in the convention area, which derive from the divergence over claims to territorial sovereignty in Antarctica. However, in a manner parallel to the Antarctic treaty, CCAMLR provides the basis for its parties to cooperate in elaborating an effective management system without prejudice to their juridical positions.

The convention also provides for the establishment of machinery necessary to carry out its objectives. This includes the Commission for the Conservation of Antarctic Marine Living Resources, with headquarters in Hobart, Tasmania; the Scientific Committee for the Conservation of Antarctic Marine Living Resources, designed to provide objective scientific assessments and recommendations to the commission; and a secretariat to serve both the commission and scientific committee.

Since the entry into force of CCAMLR, there have been annual meetings of the commission and its scientific committee. The first two annual meetings (at the headquarters site in Hobart, Tasmania) were concerned largely with start-up functions--establishment of the secretariat, elaboration of the headquarters agreement, rules and procedures, and staff regulations and financial matters including budgets and financial regulations. At the same time, the scientific committee made substantial progress in delineating its initial program of work.

The third annual meeting of the commission and the scientific committee took place in Hobart on September 3-14, 1984. The 16 members of the commission participated, as well as observers from Brazil and Sweden and six international organizations. At the 1984 annual meeting initial measures dealing with fishing activities in Antarctica were adopted. The transition from the start-up phase to the implementation stage of CCAMLR seems to have taken place. The fourth annual meeting takes place in September, 1985.

With respect to antarctic mineral resources, the Antarctic Treaty consultative parties, at the eleventh consultative meeting, held in Buenos Aires in 1981, agreed that a regime for antarctic mineral resources should be concluded "as a matter of urgency." The relevant recommendation--Recommendation XI-1--called for convening a special consultative meeting to elaborate a regime and to undertake other steps relating to the negotiations, including a decision as to the form of the regime and procedures for its adoption.

The term "regime" used in Recommendation XI-1 is understood to mean an international system for making decisions about possible mineral resource activities in Antarctica. Essentially, the system would have as its purpose determination of the acceptability of mineral resource activities in Antarctica, should interest in these develop, and the management of any such activities determined to be acceptable. Specifically, Recommendation XI-1 provided that the regime should include means for

(1) Assessing the possible impact upon the antarctic environment of antarctic mineral resource activities;
(2) Determining whether such activities are acceptable;
(3) Governing the environmental, technological, political, legal, and economic aspects of such activities as may be found to be acceptable;
(4) Establishing rules for the protection of the antarctic environment; and
(5) Ensuring that any antarctic mineral resource activities undertaken are in strict conformity with such rules.

The consultative parties reaffirmed their commitment to ensure that no exploration or development of antarctic mineral resources take place while making progress toward the timely adoption of an agreed regime. They also identified a number of principles and elements that should be reflected in the regime. These include the following:

- Maintenance of the Antarctic Treaty in its entirety,
- Ensuring protection of the unique antarctic environment and of its dependent ecosystems,
- Ensuring that the interests of all mankind in Antarctica are not prejudiced,
- Ensuring that the principles of Article IV of the Antarctic Treaty are safeguarded,
- Inclusion of procedures for adherence by states other than consultative parties,
- Application of the regime to all mineral resource activities on the Antarctic continent and its adjacent offshore areas but without encroachment on the deep seabed,
- Provision for cooperative arrangements between the regime and other relevant international organizations, and
- Promotion of the conduct of research required to make the necessary environmental- and resource-management decisions.

It is not known at this stage whether there are deposits of mineral resources in Antarctica whose development would be economically feasible. The commitment of the consultative parties to negotiate a regime rests on the belief that it is important to have an effective mechanism in place for making informed decisions about possible mineral resource activities before any specific interest in those activities might develop. The consultative parties share a commitment to ensure that no mineral resource activities take place unless it can be demonstrated that they could be undertaken in an environmentally sound fashion. Agreement to develop such a regime also reflects a determination that interest in antarctic mineral resource activities, should it develop, does not become the source of international discord or conflict.

The first session of the special consultative meeting on antarctic mineral resources convened to begin the process of elaboration of the regime took place in Wellington, New Zealand, in June 1982. This has been followed by an informal session, also in Wellington, in January 1983; a formal session in Bonn in July 1983; an informal working group in Washington in January 1984; an informal session in Tokyo in May 1984; and an informal session in Rio de Janeiro in February/March 1985. The next round takes place in Paris in September/October 1985.

THE ANTARCTIC TREATY SYSTEM--INSTITUTIONAL RESPONSE

A summary of the substantive elements of the Antarctic Treaty System reveals a complex and evolving pattern of measures for ensuring necessary coordination and management of activities in Antarctica. This pattern also represents a set of institutional responses or perhaps institutional processes.

The Antarctic Treaty itself pays scant attention to institutional elements. Article IX, as is indicated above, makes provision for consultative meetings, but, other than calling for such a meeting within two months of entry into force and "thereafter at suitable intervals and places," does not address the operation and organization of these meetings. As I have indicated elsewhere, "the lack of institutional provisions in the Treaty... appears to stem more from an intentional desire to provide flexibility in future institutional development rather than an inability to agree upon institutional mechanisms."[1] Certainly, the Antarctic Treaty System has demonstrated flexibility and pragmatism in response to new issues and circumstances.

From the outset, the Antarctic Treaty System has been science intensive. A primary impetus to conclusion of the treaty itself came from the scientific community, which had been involved in the International Geophysical Year (IGY) and which wished to perpetuate the creative arrangements for cooperative scientific research carried out during the IGY. In fact, the nongovernmental body that was formed to coordinate IGY programs in Antarctica-- SCAR--was placed on permanent footing even prior to entry into force of the treaty. SCAR, a committee of the International Council of Scientific Unions, acts as the key, though informal, vehicle through which the scientists active in Antarctica coordinate scientific activities and scientific priorities.

Equally important, SCAR has also functioned as the scientific advisory body for the Antarctic Treaty System from the outset. The linkage between SCAR and the Antarctic Treaty consultative parties is an indirect one. Recommendations to and from SCAR are conveyed through the individual national committees of SCAR rather than directly to or from governments. Nonetheless, SCAR has been instrumental in providing expert scientific advice on issues requiring common action by the consultative parties and, further, in identifying issues in need of such attention. Because of the nature of its linkage to the consultative parties, SCAR plays an important peer-review function, bringing to bear an independence in scientific advice and judgment on the operation of the consultative mechanism.

Supported from the outset by SCAR, the consultative parties, on the basis of the provision for meetings contained in Article IX, have elaborated an effective consultative mechanism. The practice has emerged of holding a regular consultative meeting at approximately two-year intervals. The site of these meetings rotates among the consultative parties based on alphabetical sequence (in English). No permanent secretariat has been established for the consultative mechanism, but the secretariat function rotates with the responsibility of the host country. The host of the next consultative meeting undertakes the necessary preparations, including organization of a meeting or meetings to prepare the agenda for the consultative meeting as well as coordination of communications and information flow among the consultative parties in the period leading up to the meeting.

Early in the development of the consultative mechanism, the need to supplement the regular biennial consultative meeting forum with techniques for directing concentrated attention to particular issues or subject areas became apparent. These techniques have included meetings of specialists or experts to bring to bear particular knowledge or experience to the resolution of particular issues. For example, there have been three Antarctic Treaty meetings on telecommunications, which provided detailed recommendations acted on at subsequent regular consultative meetings. In analogous fashion, groups of experts have been constituted at regular consultative meetings to examine and report on particular subject areas.

This trend toward development of distinct negotiating forums accountable to, but distinct from, the regular consultative meeting has led to the emergence of the special consultative meeting. The first special consultative meeting was held in London, in 1977, to address the question of Poland's participation in regular consultative meetings pursuant to Article IX of the treaty. At that meeting, the participants agreed that Poland should be seated at the next regular consultative meeting and elaborated procedures for considering such instances in the future. The procedures include provision for convening special consultative meetings. The third special consultative meeting, in 1981, confirmed the consultative status of the Federal Republic of Germany; the fifth special consultative meeting, in 1983, confirmed the consultative status of Brazil and India.

The second and fourth special consultative meetings were convened to deal with resource issues--antarctic marine living resources and antarctic mineral resources, respectively. An antecedent for the special consultative meetings is to be found in the special preparatory meeting on antarctic mineral resources, which took place in Paris in 1976 to prepare for the item on mineral resources on the agenda of the ninth consultative meeting. The second special consultative meeting resulted from Recommendation IX-2 of the ninth consultative meeting, which called for such a meeting to elaborate a draft regime on antarctic marine living resources and to determine its form as well as procedures for its adoption. The second special consultative meeting held three sessions: the first in Canberra in January/February 1978, the second in Buenos

Aires in July 1978, and the third in Canberra in May 1980, immediately prior to the diplomatic conference that concluded the CCAMLR. These formal sessions were supplemented by intersessional meetings and consultations of an informal character.

As indicated earlier, the same pattern is being followed with regard to antarctic mineral resource issues. The fourth special consultative meeting opened with a formal session in Wellington in June 1982, and has been followed by one formal session, three informal sessions and an informal working group meeting.

The institutional development of the Antarctic Treaty System has also included negotiation of separate legal instruments, specifically, the Convention for the Conservation of Antarctic Seals concluded in 1972 and the CCAMLR concluded in 1980 (whose provisions are described in the previous section). The initiative for each of these conventions--which, though separate legal instruments, are tied closely with the Antarctic Treaty-- emerged from the consultative mechanism. Each was preceded by agreement by the consultative parties to act in accordance with interim guidelines while developing the legal arrangements determined to be necessary. (It should be noted that the commitment by the consultative parties to ensure that no commercial mineral resource activities take place while making progress toward conclusion of a regime for antarctic mineral resources is another example of this practice.)

The Convention for the Conservation of Antarctic Seals foresees the establishment of permanent machinery to regulate sealing activities in Antarctica if and when commercial sealing occurs. As noted above, the need for such machinery has not arisen, and SCAR continues to perform--on an interim basis--necessary institutional functions under that convention.

With the entry into force of the CCAMLR, however, there have been created continuing institutions that are distinct from the consultative mechanism itself. Though CCAMLR represents an important component of the Antarctic Treaty System, it establishes for the first time within the system a separate regulatory mechanism commission and scientific committee--and a permanent secretariat. This development responds to the requirements of pursuing effective management of the antarctic marine ecosystem. It also represents an important stage in the evolution of institutional techniques within the Antarctic Treaty System.

The elaboration of the consultative mechanism--
including the development of separate instruments and new
institutions--has been accompanied by the development of
ties with other international bodies. Cooperative working
relationships with U.N. specialized agencies and other
international organizations having scientific or technical interest in Antarctica are foreseen in Article
III(2) of the treaty.

In addition to the long standing ties between SCAR and
the consultative mechanism, working relationships have
been developed with a number of international bodies,
including the World Meteorological Organization, the
International Telecommunications Union, the Intergovernmental Oceanographic Commission (IOC), and the International Civil Aviation Organization.

CCAMLR also makes specific provision for establishment
of such working relationships. At the most recent meeting
of the CCAMLR in September 1984, observers participated
from the U.N. Food and Agriculture Organization, the
International Whaling Commission, the International Union
for the Conservation of Nature and Natural Resources, and
the Scientific Committee on Oceanographic Research, as
well as from SCAR and the IOC.

As a conclusion to this section on the institutional
response of the Antarctic Treaty System, I would like to
repeat the assessment of the nature of the system which I
set forth in 1982:

> The system which has evolved under the Antarctic treaty appears both simple and pragmatic.
> These characteristics should not mask the fact
> that the treaty system has also been flexible and
> innovative in response to new and evolving issues.
>   Within this system, the consultative mechanism
> itself--the consultative meetings--plays the
> primary role in the identification of issues
> requiring common action or response by the
> Consultative Parties. Even with regard to subject
> areas, which eventually elicited establishment of
> new institutions such as conservation of marine
> living resources, the need for a regime and the
> purposes and principles it should incorporate were
> defined within the regular meetings of the
> Consultative Parties, drawing upon the scientific
> advice of SCAR.
>   Within the framework of this consultative
> mechanism, a wide variety of techniques have been

established for the analysis of possible responses to issues once identified. This function has been performed at the regular Consultative Meetings themselves or through more specialized meetings, including Special Consultative Meetings. Equally important, these techniques provide a means for achieving consensus among the Consultative Parties on the appropriate type of solution to the issue at hand.

Finally, the system has adopted a number of institutional means for responses to issues requiring common action. These range from the agreed measures approach, fully within the consultative mechanism, to the establishment of new instruments such as the CCAMLR which can stand independently of that mechanism. The Convention on the Conservation of Antarctic Seals represents a mid-way point on this spectrum.

In contrast to many collective international undertakings, the Antarctic Treaty system has created new institutions and techniques only as and when necessary. New institutions and new institutional techniques have been conceived and perfected in response to its well-defined need and well-defined problems. This decentralized and evolutionary approach to institutional building has permitted the institutions themselves to be tailored to the function they were designed to perform. For example, differing components of the system may involve differing types of obligations, differing participation and differing types of relationships to other international bodies. The CCAMLR, again, illustrates this point. The Convention applies to an area which is larger than that of the Antarctic Treaty area. The requirements of conserving antarctic marine living resources were determined to be controlling rather than the definition of the Antarctic Treaty area. The CCAMLR also envisages participation in the management system by a group of countries which may, and in fact does, differ distinctly from the Consultative Parties currently entitled to participate in the meetings provided for in Article IX of the Antarctic Treaty. All of the Consultative Parties are entitled to participate but additional parties engaged in research or harvesting activities are also so entitled. The concept of an activities criterion such as that included in the Antarctic Treaty has been adapted to the requirements of living resources management.

This decentralized and functionally-oriented system which has emerged over the past two decades, has played an important part in the practical realization of the obligations of the Antarctic Treaty and the unique form of international cooperation which has taken place pursuant to it. It demonstrates the will of the Consultative Parties not only to implement the provisions of the Treaty, but also to deal effectively with the new strains and challenges generated by resource issues. In fact, the emergence of resource issues has provided a major impetus to the evolution of the system and may well be the key to its future development.[2]

OPERATION OF THE ANTARCTIC TREATY SYSTEM

In the two years since reaching the above conclusion about the Antarctic Treaty System (which I consider equally valid today), there has been increased attention to the operation of the system. The subject figured prominently on the agenda of the twelfth Antarctic Treaty consultative meeting, held in Canberra in September 1983, and is likely to be an important item at the thirteenth consultative meeting in late 1985.

Antarctica and the Antarctic Treaty System have become the object of growing international attention and interest. In the past two years, 9 nations have acceded to the Antarctic Treaty, bringing the total number of parties to 32. Two parties, Brazil and India, have achieved consultative status, bringing the total number of consultative parties to 16. All of CCAMLR's original signatories are now parties to the convention, as are the European Economic Community, Spain, Sweden, Uruguay, the Republic of Korea, and India. Finally, as a reflection of the increased awareness of Antarctica among nations that are not party to the treaty, an item on Antarctica has been included on the agenda of the United Nations, and, as a result, an extensive report on Antarctica has been prepared and circulated by the U.N. Secretary-General.

This interest has had the result of illuminating the important accomplishments of the Antarctic Treaty System.

The realization of the principles and purposes of the treaty, the implementation of its specific provisions, and the measures adopted pursuant to the treaty constitute an impressive record. Concurrently, attention has been directed toward the operation of the Antarctic Treaty System--particularly to its decision-making and participation requirements and to the relationship of the system to other elements of the international system. This effort to bring the institutional perspective (broadly construed) to bear on the Antarctic Treaty System will therefore conclude by seeking to address these areas.

The operation of the Antarctic Treaty System--both in the substantive sense and in the institutional sense--rests on the juridical provisions contained in Article IV. These provisions reflect two basic conclusions:

- First, that it is not necessary to resolve differences over the legal and political status of Antarctica in order to establish the basis for managing human activities in Antarctica, and
- Second, that efforts to resolve differences over the legal and political status of Antarctica would be inconsistent with the commitment to reserve Antarctica exclusively for peaceful purposes.

Recognition of these factors permitted negotiation of the Antarctic Treaty. The framers of the treaty perceived that an effort to resolve the claims issue--either through perfection and recognition of claims to territorial sovereignty in Antarctica or through extinguishing and renunciation of such claims--would simply have resulted in discord and conflict. They also understood that it is possible, without agreeing on who, if anyone, owns Antarctica, to construct a system for applying necessary obligations and controls on activities in Antarctica.

The conclusions reached by the framers of the treaty have been corroborated in the years of operation of the treaty system. In a period of ideological competition--East and West, North and South--there is no doubt that an effort to determine the legal character of Antarctica would undercut the pattern of peaceful international cooperation which has emerged there. Equally, the development of an effective system of managing activities in Antarctica has demonstrated that there is no need for such determination. The "irresolution" of the claims question is often considered a drawback of the Antarctic Treaty. In fact, it has proved to be one of its strengths. The Article IV approach has been effective in neutralizing political and ideological impediments to dealing effectively with antarctic issues, and the operation of the treaty system stands as an unusual example of conflict management and avoidance.

The Article IV approach--to use that shorthand--has important implications for decision-making and participation within the Antarctic Treaty System as well as for the relationship of the system with the international community. First, it has substituted a functional basis for a political or ideological basis for involvement in decision-making. Commitment to a particular legal status for Antarctica does not establish eligibility to take part in decisions relating to activities in Antarctica. Demonstration of concrete interest in those activities becomes the standard. The activities criterion for achieving consultative status under the Antarctic Treaty or for membership in the Commission under CCAMLR therefore flows from the Article IV approach.

This approach has permitted an extremely diverse group of countries to cooperate effectively on Antarctic matters even during periods when relations among individual consultative parties have otherwise been hostile or even combative. It is a signal achievement that the Antarctic Treaty and all the measures and agreements that together with it constitute the Antarctic Treaty System have been negotiated by consensus.

Second, and relatedly, the Article IV approach implies a functional cast to the relationships between the Antarctic Treaty System and other components of the international system. The importance of not seeking to impose a particular legal status on Antarctica is as essential to the effective relationships among the international institutions concerned with Antarctica as it is among the states concerned.

A third implication of the Article IV approach is that the system must be open to new participants. Specifically, this approach entails that any state be able to express an interest in Antarctica by accession to the treaty or to another component of the treaty system, and that any state that gives concrete form to such interest by actual involvement in activities in Antarctica, be able to participate in decision-making regarding those activities.

Finally, the approach implies a responsibility on the part of those participating in the Antarctic Treaty System to ensure that there is available adequate information about Antarctica and the operation of the system. The Antarctic Treaty System places an emphasis on demonstration of concrete interest in Antarctica. For such a system to operate effectively, there must be sufficient information available to allow potential participants--states or organizations--to make informed judgments as to their interest in Antarctica.

CONCLUSION

At this point, it is perhaps appropriate to look for trends in the ongoing evolution of the Antarctic Treaty System. As is evident from the previous section, the catalyst to the development of the treaty system is the approach of defining and resolving issues inherent in Article IV of the treaty. This approach--with its consequent functional, activities-oriented, and consensual form of decision-making--remains and will remain the basis for the successful operation of the treaty system.

With this in mind, there are in the results of the most recent regular consultative meeting--the twelfth, held in Canberra in September 1983--indications of the direction of the evolution of the Antarctic Treaty System. First, parties to the Antarctic Treaty that are not consultative parties participated as observers, and the rules of procedure of consultative meetings were amended to provide for such participation. The question of observers had been discussed informally at previous consultative meetings and was a specific item on the agenda of the eleventh consultative meeting in 1981. The provision for participation by nonconsultative parties initiated at the twelfth consultative meeting represents a response to the growth in interest in Antarctica, which has also been reflected in increased accession to the treaty. Action at the twelfth consultative meeting was followed by similar agreement in the special consultative meeting on antarctic mineral resources; nonconsultative parties to the antarctic treaty were invited to attend as observers the antarctic mineral resource negotiations. As noted earlier, there was attendance at the 1984 CCAMLR meeting by observers from Spain and Sweden (parties to CCAMLR) and Brazil (a nonparty) as well as by interested international organizations.

Discussions at the twelfth consultative meeting also emphasized further development of working relationships between the components of the Antarctic Treaty System and other international organizations interested in Antarctica. The report of the meeting indicates agreement that consideration should be given in preparation of agendas for future consultative meetings as to whether discussion of particular agenda items would be assisted by attendance of a U.N. specialized agency or other international organization having a scientific or technical interest in Antarctica. Further, Recommendation XII-6, "Operation of the Antarctic Treaty System," calls for bringing to the attention of such organizations material in consultative meeting reports relevant to their interest in Antarctica.

With respect to the general question of information about the Antarctic Treaty System, the same recommendation of the twelfth consultative meeting suggested a number of items to ensure broader availability of consultative meeting reports, the record of agreed recommendations adopted at consultative meetings, and consultative meeting documentation. The consultative meeting also called on the depository government (United States) to undertake a general examination of this issue. While treatment of the information requirements of the Antarctic Treaty System will be a continuing item, the recommendations developed at the twelfth consultative meeting represent the fruition of discussions initiated at the ninth consultative meeting in 1977.

From these developments, there emerges a picture of the Antarctic Treaty System in the process of responding to and accommodating new circumstances and situations. The specific nature of this response involves an opportunity for greater involvement in the operation of the consultative mechanism by nonconsultative parties, an opportunity for the emergence of closer functional ties between the Antarctic Treaty System and other international bodies, including the functional components of the U.N. system, and an opportunity for greater information flow to and from the components of the Antarctic Treaty System. From a broader perspective, however, those developments fit the pattern of flexible and pragmatic institutional development that has been characteristic of the Antarctic Treaty System from its earlier days. It is the picture of this resiliency that permits confidence that our children and grandchildren will also see Antarctica as an area of the world reserved exclusively for peaceful and cooperative pursuits and, it is hoped, an example that will have been emulated elsewhere.

NOTES

1. Scully, R.T. 1983. Alternatives for Cooperation and Institutionalization in Antarctica: Outlook for the 1990's. In F. Orrego Vicuna, ed. Antarctic Resources Policy: Scientific, Legal and Political Issues, Cambridge University Press (Cambridge), 1983, p. 283.
2. Scully, R.T. 1983. Alternatives for Cooperation and Institutionalization in Antarctica: Outlook for the 1990's. In F. Orrego Vicuna, ed. Antarctic Resources Policy: Scientific, Legal and Political Issues, Cambridge University Press (Cambridge), 1983, p. 291-292.

27.

Panel Discussion on Institutions of the Antarctic Treaty System

The panel consisted of Lee A. Kimball (moderator), Cristian Maquieira, and Rolf Trolle Andersen

The summarized presentations of three participants who requested the opportunity to make statements on the theme of this section are included below. Their remarks are followed by those of the panel.

REMARKS BY BO JOHNSON THEUTENBERG

Theutenberg noted Sweden's accession to the Antarctic Treaty on April 24, 1984, and to the Convention on the Conservation of Antarctic Marine Living Resources (CCAMLR) in June 1984. He described Sweden's early involvement in both antarctic and arctic science, as had been mentioned in other workshop sessions, but noted that since the 1959 treaty had been concluded Swedish participation in Antarctic science had been manifested primarily by individual scientists.

In this context he cited the transition--once the Antarctic Treaty was concluded--from individual scientists and scientific societies to government representatives as the major determinants of antarctic developments. Since Sweden was not an original signatory to the Antarctic Treaty, its government at the time the treaty became effective was not interested in being restricted to non-consultative status, nor did the Swedish scientific society seek government support to promote its specific interests within the treaty system. So for 25 years pursuit of antarctic interests in Sweden had lapsed.

During the same period, as securing access to resources had become a critical national objective, science and technology forced their way into the lives of states. Theutenberg noted that these factors affect the formation of national foreign policies as well as principles of international law, not least those regarding protection of nature and the development of international environmental law. Sweden was no exception, and it now wished to maintain and even expand its scientific interests with regard to neighboring ocean areas, including polar research. In this regard, he stated, Sweden must seek cooperation with other interested nations. Renewed efforts by Swedish scientists in these fields have made it natural for Sweden to enter the Antarctic Treaty System (ATS) and the Scientific Committee on Antarctic Research (SCAR), and Sweden will seek to promote regional and international scientific cooperation in polar areas.

In commenting on the Antarctic Treaty, Theutenberg suggested two alternative outcomes that might have resulted had the 1959 negotiations failed: military conflict in the area or a legal/political solution to the claims based on argumentation by claimants and nonclaimants. While the first alternative might have de facto settled the territorial status of Antarctica, he doubted that a legal/political solution could have been arrived at. In either case, however, sovereignty in Antarctica would have been resolved long ago. Had that been the case, the wider circle of nations and other interested parties would never have found themselves today in the position of discussing these matters. He stated that the founding members of the treaty should be given some credit for the course of action taken in 1959, because they actually opened the way for other states to take an active interest in Antarctica and to participate in the Antarctic Treaty. If the treaty system were to fail, he believed that the alternatives for what is called "mankind" would probably be much less attractive than under the present regime.

Moreover, a common heritage regime could be established only by consensus, as had been done in the Law of the Sea conference, and it was clear that this goal could not be achieved with regard to Antarctica from either a political or a legal point of view. He did not believe that the common heritage principle contains any kind of ownership- or management-oriented elements; it is not in itself a regime for exploitation, common ownership, or joint management of resources. It does contain the legal/political basis for custodianship and has the meaning that the heritage should be passed on to future generations without destruction. He wondered whether this kind of regime will be of any help to the world community when it comes to the question of exploitation of antarctic resources.

Theutenberg stressed that his country believes that it is important that the treaty system be supported by a wider circle of nations and that in general every member of the international community have a right to follow questions of international importance and advance proposals and ideas relating to them, especially concerning environmental protection and exploitation of resources. For states affected by or believing that they have vital interests to preserve in antarctic affairs, he stated, the mechanisms for involvement should be strengthened.

He believed that it is difficult enough to solve real conflicts within the ATS as it operates today; but outside parties have no way in a legal sense to intervene in potential conflicts, although they could possibly do so by influencing world opinion. Criticisms of the ATS have therefore arisen because outsiders believe that they have no ability to influence or intervene in affairs affecting them.

Taking this into account, some promising proposals exist in the minerals regime negotiations not only for mechanisms that would allow other concerned parties a great deal of influence, but also with respect to some ideas advanced by developing nations. There are also some interesting ideas aimed at environmental protection in the minerals regime papers, which could perhaps be incorporated into the ATS itself. Theutenberg supported working in the direction of openness with regard to environmental issues, possibly by establishing some kind of information and management mechanism in which broader interests could have some influence.

Finally, he expressed the hope that the status of observer will provide the nonconsultative states with the ability to influence the important matters discussed at biennial treaty meetings and in the minerals regime negotiations.

REMARKS BY ALEXANDRE KISS

Kiss elaborated on the background and recent evolution within the international legal system of the ill-defined concept of the "common heritage of mankind," which had arisen a number of times during discussions in the workshop. He noted its appearance in the 1979 Agreement Governing the Activities of States on the Moon and Other Celestial Bodies (Moon Treaty) and in the 1982 Convention on the Law of the Sea. (The Moon Treaty entered into force on July 11, 1984, whereas the Law of the Sea Convention is not yet in force.)

Kiss recalled that as modern international law took definite form at the beginning of the nineteenth century, it resulted in a growing number of treaties whose main characteristic was that they were based on material reciprocity; that is, states accepted obligations restricting their sovereignty in order to gain perceived advantages of equivalent value.

As early as 1815, however, a few treaty rules were formulated that were not based on this criterion of material reciprocity. They proclaimed freedom of navigation, first in international rivers and later in the newly constructed interoceanic canals of Suez and Panama. Although these were agreements among a limited number of countries, they could be considered the first expression of what could be called the "common interest of mankind": In the aim of maintaining freedom of international communications, states parties accepted obligations without any immediate advantage to themselves.

Following World War I, the concept of the common interest of mankind found expression in international labor treaties drafted by the International Labor Organization: States accepted obligations to apply certain social rules to their nationals without any compensation other than the belief that other states would be encouraged to act in the same way and thus improve human social conditions worldwide. After World War II, this concept was developed in the fields of human rights and environmental protection law.

The common feature of all these international treaties, according to Kiss, is that they are not based on reciprocity, unlike most treaty rules, and in particular that they do not ensure any immediate advantage to the contracting states. On the contrary, they impose on them obligations that restrict their sovereign rights: States must adopt legislation protecting the fundamental rights and freedoms of individuals or of the environment within their territories, and they may be obliged to explain their acts before international organizations when a violation of their treaty obligations is alleged. States subordinate their immediate interest and their freedom of action in order to promote a superior objective, such as human rights or environmental protection. This works only when all states of the world—or at least those of a certain region—agree likewise to restrict their freedom of action.

Since the early nineteenth century, then, the application of treaty rules based on the concept of the common interest of mankind had been extended from international waterways to certain substances and situations within national territory. More recently, with the depletion of natural resources owing to growing demand and demographic pressures, a new perspective had appeared: That future generations should not be deprived by human activities of the benefit of natural resources, or, put another way, that this generation has inherited resources that should be managed in a way that quantitatively and qualitatively should offer the same possibilities to its children and grandchildren.

The main features of this new regime of the common heritage of mankind are the following: (1) common use of or free access by every nation, (2) prohibition of nonpeaceful use, (3) international management with a view to ensuring rational use of resources in the interest of present and future generations, and (4) access by the whole of humankind to an equitable share of the benefits, be they material, cultural, or scientific.

Some persons would add a fifth feature: the prohibition of appropriation by states, but this is an open question. It appears in the regime for deep seabed minerals in the 1982 Law of the Sea Convention and in the 1979 Moon Treaty, and it is implicit in the regime for the geostationary orbit. Kiss believed that it is questionable whether it could be applied to the radio spectrum or to Antarctica. He added that it should be borne in mind that there are various nonappropriable elements that are not necessarily part of the common heritage of humankind. This is the case with the high seas and the airspace above them as well as with outer space on the whole (though not of celestial bodies). In these areas, which are nonappropriable and should be used for peaceful purposes, there is neither a prohibition on military use nor, as yet, a prescription for international management and rational use.

Kiss added that another aspect of the common heritage of mankind was illustrated by the 1972 United Nations Economic, Social, and Cultural Organization (UNESCO) World Heritage Convention. While monuments, historic cities, parks, etc. may remain under national sovereignty, or even as private property, the convention provides that their conservation is in a way under international scrutiny to ensure that they benefit future generations. Several international conventions on the conservation of wild fauna and flora have the same effect. Thus, while no one contests Italian sovereignty over Venice or the sovereign rights of Kenya to use its territory, the conscience of mankind would hardly accept the destruction of Venice in favor of an industrial complex or the destruction of all wildlife in Kenya. He saw the growth of a sense of moral rights shared by all humanity with respect to cultural and natural components of an increasingly universal civilization.

In conclusion, Kiss stressed the important concept of trust: Individual states, groups of states, or in some cases the whole international community are vested with the task of managing, conserving, and transmitting to future generations some essential components of our planet or even of bodies in outer space. The common heritage of humankind could be organized according to different legal patterns, but these patterns all share the four criteria identified earlier. In addition to the resources of the deep seabed and those of celestial bodies, other items falling into this category include the geostationary orbit, the radio spectrum, Antarctica, components of our cultural and natural heritage, and certain wild fauna and flora. In the future the ozone layer of the Earth's atmosphere, the global climate, and our genetic heritage could be added. In all cases where the concept of the common heritage of humankind has already been applied, the common interest of humankind materialized in a legal instrument without any commensurate material interest being granted to contracting states.

In his view, whatever the territorial status of Antarctica might be, the area satisfies the criteria of the common heritage of humankind: All nations have access to it, nonpeaceful and even military uses are prohibited, it is managed by international bodies, and the benefits--that is, the scientific knowledge acquired in Antarctica--are largely shared with the rest of the world community. He believed that this is how the ATS serves the common interest of humankind.

REMARKS BY JOSE SORZANO

Requesting an opportunity to "critique Zain's critique" of the ATS (see Chapter 21 by Zain Azraai), Sorzano affirmed Zain's impressions of the "reaction of extreme sensitivity and resentment" on the part of the Antarctic Treaty consultative parties (ATCPs) at the United Nations regarding any questioning of the current antarctic system and Malaysia's initiative at the United Nations. He took note of the reassurances obtained from repeated statements by Malaysia that it does not wish to undermine the ATS, preferring consensus to confrontation, and reported that the ATCPs have reciprocated by showing flexibility and not opposing inscription of the antarctic issue on the United Nations General Assembly agenda nor the production of the United Nations Secretary-General's study of Antarctica.

He pointed out, however, that there have been other Malaysian statements of intention that point in different directions and therefore seem incompatible with the reassurances previously articulated by Malaysia. These divergent statements have created a great deal of speculation concerning Malaysia's objectives and intentions with regard to the ATS.

Sorzano believed that Zain's presentation as set forth in Chapter 21 illustrates this point: Zain examined the possible justifications for the ATS but did not find them persuasive. He argued that a justification based on know-how and expertise is insufficient because it creates a two-tiered, nonuniversal body, which does not meet the criteria of "natural justice" and "interest."

The ATCPs had found Malaysia's intentions confusing, because Malaysia was simultaneously saying that it did not want to destroy the treaty and then challenging the treaty's legitimacy; it appears Malaysia seeks consensus and compromise only to raise an issue that cannot be negotiated or compromised. In Sorzano's view, questions of legitimacy <u>by their very nature</u> cannot be compromised; they are zero sum: the treaty is either legitimate or it is not; it cannot be something between.

In examining Zain's arguments, Sorzano first wished to distinguish between the basis for territorial claims and the basis for justification of the treaty. While he found it doubtful that the basis for territorial claims could be set aside as a historical anachronism, as Zain suggested, nevertheless, even if the claims were set aside, that fact would not impinge on the basic justification and validity of the treaty. He believed that the treaty stands on its own, independent of the claims and their validity or lack thereof, so that it is not enough to state that territorial claims are not universally recognized to challenge the validity of the treaty.

Turning to expertise and know-how as the basis for justification of the treaty, Sorzano noted that this is neither a new nor a negligible justification for authority or for sharing in authority (e.g., Plato's Philospher King) and that it is in widespread use today in both international and domestic politics. Even the most egalitarian democracies restrict voting rights to those who could exercise them rationally; for example, minors and lunatics are excluded.

So, too, are two-tiered political systems extremely common. The United Nations Charter created two-tiered systems both in granting veto power to the permanent members of the Security Council and in its grant of powers to the council as compared with the powers of the General Assembly. He believed that Malaysia's own constitutional arrangements are a form of two-tiered bicameral parliamentary system, and noted that Malaysia was a founding member and driving force behind two-tiered agreements governing tin and rubber production. He therefore did not believe that "two-tieredness" could be a disqualifying ground for the legitimacy of the Antarctic Treaty.

Nor did Sorzano accept that nonuniversality could be used to challenge the legitimacy of the treaty, since there are multiple examples of nonuniversal international institutions, such as the European Economic Community, the United Nations Food and Agriculture Organization, the International Monetary Fund (IMF), the International Bank for Reconstruction and Development (IBRD), and UNESCO, whose legitimacy nobody would even dream of challenging. Even the UN General Assembly is not a universal body, since some states, such as Switzerland, are not members.

Sorzano concluded by suggesting a political, as opposed to a legal, justification for the Antarctic Treaty. Noting that much has been heard of the uniqueness of Antarctica, he agreed with this assessment in most respects but emphasized that its situation is not unprecedented in history: Following the age of exploration and discovery in the sixteenth, seventeenth and eighteenth centuries, there was a need to elaborate a political theory to justify the new governments that were set up in the newly discovered lands. The theoretical work of Hobbes, Locke, and Rousseau postulated a situation of a "state of nature," in which there was no governmental authority, individuals behaved with unrestrained self-interest, and thus conflict was ever-present. This undesirable state of affairs was ended by a "social contract," in which previously unrelated individuals voluntarily joined in forming a governing structure based on the consent of its members.

In Sorzano's view, the situation before and after the treaty became effective bears a strong similarity first to the "state of nature": no governing authority, unrestrained self-interest, and conflict, and then to the "social contract": a treaty based on consent, setting aside self-interest and creating a governing structure that, like any government's, including Malaysia's, is exclusive, total, and unaccountable.

REMARKS BY CRISTIAN MAQUIEIRA

Maquieira commented that, having visited Antarctica, he now understood the twinkle in the eye of all those who had come before. This feeling complemented his earlier understanding of the great respect for the treaty system demonstrated by those who had worked with it.

Referring to Zain's point about the "totality" of ATCP rights with respect to Antarctica, Maquieira did not believe that it is possible to divorce demilitarization and cooperation in antarctic science from cooperation with respect to resources under the Antarctic Treaty. He described the system as placing equal limitations on all parties to it, and believed that it would upset the balance among parties if some of these limitations were incorporated into a resources regime and others were rejected.

On the other hand, he believed that the system must prove itself adaptable to states and organizations newly interested in Antarctica in order to make it attractive and acceptable to them. He noted that a start in this direction has already been made with the invitation to the nonconsultative states to attend treaty meetings as observers and with the increased distribution of documentation from treaty meetings, including publication of the Handbook of the Antarctic Treaty System. He also introduced the idea of creating a fund to help developing countries that wished to join the treaty and take part in scientific research. He stated that there is a need to continue and expand scientific cooperation within the ATS and that the interests of the international community require more extensive participation by states in Antarctic science. These two elements constitute the basis of the fund idea that had been presented in the minerals regime negotiations. (For additional discussion of the fund see chapter 20.) He added that all these adaptations must be initiated by the parties to the treaty.

Maquieira brought up the fact that various proposals for the evolution of the treaty system have been made elsewhere. For instance, he noted from the report of the United Nations Secretary-General that Zimbabwe has proposed the establishment of an international scientific research station.The possibility of establishing a joint scientific venture among groups of developing countries and consultative party states has also been mentioned. It had been noted earlier that SCAR is contemplating establishing a category of "associate state" membership. In addition, various ideas have been proposed to extend the cooperative working relationships between the ATS and other international organizations. Maquieira suggested that a group of developing countries could establish a program of technical cooperation among themselves and then seek, say, United Nations Development Program funding to carry out activities in Antarctica. He cautioned, however, that all these possibilities presuppose a legitimate and sustained interest in Antarctica.

REMARKS BY ROLF TROLLE ANDERSEN

Andersen likewise stressed the ability of the ATS to adapt, as demonstrated by the invitation in 1983 to the nonconsultative states to attend treaty meetings. He stated that while consultative status requires a concrete demonstration of interest, there are other ways of participating in the system, such as through the newly created observer role. He acknowledged that this development will create some difficulties as well, since the sheer number of states involved will of its nature complicate a system that operates by consensus.

Andersen believed that further evolution will take place in the system and is desirable. This is a matter that naturally belongs on the agenda of regular consultative meetings. During the twelfth consultative meeting, in Canberra in September 1983, an agenda item entitled "The Operation of the Antarctic Treaty System" had resulted in a recommendation on a number of measures to be taken in order further to open up the ATS.

With respect to developments in the United Nations, he expressed the view that establishing a United Nations committee on Antarctica is unnecessary and undesirable; a forum already exists within which antarctic matters are discussed. He urged states interested in antarctic questions to accede to the Antarctic Treaty and express their views within that body. He disagreed with the Malaysian proposal to call for specific comments on the 1984 United Nations study in preparation for the 1985 United Nations General Assembly. It is up to individual governments to review the study, which they are doubtless already doing.

Elaborating on the question of the common heritage of humankind, he did not find the principle applicable to Antarctica, inter alia, because (1) there has been human activity in Antarctica for most of the twentieth century, (2) claims to sovereignty exist there, and (3) an existing legal regime applies there.

In response to Zain's comments that the parties to the Antarctic Treaty always act in concert, he viewed this coherence as necessary and desirable: It is an impressive demonstration of the importance attached by all the consultative parties to the maintenance and strengthening of the ATS.

SUMMARY

The final session's discussion encompassed a number of topics raised in earlier sessions that in one way or another relate to the institutional evolution of the ATS. On one level, the legitimacy of the ATS, as opposed to a more universally based system of governance for Antarctica, was debated at greater length and depth than in the discussions on this point summarized in Chapters 7 and 20. On a second level, for those who not only endorse the legitimacy of the ATS but believe that the ATS has proved to be a most workable system of governance and appropriately responsive to changing interests and requirements in Antarctica, the key question was to identify the directions in which the ATS should continue to adapt to changing circumstances.

The following discussion is divided into sections on legitimacy and the evolution of the ATS, followed by concluding remarks from three of the primary speakers.

LEGITIMACY

The discussion on legitimacy covered some of the same ground as that outlined in Chapters 7 and 20. Although neither those challenging the legitimacy of the ATS nor those defending it were about to concede defeat, the conclusion of the workshop produced a mood that may be best summed up as follows: There is nothing to prevent a review of the Antarctic Treaty after 1991, and it is clearly in the interest both of the parties to the treaty and those outside to take their time and allow the treaty to evolve and adapt rather than to maintain extreme positions opposing its review in any form or to seek its modification or replacement. All sides will have to bridge their differences. The corollary is that the extent of the evolution and adaptation within the system will clearly influence the possibility and/or the extent of any formal review; there is a need for a forum, informal or otherwise, to continue the dialogue.

The main points aired in the legitimacy debate were as follows:

Many speakers stated that they did not believe that the legitimacy of the Antarctic Treaty could be called into question. It is not opposed to any rule of jus cogens, nor does any rule of general international law disqualify the treaty or the ATS as illegal. The Antarctic Treaty had been negotiated and entered into by 12 states operating in accordance with Article II(3) of the United Nations Charter, which calls on all members to settle their international disputes by peaceful means.

The treaty was not meant to be a universal regime but rather a set of rules negotiated among states parties to regulate their own activities in Antarctica. It created rights only as between parties, as is permissible under the international law of treaties. These rights are exclusive only as between parties and do not generate rights or obligate third parties without their consent.

The legitimacy of the Antarctic Treaty on the basis of its standing under international law, however, was not so much challenged as was the fact that its administration is undemocratic because of the two-tier status of contracting parties and the costs of achieving consultative status. Yet various speakers denied the "exclusivity" charge leveled by Zain, noting both that the Antarctic Treaty is open to accession by any country and that the flexible criteria governing consultative party status allow any adhering state meeting these criteria to qualify for consultative status. The only way to make the Antarctic Treaty more universal is for more states to adhere to it.

If the criticism of lack of democracy was based on the

two-tier status of consultative party and nonconsultative party members, as opposed to a one-nation/one-vote rule for all states parties, it must be noted that there are numerous international organizations that do not operate on the latter principle. Additional examples exist of limited-membership agreements or regional treaties and agreements that do not include the whole of the international community yet have never been challenged as illegitimate. Moreover, the two-tier system permits states to participate up to their level of interest or willingness to take on the costs and responsibilities of carrying out activities in Antarctica; it does not unduly burden countries less actively involved.

The three basic questions posed in exploring the legitimacy of the ATS were the following:

(1) Whether it is possible to pick and choose among the fundamental elements of the ATS, rejecting the position of the claimant states by applying the common heritage of humankind approach, while assuming that peaceful, international cooperation in scientific research throughout the Antarctic Treaty area will prevail (see Chapter 7);
(2) Whether, in effect, Article IV of the treaty could remain operative with respect to scientific research in Antarctica and the objectives of the Antarctic Treaty, as well as with respect to the marine living resources regime, while a totally different foundation could be constructed as the basis for the regime governing potential mineral resources development in Antarctica; and
(3) Whether and how it would be possible to reconcile different views of "rights" and "interests" in Antarctica in a forum where all states were comfortable with their ability to contribute to and influence the dialogue.

On the first question, Article IV was described as a "Pandora's box," sealed in 1959 by the 12 original signatories to the treaty and since endorsed by 20 additional states. Unless the claimant states renounced their claims, to attempt to internationalize Antarctica or apply to it the common heritage of humankind principle will reopen the issues dealt with in Article IV and release the plagues, sorrow, and mischief long contained in this Pandora's box. Both claimant and nonclaimant states had agreed to restrict their perceived freedom of

action in Antarctica existing prior to completion of the
Antarctic Treaty, and they protected their positions--
maintained under Article IV--through the consensus
decision-making procedure.

Nonetheless, some participants maintained that the
claimant/nonclaimant positions in Antarctica contain the
seeds of latent conflict and that internationalizing
Antarctica might bring about the renunciation of the
claims--which in some eyes are an anachronism that newly
independent states neither recognize nor acquiesce.
Internationalization does not mean that it is essential
to create new management institutions or to base them on
a one-nation/one-vote decison-making process; the
countries nearest Antarctica might be assigned to manage
the area, or a condominium system of management could be
established. In the view of these participants, United
Nations interest does not necessarily mean United Nations
control, nor does it mean the destruction of the
Antarctic Treaty. The "sacred cow" attitude of the
consultative parties that oppose all challenges to the
treaty seemed to be a violation of the scientific
approach, which was based on questioning the status quo
and was enshrined in the purposes of the Antarctic Treaty.

While the concept of the common heritage of humankind
was deemed appropriate for outer space and the deep
seabed, it did not take account of the existing legal
situation in Antarctica. The common-heritage principle
had been applied to outer space and the deep seabed only
by means of consensus. No such consensus exists with
respect to Antarctica. Therefore, the principle is
clearly not applicable. Moreover, the common heritage
was meant only to describe a framework for a future
regime. Part XI of the Law of the Sea (LOS) Convention
is not the mandatory interpretation of the principle.

A regime could be envisaged that, being in full harmony
with the common heritage principle, nevertheless would
entrust certain states with obligations and rights to
ensure the use of Antarctica for the benefit of humankind
as a whole. But if the common heritage concept were to
be applied in the meaning of Part XI of the 1982 LOS
Convention, it would mean the collapse of the treaty
system, and this would not be in the interest of
humankind. (There was further debate about whether the
common heritage concept is an exploitative one. See
chapters 19, 20 and 25.)

The second question related to Zain's characterization
of the rights of the consultative parties as "total,"

that is, bearing on all activities in Antarctica. Divorcing the issue of minerals development from the framework of the ATS, however, was seen by some as undermining the system; it would also remove potential minerals activities from the umbrella of the Antarctic Treaty's emphasis on environmental protection. Others viewed this "totality" as part of a natural evolutionary process under the Antarctic Treaty as new interests and activities arose in Antarctica. The ATS did not ab initio govern all activities in Antarctica, but it demonstrated its flexibility and responsiveness to new circumstances with the adoption of additional legal instruments and recommendations.

The third question, that of a forum fully representative of all states' views on antarctic questions, was left to future antarctic symposia and meetings. Its answer is contingent on (1) evolution within the ATS and whether this will be sufficient to provide the forum sought (see the discussion below) and (2) the approach taken in the United Nations and the feasibility of agreement there on the future system of governance for Antarctica.

EVOLUTION OF THE ANTARCTIC TREATY SYSTEM

Drawing on the speakers' presentations that described the evolution of the ATS to date, the Workshop on the Antarctic Treaty System explored again, in the discussion of institutional changes within the ATS, possible directions for continued adaptation of the ATS that had arisen throughout the workshop.

(1) The question of participation in the ATS was addressed again in Chapters 21-26 with respect to nonconsultative parties and international organizations, both governmental and nongovernmental.[1] It was also raised again with respect to non-United Nations-member states, such as the Republic of Korea, a topic addressed in Chapter 6, and with respect to helping developing countries to take part in antarctic scientific research projects and share in their results.

This topic encompasses in addition the question asked in Chapters 13 and 14--how to institutionalize input from the concerned public in a fruitful manner--and the possibility noted in Chapters 19-20 of international comment on decisions pursuant to the minerals regime.

Many speakers acknowledged that there is room to perfect the evolution of the ATS in cooperating with non-treaty states and international organizations to the benefit of all. Bruckner's presentation in Chapter 22 examined the role of NCPs in the ATS and the criteria governing CP status, in particular the standing of joint scientific research programs in this regard (see Chapter 7).

It was noted that increasing cooperation with international organizations could also improve the achievements in Antarctica in fields where these organizations exercise competence. Further study of the types of relationships described by Woolcott (Chapter 25), in order to develop the relationships between the ATS and the United Nations system, was envisaged.

Spokesmen for the International Union for the Conservation of Nature and Natural Resources (IUCN) and ASOC made cases for the involvement of these two organizations in the ATS as a measure of the flexibility and institutional development of the system. It was noted that IUCN has expertise relevant to Antarctica, interests in the area, and a broad-based constituency. It could also contribute to worldwide understanding of antarctic issues and is already involved in sponsoring with SCAR a meeting on scientific requirements for antarctic conservation (see Chapter 14).

ASOC, founded in 1977, represents 150 member organizations in 35 countries, drawn together by their goal of implementing the World Conservation Strategy developed in 1980 by IUCN, the United Nations Environment Program, and the World Wildlife Fund International. ASOC hoped for more activity with respect to the concept in the Agreed Measures for the Conservation of Antarctic Fauna and Flora, adopted in 1964, which state that the treaty parties consider the treaty area a special conservation area. ASOC believed that Antarctica was legitimately a concern of the international community as a whole.

One speaker urged elimination of the "discrimination" against non-United Nations-member states' ability to adhere to the treaty.

With respect to participation by developing countries in antarctic science, the point made in the section on Antarctic Science and Chapters 17-20, on the lack of immediacy with respect to any foreseeable benefits from antarctic minerals development and potential sharing of the revenues therefrom, was repeated, emphasizing instead the already extensive benefits derived from scientific and environmental knowledge of Antarctica. It was suggested that these benefits could be expressly extended to developing countries through SCAR programs for scientific activities and could be underwritten financially by the consultative parties or by some other groups in order to ensure more widespread distribution of the existing benefits of the ATS.

Another speaker added that for antarctic science to contribute most effectively to scientific knowledge and associated benefits, it should not be conducted in isolation but should be integrated into domestic science programs and form part of the global research agenda. Otherwise its potential for addressing domestic problems and priorities in many countries would be lost. He believed that the increasing sophistication of science must be used for the benefit of all countries. Perhaps the best way for countries to become influential in the ATS would be to encourage their scientists to cooperate with the countries already active in antarctic science.

He added that while his country had never become a member of the Antarctic Treaty, it had taken part in science and supporting activities since 1914. Yet, by providing supplies and commercial equipment for antarctic activities, his country had probably made more money out of Antarctica than any other--an analogy to the saloon-keeper at a gold rush camp: while others made headlines and won or lost fortunes, the saloonkeeper made money. He posed two questions: whether it is necessary to join the ATS to benefit from it, which his country's experience would indicate was not necessarily the case, and whether there is a responsibility on the part of nations benefitting from the ATS to join it and to join in the collective responsibility that membership entails.

(2) The role of consensus in the ATS and its continuing viability as membership in the ATS expanded and as new legal instruments were articulated to govern antarctic activities received some attention. (See section on Legal and Political Background and Christopher Beeby's presentation on the minerals regime in Chapter 19.)

(3) The coordination, integration, or consolidation of the ATS was touched on briefly again in relation to competing uses in Antarctica and the preservation of the antarctic environment. This topic is on the agenda of the XIII consultative meeting in October, 1985. (See the proposals for a continental conservation strategy for Antarctica and for an environmental protection agency to independently monitor environmental protection policies and compliance with them in Chapters 13 and 14). So too

was the relationship between adherence to the minerals regime and accepting the Antarctic Treaty in its entirety (see Chapters 19 and 20).

The question of whether growth within the ATS might require some form of permanent infrastructure or secretariat was also raised.

In addition, one participant suggested amending the World Heritage Convention, cited by Kiss earlier in the discussion, so that it could apply to the special circumstances in Antarctica created by the status of claims and Article IV. Additional comments were made about the paramount value of scientific research in Antarctica (see section on Antarctic Science), the possibility of a 30-year moratorium on minerals development activities (see Chapter 20), and the possibility of extending the antarctic zone of peace north by 10 degrees every year.

(4) The possibilities of renewed conflict and militarization of Antarctica, the role of effective enforcement, the viability of inspection under the antarctic minerals regime and within the ATS as a whole, and the relationship of enforcement and inspection to the accountability of the ATS were considered as well.

Some fears were expressed that the minerals regime might undermine the ATS, because developers would oppose inspection of their operations in order to protect proprietary information, and this could compromise on-site inspection. Such possible deterioration of inspection might in turn undermine the demilitarization of Antarctica. One participant suggested that the fact that Antarctica was not militarized might simply be due to the fact that appropriate technology did not exist at the time of the negotiation of the treaty; new technologies might produce a situation in which Antarctica would be militarized. He claimed that some governments questioned the treaty because it prevented them from developing military options there.

Other speakers countered the possibility of the militarization of Antarctica by noting that both the United States and the USSR have a strong interest in avoiding that eventuality. Moreover, Antarctica was the only place where inspection works, and there is no reason to presume that it will not continue to do so.

(5) With respect to the information policies of the ATS, all participants emphasized the importance of accessibility, continuity, regularity, and sufficiency of information—whether dealing with activities in Antarctica, the results of scientific research projects there, or proceedings in ATS forums. They took note of progress

made in the area over the past 18 months and acknowledged
the relationship between information policies and building
confidence in the ATS and making it more accountable.
This issue is to be considered again at the XIII consultative meeting in October 1985.

Some speakers noted, contrary to Zain's view, that
there is accountability under the treaty among parties as
well as accountability to third states under general
principles of international law. Nor does the Antarctic
Treaty depart from other general international law
governing state responsibility and liability.

CONCLUDING REMARKS

R. Tucker Scully

Scully stated that the ATS has evolved and would continue
to evolve. There is no alternative and this represents a
strength of the system. He believed that there is a
logical inconsistency in the position, articulated by
countries such as Malaysia, that argues that the system
should be changed but at the same time does not define
the country's interest or interests with respect to
Antarctica. Nor is there any justification in calling
for the renegotiation of the Antarctic Treaty based upon
fears or speculation as to what might happen. In his
view, the treaty system had seen an oustanding example of
international cooperation--in arms control, in scientific
research, in environmental protection, and in resources
management. To achieve these objectives the system had
demonstrated that when legitimate (and concrete) interests
arose in these areas, they would be effectively
accommodated.

Richard Woolcott

Woolcott acknowledged that the Malaysian initiative had
served a useful purpose in assisting in the further
opening up of the ATS and in the production of the
comprehensive United Nations study on Antarctica. But he
did not agree with the impression left by Malaysia that
its proposal for a United Nations special committee had
been widely supported among non-treaty states. Various
UN member states, including some members of the Group of
77, had not favored the committee. To the two categories
of nontreaty states identified by Zain, he would add the

two-thirds of the United Nations member states which had shown very little interest in the subject and which did not contribute to the United Nations study and also those nontreaty states which were actively considering acceding to the treaty. Even of the 54 countries that did contribute to the United Nations study, the majority were parties to the treaty. Antarctica simply was not an issue about which the United Nations was, or needed to be, concerned. The issue had been artificially stimulated by Malaysia.

He believed that countries outside the treaty system were able to influence decisions taken by the parties, and noted that a measure of accountability was generated through the pressure of public opinion and indeed through an institution like the United Nations, should some real problem arise in respect of Antarctica. He wished to revise his introductory presentation to indicate that the United Nations did not really create problems. Rather, what he had had in mind was that, in the case of Antarctica, an issue had been artificially promoted when there was, in fact, no contention in Antarctica. In this case, United Nations consideration had the capacity to exacerbate the matter and bring about a confrontation between parties to the Antarctic Treaty and some nontreaty states. In his view, this would create an unnecessary problem. Too often the United Nations General Assembly becomes an arena for confrontation rather than a forum for conciliation.

He believed that certain political realities should be taken into account in considering any United Nations involvement with Antarctica, because there is a very broad base of support for the ATS. For instance, the two superpowers, the five permanent members of the UN Security Council, all the present nuclear powers, six of the seven most populous nations on Earth, and the current chairman of the nonaligned movement are members of the treaty system and oppose institutionalizing the involvement of the United Nations in Antarctica. The antarctic issue could not be presented as either an east/west issue, because of the membership of the United States and the USSR, or a north/south issue because of the membership of countries like India, China, Brazil, and Argentina.

He confirmed Zain's impression that the Antarctic Treaty consultative parties believed that the less the United Nations is involved, the better. This was because the parties do not want any actions taken that could lead to undermining a successful and valuable treaty. His own consultations indicated that a number of nonaligned and nontreaty countries are in fact confused about why Malaysia has raised the issue in the first place and simply hope that it will fade from the United Nations agenda.

His reasons for stating that the United Nations will not be able to manage effectively some new antarctic regime were the following: the United Nations is not a world parliament but an assembly of sovereign states; these states are divided or disinterested in the issue; United Nations members do not easily reach agreement on any major political, economic, or social issue especially when deep divisions are involved. In addition, the group system has resulted, on occasion, in support for various issues being horsetraded independently of the merits of the issue. The United Nations is a politicized body, tending toward preoccupation with essentially political conflicts: Arab versus Jew in the Middle East, white versus black in southern Africa, East versus West in a number of areas, and North versus South on economic questions. For instance, Pakistan has criticized the Antarctic Treaty not, in his view, because it has made a considered assessment of its value but probably because India has joined it; similarly, Ghana and some black African countries have attacked the treaty not on the basis of disagreements with its objectives but because South Africa is a member. South Africa had joined the Law of the Sea Convention without attracting African opposition.

Zain Azraai

Zain agreed that clearly the countries outside the ATS must identify answers of their own but stated that their first efforts have been directed toward understanding the position of the parties to the treaty. He reiterated that those outside do not necessarily wish to replace the ATS. On the other hand, he saw joining the Antarctic Treaty as a "catch-22" situation: once a country joins, it can no longer challenge the system, but it was not yet clear to him to what extent the country would be able to influence determinations within the system.

He did not see why the exclusive rights of the consultative parties with respect to decision-making should be extended to _all_ activities in Antarctica nor why the consultative parties are unwilling to be accountable in some forum to the wider international community. He questioned the assertion that by not objecting to Antarctic matters being dealt with in some other forum the Antarctic Treaty parties do not restrict the rights of others; he recalled earlier efforts by the consultative parties to restrict possible involvement in antarctic affairs by the United Nations Committee on Natural Resources, the United Nations Environment Program, and the United Nations Food and Agriculture Organization.

Zain also wished to get a clearer definition of the special rights of the consultative parties and whether those rights derive merely from spending money in Antarctica. Nor was he yet certain that he understood why the Antarctic Treaty creates only obligations but no rights, as several speakers had indicated, because while regulations by definition create restrictions or obligations, the determination or decision as to what those regulations should be is the exercise of a right.

He doubted the relevance of attempting to justify the Antarctic Treaty's status by citing various limited-membership treaties, because the mandate for the establishment of these other limited-membership bodies came from the United Nations or was negotiated in the United Nations or otherwise freely negotiated by the parties concerned. He also believed that it is unhelpful to ascribe extraneous motives to the positions of member states, as had earlier been suggested in the case of Pakistan and Ghana. Members states' views should be ascertained from their own statements rather than asserted by individuals on the basis of their perception as a result of any consultation in which they may have engaged.

He concluded that, as a representative from a country outside the ATS, he would follow carefully the evolving role of the nonconsultative states and was interested in the role of the United Nations specialized agencies and other international organizations vis a vis the ATS. He added that he did not like to see the latter involvement pursued only as a reluctant concession on the part of the consultative parties to the treaty. He supported the development of extensive working relationships with these bodies, by which he meant that Antarctica should be discussed in the governing councils of the relevant specialized agencies; he did not mean simple attendance by an individual from a specialized agency at Antarctic Treaty meetings. These developments would help create confidence among members of the international community that Antarctica is being managed in their interest and for their benefit.

His fundamental questions remained: Who should reconcile differing views on Antarctica, in what forum, and how? He noted that many speakers, including those who were supportive of the present ATS, had stated that the system could be improved. It was Malaysia's position that discussion of these issues should take place in a forum where member states would be on an equal footing, without prejudice to any position that they may take regarding the ATS.